Analysis and Synthesis of Fuzzy Control Systems

A Model-Based Approach

AUTOMATION AND CONTROL ENGINEERING

A Series of Reference Books and Textbooks

Series Editors

FRANK L. LEWIS, Ph.D.,
Fellow IEEE, Fellow IFAC

Professor
Automation and Robotics Research Institute
The University of Texas at Arlington

SHUZHI SAM GE, Ph.D.,
Fellow IEEE

Professor
Interactive Digital Media Institute
The National University of Singapore

Analysis and Synthesis of Fuzzy Control Systems

A Model-Based Approach

Gang Feng

City University of Hong Kong
Hong Kong SAR

CRC Press
Taylor & Francis Group
Boca Raton   London   New York

CRC Press is an imprint of the
Taylor & Francis Group, an **informa** business

CRC Press
Taylor & Francis Group
6000 Broken Sound Parkway NW, Suite 300
Boca Raton, FL 33487-2742

First issued in paperback 2017

© 2010 by Taylor and Francis Group, LLC
CRC Press is an imprint of Taylor & Francis Group, an Informa business

No claim to original U.S. Government works

ISBN 13: 978-1-138-11424-1 (pbk)
ISBN 13: 978-1-4200-9264-6 (hbk)

Library of Congress Cataloging-in-Publication Data

Feng, Gang.
    Analysis and synthesis of fuzzy control systems : a model-based approach / Gang Feng.
       p. cm. -- (Automation and control engineering)
    Includes bibliographical references and index.
    ISBN 978-1-4200-9264-6 (hardcover : alk. paper)
    1. Automatic control--Mathematical models. 2. Fuzzy systems. I. Title. II. Series.

TJ213.F392 2010
629.801'511313--dc22   2009049468

Visit the Taylor & Francis Web site at
http://www.taylorandfrancis.com

and the CRC Press Web site at
http://www.crcpress.com

Contents

Preface

Fuzzy sets and systems have gone through significant development since the introduction of fuzzy set theory by L. A. Zadeh about four decades ago. They have found a great variety of applications ranging from control engineering, pattern recognition, signal processing, information processing, and machine intelligence to decision making, management, finance, medicine, robotics, and so on. In particular, fuzzy logic control (FLC) has become one of the most successful applications and has proven to be a popular control approach for many complex nonlinear systems or even nonanalytic systems. In many cases it has been suggested as an alternative approach to conventional control techniques.

Conventional fuzzy logic control developed in the 1970s to 1980s unfortunately suffers from a lack of tools for systematic stability analysis and controller design even though it has had great success in industrial applications. It has been recognized that conventional fuzzy logic control is fundamentally model free, which more or less is the essential cause of the problem. To address the difficulties encountered by conventional fuzzy logic control, many model-based fuzzy control approaches have been developed in the past decade.

In particular, the fuzzy dynamic model or the Takagi and Sugeno (T–S) fuzzy model-based approaches have been proposed and widely studied. This model is based on using a set of fuzzy rules to describe a nonlinear system in terms of a set of local linear models that are smoothly connected by fuzzy membership functions. This fuzzy modeling method offers an alternative approach to describing complex nonlinear dynamic systems. One of the most appealing features of T–S fuzzy models is that two kinds of knowledge, one the qualitative knowledge represented by fuzzy IF–THEN rules and the other the quantitative knowledge represented by local linear models, can be embedded into one model. In addition, T–S fuzzy models have a compatible structure with a two-level control system with the lower level providing basic feedback control and the higher level providing supervisory control or scheduling. By using T–S fuzzy models one can formulate these two kinds of knowledge or two-level systems into a unified mathematical framework. And more important, T–S fuzzy models provide a basis for the development of systematic approaches to stability analysis and controller design of fuzzy control systems in view of powerful conventional control theory and techniques.

Many fundamental theoretical results on function approximation, stability analysis, and controller synthesis of T–S fuzzy systems have been developed during the past 10 years or so. In fact, it has been shown that T–S fuzzy models are universal function approximators in the sense that they are able to approximate any smooth nonlinear functions to any degree of accuracy in any convex compact region. This result provides a theoretical foundation for using T–S fuzzy models to represent complex nonlinear systems approximately. Based on T–S fuzzy models many approaches to systematic analysis and synthesis of fuzzy control systems have been developed. Moreover, it has been successfully demonstrated that conventional control technology and fuzzy logic control can be elegantly combined and further developed so that the disadvantages of conventional fuzzy logic control can be avoided and the horizon of conventional control technology can be greatly extended.

This book is devoted to systematic analysis and synthesis of model-based fuzzy control systems. It mainly consists of two parts: fuzzy system modeling and identification, describing how a nonlinear system can be represented by a T–S fuzzy model; and analysis and synthesis of T–S fuzzy model-based control systems, which is the main focus of the book. In addition, many chapters of the book also feature application simulation examples and numerical examples based on the software platform MATLAB®. The introduction in Chapter 1 gives a brief review of the varieties of fuzzy logic control and in particular T–S fuzzy model-based fuzzy logic control. Chapter 2 introduces the fundamental concepts of fuzzy sets, fuzzy logic, and fuzzy systems that enable the book to be self-contained and provide a basis for the developments in the later chapters. T–S fuzzy modeling and identification via nonlinear models or data are studied in Chapter 3, which is followed by stability analysis of T–S fuzzy systems in Chapter 4, stabilization controller synthesis in Chapter 5, robust H_∞ controller synthesis in Chapter 6, and observer and output feedback controller synthesis in Chapter 7. Robust controller synthesis of uncertain T–S fuzzy systems and time-delay T–S fuzzy systems are studied in Chapters 8 and 9, respectively, followed by fuzzy model predictive control in Chapter 10, robust fuzzy filtering in Chapter 11, and finally adaptive control of T–S fuzzy systems in Chapter 12.

The book can be used as a textbook for graduate students in a course of intelligent or fuzzy control, and also as a reference for scientists and engineers in the area of systems and control. The prerequisite knowledge required of readers includes basic theories of linear algebra, advanced calculus, control systems, and linear systems.

The completion of this project would have been impossible without the generous support of the many people involved. In particular, Professor Frank Lewis inspired the author to initiate this project. Many colleagues or students of the author, including N. W. Rees, Shuguang Cao, Zhixiu Han, C. K. Chak, Louis Wang, Cailian Chen, Tiejun Zhang, Bo Yang, Jianbin Qiu, He Huang, Yonghui Sun, Yuan Fan, and Cheng Song, have made various and significant contributions. In fact, some chapters in this book are adapted from joint publications of the author and his colleagues or students, and the author would like to take this opportunity to acknowledge their contributions. In addition, the author would like to acknowledge the contribution of Yanyan Shen and Jianbin Qiu who have carried out all the simulation studies contained in this book. The author is also grateful to LiMing Leong, Acquiring Editor; Yong Ling Lam, Assistant Editor; Iris Fahrer, Project Editor, and Jennifer Ahringer, Project Coordinator at Taylor & Francis for their assistance and generous support during the long project period.

The financial support under a number of research grants from the Australian Research Council, Research Grants Council of Hong Kong, and City University of Hong Kong are also acknowledged for many of the research results presented in this book.

MATLAB® is a registered trademark of The MathWorks, Inc. For product information, please contact:

The MathWorks, Inc.

3 Apple Hill Drive

Natick, MA 01760-2098 USA

Tel: 508-647-7000

Fax: 508-647-7001

E-mail: info@mathworks.com

Web: www.mathworks.com

Author

Gang Feng, Ph.D., received his B.Eng. and M.Eng. in automatic control (of electrical engineering) from Nanjing Aeronautical Institute, China in 1982 and 1984, respectively, and his Ph.D. in electrical engineering from the University of Melbourne, Australia, in 1992. He was with the School of Electrical Engineering, the University of New South Wales, Australia, from 1992 to 1999, first as a lecturer and then as a senior lecturer. He was a visiting fellow at the National University of Singapore and Aachen Technology University, Germany, in 1996 and 1997, respectively. Since 2000, he has been with City University of Hong Kong where he is currently a chair professor of mechatronic engineering and also serves as the associate provost. He is also the Cheung Kong Chair Professor at Nanjing University of Science and Technology conferred by the Education Ministry of China.

Dr. Feng has done research work in adaptive control, robot control, intelligent control, and, more recently, control of piecewise linear systems and switched systems. His research work has led to the publication of one edited book, seven invited book chapters, and over 180 international journal papers and numerous international conference papers. He has received the *IEEE Transactions on Fuzzy Systems* Outstanding Paper Award (2007), an Alexander von Humboldt Research Fellowship of Germany (1997), and a number of best conference paper awards. He has been invited to give plenary/special lectures at a number of international conferences.

Dr. Feng is an IEEE Fellow and an associate editor of *IEEE Transactions on Automatic Control, IEEE Transactions on Fuzzy Systems,* and *Mechatronics.* He has served as an associate editor of *IEEE Transactions on Systems, Man, & Cybernetics, Part C, Control Theory and Applications,* and on the Conference Editorial Board, IEEE Control Systems Society.

1 Introduction to Fuzzy Logic Control

1.1 INTRODUCTION

Since their introduction by Professor L. A. Zadeh in his seminal works (Zadeh, 1965, 1968, 1971, 1973, 1975) in the early 1970s, fuzzy sets and fuzzy logic theory have found a great variety of applications in control engineering, power systems, telecommunication, consumer electronics, information processing, pattern recognition, signal processing, machine intelligence, qualitative modeling, decision making, management, finance, medicine, the chemical industry, motor industry, robotics, and so on (see, e.g., Babuska, 1998; Bellman and Zadeh, 1970; Bezdek et al., 1999; Bonissone et al., 1995; Koska, 1992; Lee, 1990a,b; Mendel, 1995; Pedrycs, 1993; Sugeno, 1985; Teodorescu, Jain, and Kandel, 1998; Zimmermann, 1991). In particular, fuzzy logic control, or fuzzy control for short, as one of the earliest applications of fuzzy sets and fuzzy logic theory, has become one of the most notable applications. In fact, fuzzy logic control has proven to be a successful control approach to many practical and industrial systems, especially to complex nonlinear systems or even nonanalytic systems, and has been widely accepted as an alternative or complementary approach to conventional control techniques in many engineering applications.

It is generally believed that the first real fuzzy logic control system was developed by Mamdani and Assilian (Mamdani, 1974; Mamdani and Assilian, 1975) and called a Mamdani-type or conventional fuzzy controller. Since its inception, fuzzy logic control has attracted great attention from both the academic and industrial communities. Many people have devoted considerable effort to both theoretical research and application techniques of fuzzy logic controllers. In particular, fuzzy logic control has been developed substantially since the 1980s and put to use in a variety of applications in industry. This widespread research and development is evidenced by a number of excellent books and tutorial articles on the topic; see, for example, Babuska (1998), Lee (1990a,b), Feng (2006), Mendel (1995), Palm, Driankov, and Hellendoorn (1996), Passino and Yurkovich (1998), Pedrycs (1993), de Silva (1995), Bezdek et al. (1999), Bonissone et al. (1995), Sala, Guerra, and Babuska (2005), Tanaka and Wang (2001), Wang (1994, 1997), Sugeno (1985), and Yager and Filev (1994).

Successful engineering application areas include, but are not limited to, power systems (Abdelazim and Malik, 2003; Flores et al., 2005; Guesmi, Adballah, and Toumi, 2004; Ko and Niimura, 2002); telecommunications (Aoul et al., 2004; Chen, Lee, and Guo, 2003a,b; Chen, Tsai, and Chen, 2003; Kandel et al., 1999; Lee and Lim, 2001; Zhang and Phillis, 1999); mechanical/robotic systems (Bai, Zhuang, and Roth, 2005; Baturone et al., 2004; Boukezzoula, Galichet, and Foulloy, 2004; Chang and Chen, 2005; Hagras, 2004; Hwang and Kuo, 2001; Li, Lee, and Cheng, 2004;

Santibanez, Kelly, and Llama, 2005; Sun and Er, 2004; Yang et al., 2004); automobiles (Bonissone et al., 1995; Huang and Lin, 2003; Lin and Hsu, 2004; Mar and Lin, 2001; Murakami and Maeda, 1985; Niasar, Moghbeli, and Kazemi, 2003; Sugeno and Nishida, 1985); industrial/chemical processes (Chen and Liu, 2005a; Frey and Kuntze, 2001; Horiuchi and Kishimoto, 2002; Juang and Hsu, 2005; King and Mamdani, 1977; Larsen, 1980; Ostergaard, 1977; Sugeno, 1985; Tani, Murakoshi, and Umano, 1996; Tong, Beck, and Latten, 1980; Umbers and King, 1980); aircraft (Chiu et al., 1991; Farinwata, Pirovolou, and Vachtsevanos, 1994; Kadmiry and Driankov, 2004; Larkin, 1985); motors (Barrero et al., 2002; Guillemin, 1996; Kim and Lee, 2000); medical services (Kwok et al., 2004; Seker et al., 2003; Zheng and Zhu, 2004); consumer electronics (Haruki and Kikuchi, 1992; Kumar, 2005; Lee and Bien, 1994; Lee et al., 1994; Nakagaki et al., 1994; Smith, 1994; Takagi, 1992; Wu and Sung, 1994); and other areas such as chaos control (S. S. Chen et al., 2005; Lian et al., 2001) and nuclear reactors (Boroushaki et al., 2003; Munasinghe, Kim, and Lee, 2005).

The literature on the topic of fuzzy logic control is vast and many different approaches have been developed. Based on the features of fuzzy control rules, these approaches can be roughly classified into the following categories:

1. Conventional fuzzy control or Mamdani fuzzy control
2. Fuzzy proportional-integral-derivative (PID) control
3. Neuro–fuzzy control or fuzzy–neuro control
4. Fuzzy sliding mode control
5. Adaptive fuzzy control
6. Takagi–Sugeno model-based fuzzy control

However, this classification is neither unique nor exhaustive, and many other different classifications can also be employed. It should also be noted that overlapping among these classification categories is inevitable. For example, conventional fuzzy control can be adaptive, fuzzy PID control can be tuned by neuro–fuzzy systems, or neuro–fuzzy control can be adaptive in nature in many cases. A brief review of all these categories of fuzzy logic control approaches is presented next.

1.2   BRIEF REVIEW OF FUZZY LOGIC CONTROL

1.2.1   CONVENTIONAL FUZZY CONTROL (MAMDANI-TYPE FUZZY CONTROL)

Mamdani and Assilian developed the first real fuzzy control algorithm in the early 1970s (Mamdani, 1974; Mamdani and Assilian, 1975), where control of a small steam engine was considered. The fuzzy control algorithm basically consisted of a set of heuristic control rules, and fuzzy sets and fuzzy logic were used, respectively, to represent linguistic terms and to evaluate the control rules, called conventional fuzzy control or Mamdani-type fuzzy control. Since then similar fuzzy logic control algorithms have been developed for many different engineering processes. For example, Kickert and Lemke (1976) developed a fuzzy control algorithm for a warm-water plant, and Ostergaard (1977) presented another fuzzy control algorithm for a small-scale heat exchanger.

There were many other applications of conventional fuzzy control, including robot (Baturone et al., 2004; Uragami, Mizumoto, and Tanaka, 1976; Xiao et al., 2004; Yang et al., 2004), stirred tank reactor (King and Mamdani, 1977), traffic junction (Pappis and Mamdani, 1977), steel furnace (Kornblum and Tribus, 1970), cement kilns (Umbers and King, 1980), automobile (Bonissone et al., 1995; Murakami and Maeda, 1985; Sugeno and Nishida, 1985), wastewater treatment (Tong, Beck, and Latten, 1980), aircraft (Chiu et al., 1991; Larkin, 1985), missile autopilot (Farinwata, Pirovolou, and Vachtsevanos, 1994), motor (Guillemin, 1996), network traffic management and congestion control (Kandel et al., 1999; Lee and Lim, 2001), bioprocesses (Horiuchi and Kishimoto, 2002), fusion welding (Bingul, Cook, and Strauss, 2000), and so on. In addition, conventional fuzzy control was also widely used in various consumer electronic devices such as video cameras, washing machines, televisions, and sound systems in the late 1980s and early 1990s (Haruki and Kikuchi, 1992; Kumar, 2005; Lee and Bien, 1994; Lee et al., 1994; Nakagaki et al., 1994; Smith, 1994; Takagi, 1992; Wu and Sung, 1994).

These methods of conventional fuzzy control are essentially heuristic and model free. Conventional fuzzy logic control has been widely accepted by engineers in industry due to its simplicity and ease in understanding and implementation. The fuzzy control "IF–THEN" rules are often obtained based on an operator's control action or knowledge. It is thus obvious that the design method works well only in the cases where operators or their knowledge play an important or critical role in controlling the system. Even though the performance of such control schemes might be generally satisfactory in practice, the stability issue of the closed-loop fuzzy control systems is often not addressed and thus often criticized in the earlier development of these methods. Moreover, design of conventional fuzzy control systems suffers from a lack of systematic tools, which often leads to difficulty in controller design and inconsistency in performance of the closed-loop control systems. Thus great efforts have been devoted to issues of stability analysis and controller design of conventional fuzzy control systems, and various approaches have been developed.

In addition to a heuristic approach provided by Braae and Rutherford (1979) for stability analysis of fuzzy control systems, a key idea of most approaches is to consider a fuzzy controller as a nonlinear controller and embed the stability or control design problem of fuzzy control systems into conventional nonlinear system stability theory. The typical approaches include describing function approach (Kickert and Mamdani, 1978), cell-state transition (Kang, 1993), Lure's system approach (Cho, Kim, and Lim 1993; Melin and Vidolov, 1994), Popov's theorem (Furutani, Saeki, and Araki, 1992), circle criterion (Opitz, 1993; Ray and Majumder, 1984; Singh, 1992), conicity criterion (Espada and Barreiro, 1999), sliding mode control (Hwang and Lin, 1992), and hyperstability (Calcev, Gorez, and De Neyer, 1998; Opitz, 1993), among others. However, a general theory for systematic stability analysis and controller synthesis of conventional fuzzy control systems is still out of reach and remains a challenging task for the fuzzy control community.

1.2.2 FUZZY PID CONTROL

It is well known that conventional proportional-integral-derivative controllers are still one of the most widely adopted methods in industry for various control applications. PID controllers have a number of distinctive advantages such as their simple structure, ease of design, and low cost of implementation in comparison to many other control methods. However, PID controllers might not perform satisfactorily or as desired if the system to be controlled is highly nonlinear or uncertain, or if the control performance specification is very demanding. On the other hand, fuzzy control has long been known for its ability to handle nonlinearities and uncertainties through the use of fuzzy set theory. It is thus believed that a better control system to take advantage of PID control and fuzzy control can be achieved by adequately integrating these two techniques. Considerable effort has been devoted to the course of this development and many significant results have been obtained. The resulting controllers are usually called fuzzy PID controllers.

It is noted that the name "fuzzy PID control" has been widely used in the literature with a variety of meanings. A particular example is the definition by Hu, Mann, and Gosine (2001). It is suggested that if the resulting control action from a fuzzy controller is in the scope of proportional-integral-derivative concepts like a conventional PID controller, then the fuzzy controller is called a fuzzy PID controller. Generally, this type of fuzzy PID controller is able to perform as well as conventional PID controllers if designed properly. However, the relative high cost of setting up a fuzzy control system would usually keep one from replacing conventional PID controllers with this type of controller. It should be noted that with this definition, the conventional fuzzy controller developed by Mamdani and Assilian (1975) would be classified as a two-input fuzzy PI controller, and this would not be in accordance with the original idea of fuzzy PID control. Instead, a more generally accepted definition is more or less relevant to integrating fuzzy logic control and PID control techniques. A typical example is that the controllers have the structure of PID controllers but with the gains implemented by fuzzy logic theory. This type of fuzzy PID controller is generally classified as "gain-scheduling" (He et al., 1993a,b; Zhao, Tomizuka, and Isaka, 1993) type fuzzy PID controllers, for the reason that controller gains change as the operating condition or dynamics of a system varies.

These gain-scheduling fuzzy PID controllers have gained wide acceptance by industry because they take advantage of both conventional PID control and fuzzy control techniques. In fact, this type of fuzzy PID control has indeed received much more attention and has had great success in industrial applications. It has also been shown that many fuzzy PID controllers are nonlinear PID controllers and perform much better than conventional PID controllers in most cases.

Other topics of interest relevant to fuzzy PID control include tuning of fuzzy PID parameters (Mann, Hu, and Gosine, 2001; Mudi and Pal, 1999; Woo, Chung, and Lin, 2000; Xu, Hang, and Liu, 2000), optimal fuzzy PID controller based on a genetic algorithm (Hu, Mann, and Gosine, 1999; Tang et al., 2001), realization of conventional PID controllers by fuzzy control methods (Mizumoto, 1995), improved robust fuzzy PID controller with optimal fuzzy reasoning (Li et al., 2005), and stability analysis of fuzzy PID controllers (Chen and Ying, 1993; Sio and Lee, 1998). Chen (1996) presented an excellent overview on fuzzy PID controllers.

Fuzzy PID control has a number of significant advantages; however, it also suffers from one major limitation similar to conventional PID or conventional fuzzy control. That is, the tools for its systematic design with consistent and guaranteed performance are far from satisfactory and still out of reach, especially when demanding control performance is desired.

1.2.3 NEURO–FUZZY CONTROL OR FUZZY–NEURO CONTROL

Neural network control, or neuro control for short, and fuzzy control are two of the most popular intelligent control techniques. They have received great attention and undergone substantial development in the past decades. These two control techniques share a number of common features. For example, both of them are basically model-free control techniques, both are able to store knowledge and utilize it to deduce control actions, and both are able to provide robustness of control to a certain extent with respect to system variations and uncertainties. On the other hand, these two techniques differ from each other in many ways as well. For example, they have distinctive ways of obtaining knowledge. Neuro control acquires knowledge mainly through data training (or learning). This could be an advantage as it lets the data set "speak" for itself, but it is sometimes a disadvantage if the training data set does not fully represent the domain of interest. Fuzzy control, in particular conventional fuzzy control, in contrast, mainly obtains knowledge via an operator or expert's perspective, which is often qualitative and sometimes even imprecise.

It has been recognized that these two control techniques complement each other in the sense that neuro control provides learning capabilities and high computational efficiency in parallel implementation, whereas fuzzy control provides a powerful framework for expert knowledge representation. The combination or integration of these two techniques is expected to result in better control approaches and thus has attracted a lot of attention from the control community. A typical combination of these two techniques is the so-called neuro–fuzzy control, which is basically a fuzzy control augmented by neural networks to enhance its characteristics such as flexibility, data processing capability, and adaptability (Da and Song, 2003; Farag, Quintana, and Lambert-Torres, 1998; Frey and Kuntze, 2001; Jang, 1993; Jang and Sun, 1995; Lazzerini, Reyneri, and Chiaberge, 1999; Li and Lee, 2003; Li, Lee, and Cheng, 2004; Lin and Hsu, 2004; Lin and Shen, 2006; Liu, Lara-Rosano, and Chan, 2004; Wai and Chang, 2006; Wang and Lee, 2003; Wang, Cheng and Leu, 2005, 2002). The process of fuzzy reasoning is realized by neural networks whose connection weights correspond to the parameters of fuzzy reasoning. Using back-propagation-type, reinforcement-type, or any other type of neuro network learning algorithms, a neuro–fuzzy control system can identify fuzzy control rules and learn membership functions of fuzzy reasoning, and thus realize neuro–fuzzy control. An excellent survey was given in Mitra and Hayashi (2000) for neuro–fuzzy rule generation in a more general setting of soft computation.

One of the main advantages of neuro–fuzzy control is that it does not require information on the mathematical model of a system to be controlled. Thus this class of fuzzy control offers a new avenue to solving many difficult control problems in real life where the mathematical model of a system might be hard, if not impossible, to obtain.

Other topics of interest related to this class of control schemes include tuning parameters in neuro–fuzzy controllers via genetic algorithm (Melin and Castillo, 2001; Seng, Bin Khalid, and Yusof, 1999; Wang, Cheng, and Leu, 2004), self-organizing or adaptive neuro–fuzzy control (Da and Song, 2003; Li and Lee, 2003; Munasinghe, Kim, and Lee, 2005; Wang and Lee, 2003), and input–output stability analysis based on the small gain theorem (French and Rogers, 1998). It should also be noted that the T–S fuzzy model is one of the general fuzzy systems that can be used to realize neuro–fuzzy control in this category; for example, see Juang and Hsu (2005), Tzafestas and Zikidis (2001), Wai and Chen (2004), and Yu and Li (2004).

Neuro–fuzzy control or fuzzy–neuro control has achieved great success in control applications; however, among its major limitations are that the systematic neuro–fuzzy control design, stability analysis of the resulting complex closed-loop neuro–fuzzy control systems, and convergence analysis of the learning algorithms in the context of the closed-loop control systems, in general, are very difficult and remain challenging.

1.2.4 FUZZY SLIDING MODE CONTROL

It is well known that sliding mode control is an effective robust approach to controlling nonlinear systems with various uncertainties and disturbances (Utkin, 1992; Zinober, 1994). Its salient features include good control performance for nonlinear systems, and most important, robustness to system parameter variations or external disturbances. However, it often suffers from chattering phenomena due to the discontinuous switching arising from its digital implementation. The chattering phenomena might excite the high-frequency unmodeled dynamics of systems and thus lead to degradation of control performance, or even instability of the closed-loop control systems.

Although a fuzzy controller is shown to be similar to a variant sliding mode controller (Palm, 1993), the key idea of fuzzy sliding mode control is to combine or integrate fuzzy control and sliding mode control in such a way that the advantages of both control techniques can be realized. One typical approach is to equip a sliding mode controller with the capability of handling fuzzy linguistic qualitative information (Chen and Chang, 1998; Glower and Munighan, 1997; Palm, 1994), and a direct benefit of the resulting control scheme is that the chattering often existing in conventional sliding mode control can be effectively eliminated through construction of fuzzy boundary layers instead of crisp switching surfaces (Glower and Munighan, 1997; Ha et al., 2001; Hwang and Tomizuka, 1994). Another approach is to design fuzzy control systems in a normal way for conventional fuzzy control, fuzzy PID control, or model based fuzzy control, and then to add a supervisory sliding mode controller so that not only stability of the closed-loop control systems can be guaranteed but also their robust performance against parameter uncertainties and disturbances can be greatly improved (Feng et al., 1997; Meda-Campana and Castillo-Toledo, 2005; Wang, 1993).

One important advantage of fuzzy sliding mode control is that stability analysis and controller design issues can be addressed within the framework of sliding mode control (Chen and Chang, 1998; Chen, 2001; Hwang and Lin, 1992; Palm, 1994; Su, Chen, and Wang, 2001), and the well-developed techniques of sliding mode control can be effectively applied (Utkin, 1992; Zinober, 1994). Some other topics of interest include using genetic algorithms to tune fuzzy membership functions of fuzzy sliding mode controllers (Lin and Chen, 1997), decoupling of high-dimensional systems into subsystems with lower dimensionality (Lo and Kuo, 1998), use of adaptive fuzzy systems in parameter tuning of sliding mode controllers (Erbatur and Kaynak, 2001), and adaptive fuzzy sliding mode control (Berstecher, Palm, and Unbehauen, 2001; Chen, 2001; Elshafei, 2002; Hwang, 2004; Tao, Taur, and Chan, 2004; Tong and Li, 2003). Kaynak, Erbatur, and Ertugnrl (2001) presented an excellent survey on the fusion of computationally intelligent methodologies, including fuzzy logic and sliding mode control.

1.2.5 ADAPTIVE FUZZY CONTROL

Adaptive control has been one of the most active research areas of the control community in the 1980s and 1990s. It generally refers to the control of partially known systems with some kind of adaptation mechanism, either with respect to unknown system parameters, which leads to so-called indirect adaptive control, or with respect to direct controller parameters, which is referred to as direct adaptive control. Most works in adaptive control are based on the assumption of linear or simplified nonlinear mathematical models of systems to be controlled. In fact, adaptive control of linear systems and certain special classes of nonlinear systems was well developed during the late 1970s to the 1990s (Goodwin and Sin, 1984; Ioannou and Sun, 1995; Krstic, Kanellakopoulos, and Kokotovic 1995), whereas adaptive control of general nonlinear systems still presents a challenge to the control community although great effort has been witnessed in the past decade or so. Nevertheless, mathematical models might not be available for many complex systems in practice, and the adaptive control problem of these systems is far from being satisfactorily resolved.

Following a similar idea in neural networks (Sanner and Slotine, 1992) for their universal function approximation capability, it was shown (Wang and Mendel, 1992) that a fuzzy system is capable of approximating any smooth nonlinear function over a convex compact region. Other excellent works on the topic of function approximation of fuzzy systems can be found in Ying (1998), Zeng and Singh (1994–1996), and Zeng, Zhang, and Xu (2000). Based on this function approximation capability of fuzzy systems, Wang (1993) presented an adaptive fuzzy controller for affine nonlinear systems with unknown functions. Fuzzy basis-function-based fuzzy systems were used to represent those unknown nonlinear functions. The parameters of fuzzy systems including membership functions characterizing linguistic terms in fuzzy rules were updated according to some adaptive laws based on Lyapunov stability theory. Since then, a large volume of work on adaptive fuzzy control has been reported (Andersen, Lotfi, and Tsoi, 1997; Boukezzoula, Galichet, and Foulloy, 2004; Campos and Lewis, 1999; Chen, Lee, and Chang, 1996; Fischle and Schroder,

1999; Guan and Chen, 2003; Han, Su, and Stepanenko, 2001; Koo, 2001; Lee and Tomizuka, 2000; Lee and Zak, 2004; Su, Oya, and Hong, 2003; Velez-Diaz and Tang, 2004; Yang and Ren, 2003; Yang and Zhou, 2005; Zhang and Cai, 2002; Zhang, Cai, and Bien, 2000; Zhou, Feng, and Feng, 2005; Zou, Hou, and Tan, 2008).

A key idea of these works is to use fuzzy systems to approximate unknown non-linear functions in nonlinear systems, to describe the fuzzy systems in the form of linear regression with respect to unknown parameters, and then to apply well-developed adaptive control techniques. However, it should be noted that some kinds of robust approaches in adaptive control such as dead-zone or σ-modification techniques have to be adopted to guarantee the stability of the closed-loop adaptive fuzzy control system due to the inherent approximation errors between the approximating fuzzy models and the original nonlinear functions. It should also be noted that due to the requirement of a compact set for nonlinear function approximation, most likely only semi-global stabilization can be achieved if no supplementary control strategy is employed. Unfortunately this issue has been somehow neglected or mistaken in much of the existing literature.

Other topics of interest include improved adaptive fuzzy control schemes with a smaller number of tuning parameters or better performance (Fischle and Schroder, 1999; Yang and Ren, 2003; Yang and Zhou, 2005), a robust adaptive fuzzy controller with various kinds of performance with respect to external disturbances (Chang, 2001; Chen, Lee, and Chang, 1996), fuzzy model reference adaptive control (Golea, Golea, and Benmahammed, 2002; Koo, 2001; Yin and Lee, 1995), using genetic algorithms to adaptively tune membership functions (Liu, Chen, and Tsao, 2001), and self-organizing schemes to tune fuzzy membership functions (Andersen, Lotfi, and Tsoi, 1997; Lin and Tsai, 2001). Fusion of adaptive techniques and sliding mode control techniques were presented in Berstecher, Palm, and Unbehauen (2001), Chen (2001), Da and Song (2003), Elshafei (2002), and Tao, Taur, and Chan (2004). Comparison of adaptive fuzzy control to conventional adaptive control was reported in Ordonez et al. (1997).

1.2.6 TAKAGI–SUGENO MODEL-BASED FUZZY CONTROL

The Takagi and Sugeno (T–S) fuzzy model, proposed in Takagi and Sugeno (1985) for the purpose of function approximation, can be reformulated as a fuzzy dynamic model to represent nonlinear dynamic systems (Cao, Rees, and Feng, 1995, 1997a,b). This model consists of a set of local linear models that are smoothly connected by nonlinear fuzzy membership functions. This fuzzy modeling method provides an alternative approach to describing complex nonlinear systems (Cao, Rees, and Feng, 1997a; Fantuzzi and Rovatti, 1996; Johansen, Shorten, and Murray-Smith, 2000; Tanaka and Wang, 2001; Ying, 1998; Zeng, Zhang, and Xu, 2000), and drastically reduces the number of rules in modeling higher-order nonlinear systems (Sugeno, 1999). Consequently, T–S fuzzy models are less prone to the curse of dimensionality than other fuzzy models, in particular conventional fuzzy systems. And more important, T–S fuzzy models provide a fundamental basis for development of systematic tools for stability analysis and controller design of fuzzy control systems in view of powerful conventional control theory and techniques.

A great number of theoretical results on function approximation, stability analysis, and controller synthesis have been developed for Takagi–Sugeno fuzzy models during the last 10 years or so. It has been shown that T–S fuzzy models are universal function approximators in the sense that they are capable of approximating any smooth nonlinear functions to any degree of accuracy in any convex compact region (Cao, Rees, and Feng, 1997a; Fantuzzi and Rovatti, 1996; Johansen, Shorten, and Murray-Smith 2000; Tanaka and Wang, 2001; Ying, 1998; Zeng, Zhang, and Xu, 2000). This result provides a theoretical foundation for using T–S fuzzy models to represent complex nonlinear systems approximately, or quite often to represent complex nonlinear systems in practice most accurately. On the basis of this function approximation capability of T–S fuzzy models, a variety of stability analysis and controller design approaches has been developed for T–S fuzzy systems by exploiting the particular structure of these models and conventional control techniques such as Lyapunov stability theory. Based on their fundamental features, the approaches for stability analysis and control synthesis of T–S fuzzy systems can be generally classified into the following six categories:

1. Simple local controller design and stability checking
2. Stability analysis and controller synthesis with or without various performance indices such as H_∞ and H_2 based on a nominal linear model and a single quadratic Lyapunov function
3. Stability analysis and controller synthesis with or without various performance indices based on a common quadratic Lyapunov function
4. Stability analysis and controller synthesis with or without various performance indices based on a piecewise quadratic Lyapunov function
5. Stability analysis and controller synthesis with or without various performance indices based on a fuzzy Lyapunov function
6. Adaptive control when parameters of T–S fuzzy models are unknown

The first category of methods was proposed in the earlier stage of development for T–S fuzzy control systems (Cao, Rees, and Feng, 1995, 1996a, 1999; Feng et al., 1997; Kang and Lee, 1995). The basic idea of this class of methods is to design a feedback controller for each local model, to obtain a global controller by combining all the local controllers in a meaningful way, and then to use some stability criteria to check stability of the resulting closed-loop fuzzy control system. Unfortunately, in this type of method, the design process is not constructive, necessitating many attempts before an acceptable controller design is derived. Thus this kind of method has been out of the mainstream for some time.

The key idea of the second category of methods is to represent a T–S fuzzy model as a nominal linear model with uncertainties, which includes all the nonlinearities of the T–S fuzzy model, around the equilibrium or the operating point of the system, and then to recast the control problem as a robust linear control problem with uncertainties (Feng, 2001; Feng and Ma, 2001; Kim, Ahn, and Kwon, 1995). In this way, many existing robust stability analysis and control synthesis approaches can be directly applied to or further developed for T–S fuzzy systems. Control design for this class of methods is constructive; however, this kind of method tends to be conservative because the existence of one nominal model has to be assumed which might not be the case for many highly complex nonlinear systems, thus it has not become a popular research direction in model based fuzzy control.

The basic idea of the third to the fifth categories of methods is to design a feedback controller for each local model and to construct a global controller from the local controllers in such a way that global stability with or without various performance indices of the closed-loop fuzzy control system is guaranteed. One of the advantages of this class of approach is that control design is constructive. The major techniques that have been used in this category include quadratic stabilization, linear matrix inequalities (LMIs), Lyapunov stability theory, and bilinear matrix inequalities, among others.

To date the third category of methods is the most popular (Akar and Ozguner, 2000; Assawinchaichote, Nguang, and Shi, 2004; Bergsten, Palm, and Driankov, 2002; Cao and Frank, 2000a,b; Cao and Lin, 2003; Chadli, Maquin, and Ragot, 2004; Chang and Chen, 2005; Chen and Liu, 2005; Chen et al., 2008; Chen, Tseng, and Uang, 1999, 2000; S. S. Chen et al., 2005; Choi, 2007; Cuesta et al., 1999; Delmotte, Guerra, and Ksantini, 2007; Dong, Wang, and Yang, 2009; Hong and Langari, 2000a,b; Hsiao et al., 2005; Hu and Huang, 2005; Joh, Chen, and Langari, 1998; Kim and Kim, 2001, 2002; Kim and Lee, 2000, 2005; Kim and Park, 2009; Kiriakidis, 2001; Kiriakos, 1998; Korba et al., 2003; Lam, Leung, and Tam, 2001; Lee, Park, and Chen, 2001; Lee, Kim, and Jeung, 2000; Li and Lin, 2004; Lian et al., 2001; Lian and Liou, 2006; Liang et al., 2008; Liu, Sun, and Sun, 2005; Liu and Zhang, 2003; Lo and Chen, 1999; Lo and Lin, 2004; Ma, Sun, and He, 1998; Nguang and Assawinchaichote, 2003; Tanaka and Sugeno, 1992; Tanaka, Ikeda, and Wang, 1998; Tanaka, Hori, and Wang, 2003; Tao and Taur, 2005; Teixeira, Assuncao, and Avellar, 2003; Teixeira and Zak, 1999; Tseng and Chen, 2001; Tseng, Chen, and Uang, 2001; Tuan et al., 2001, 2004; Wang, Tanaka, and Griffin, 1996; Wang and Lin, 2005; Wang and Luoh, 2004; Wang and Sun, 2005; Wu and Zhang, 2005; Xu and Lam, 2005; Yeh, 1999; Yi and Heng, 2002; Yuan, Li, and Cao, 2008). However, it requires a common quadratic Lyapunov function to be found for all the local models, and this proves to be conservative, even infeasible, in many cases.

As a less conservative alternative, the fourth category, which is based on more general piecewise quadratic Lyapunov functions, has also been well developed and accepted (Cao, Rees, and Feng, 1996b,c, 1997b; Chen, Feng, and Guan, 2005a: Chen et al., 2005b; Feng 2001, 2003, 2004a,b; Feng et al., 2005; Feng and Ma, 2001; Feng and Sun, 2002; Feng and Harris, 2001a,b; Han et al., 2000; Johansson, Rantzer, and Arzen, 1999; Kung, Chen, and Chen, 2005; Ohtake, Tanaka, and Wang, 2003; Taniguchi and Sugeno, 2004; Wang and Feng, 2004; Wang, Feng, and Hesketh, 2004; Zhang, Feng, and Lu, 2007). This category, however, demands more sophisticated techniques and presents more challenges in development and applications.

The fifth category of methods, which is based on fuzzy quadratic Lyapunov functions, has attracted more attention recently but it presents more challenges or difficulties (Choi and Park, 2003; Guerra and Vermeiren, 2004; Tanaka, Hori, and Wang, 2003; Wang, Feng, and Hesketh, 2004; Zhou et al., 2005; Zhou, Lam, and Zheng, 2007), especially for continuous time T–S fuzzy systems.

The sixth category deals with control of T–S fuzzy systems when the model parameters are unknown. Most works to date, however, are quite preliminary in the sense that they only consider unknown parameters in local linear models by assuming that the number of fuzzy rules and membership functions are all known a priori (Feng, 2002; Feng, Cao, and Rees, 2002; Johansen, 1994; Kim, Cho, and Park, 1996; Park and Cho, 2004).

1.3   SUMMARY

A brief review of fuzzy logic control is given in this chapter. Fuzzy logic control can be roughly classified into six categories: (1) conventional fuzzy control, (2) fuzzy proportional-integral-derivative (PID) control, (3) neuro–fuzzy control, (4) fuzzy sliding mode control, (5) adaptive fuzzy control, and (6) Takagi–Sugeno model based fuzzy control. Each category of fuzzy control is then discussed in terms of its development, basic features, advantages, and disadvantages. It is noted that various approaches, in particular, approaches to T–S model-based fuzzy control, provide the possibility of developing systematic tools for the analysis and design of fuzzy control systems, and also that conventional control system theories can be suitably utilized and further developed for analysis and design of such systems.

2 Fuzzy Sets and Fuzzy Systems

2.1 INTRODUCTION

Fuzzy set theory was initiated by L. A. Zadeh with his seminal papers (Zadeh 1965, 1968) in the 1960s. It was introduced as a means of modeling and representing with a precise mathematical theory imprecise uncertainties of the class of real-world objects. This new concept allows one to express imprecise and qualitative information in a precise and quantitative way. One good example of applying fuzzy set theory is to formulate human knowledge, often in the form of natural language, in a systematic and mathematically rigorous manner.

The basic concepts and terminology central to the study of fuzzy set theory and fuzzy systems include fuzzy sets, fuzzy rules, fuzzy relations, fuzzy reasoning/ fuzzy inference, fuzzifiers, defuzzifiers, fuzzy models, fuzzy systems, and so on. For easy reference, those concepts and terminology, and their brief descriptions adapted from Zadeh (1965, 1968, 1973, 1975), Dubois and Prade (1980), Lee (1990a), and Zimmermann (1991), are presented in this chapter. The rest of the chapter is organized as follows. Section 2.2 introduces the basic definitions of fuzzy sets and related concepts. Fuzzy relations and fuzzy IF–THEN rules are discussed in Section 2.3. The issues of fuzzy reasoning and fuzzy models and fuzzy systems are addressed in Sections 2.4 and 2.5, respectively, followed by some conclusions in Section 2.6.

2.2 FUZZY SETS AND RELATED CONCEPTS

A classical set is a set of objects with a crisp boundary. For example, let $\Re$ be the set of real numbers; a classical set S can be expressed as

$$S = \{x \mid x \geq 8, \ x \in \Re\}, \tag{2.1}$$

where 8 is a crisp boundary point such that if x is greater than or equal to this value, then x belongs to the set S, otherwise x does not belong to the set. In contrast to a classical set, a *fuzzy set* is a set without a crisp boundary. In other words, the transition from "belonging to a set" to "not belonging to a set" is gradual, and this gradual transition is characterized by membership functions. This notion of fuzzy sets is very useful and often critical to model commonly used linguistic expressions, such as "the car runs *fast*" or "the temperature is *very high*." Such sets play an important role in human expression and thinking, and offer a mathematically rigorous tool for representing and manipulating qualitative information originally often expressed in a linguistic form.

Definition 2.1 (Fuzzy Sets and Membership Functions)

Let X be a collection of objects, called the universe, and denote its generic element as x. A fuzzy set A in X is defined as a set of ordered pairs:

$$A = \{(x, \mu_A(x)) \mid x \in X\}, \tag{2.2}$$

where $\mu_A(x)$ is called the membership function of x in A. The membership function maps each element of X to a continuous membership value between 0 and 1.

A more convenient notation for a fuzzy set A on X is expressed as

$$A = \sum_{x_i \in X} \mu_A(x_i)/x_i \tag{2.3}$$

if X is discrete, or

$$A = \int_X \mu_A(x)/x \tag{2.4}$$

if X is continuous. The summation and integration signs in Equations (2.3) and (2.4), respectively, stand here for the union of $(x, \mu_A(x))$ pairs. They do not indicate summation or integration. Similarly, "/" is only a marker and does not imply division.

The definition of a fuzzy set can be generally recognized as an extension of the definition of a classical set in the sense that the membership function is allowed to have continuous values between 0 and 1. On the contrary, if the value of the membership function $\mu_A(x)$ is restricted to either 0 or 1, then the fuzzy set reduces to a classical set.

Example 2.1 (Fuzzy Sets with Discrete Universe X)

$$X = \{\text{positive integers}\}.$$

Let

$$A = 0.1/3 + 0.5/6 + 1.0/8 + 0.5/10 + 0.1/13.$$

Then A is a fuzzy set of "*integers close to 8*".

Example 2.2 (Fuzzy Sets with Continuous Universe X)

$$X = \{\text{real numbers}\}.$$

Let

$$A = \int_X \frac{1}{1 + \left[\frac{1}{2}(x - 8)\right]^2}/x.$$

Then A is a fuzzy set of "*real numbers clustered around 8*".

Remark 2.1

The object that a fuzzy set is used to characterize is usually "fuzzy" in the sense that the description is not precise, for example, *"integers close to 8"* or *"real numbers clustered around 8."* However, with the definition of a fuzzy set with a particular fuzzy membership function, which is a precise mathematical function and not fuzzy at all, then the described fuzzy object becomes precise. In other words, once a fuzzy object is represented by a fuzzy membership function, nothing will be fuzzy anymore. Instead, everything will be precise.

Remark 2.2

It can be easily seen from the definition of a fuzzy set that the specification of an appropriate membership function is essential in the determination of a fuzzy set. However, it should be noted that the specification of an appropriate membership function is quite subjective in the sense that the membership function specified for the same concept, for example, *"fast"* or *"high,"* by different people may vary considerably. This subjectivity originates from the indefinite nature of abstract concepts rather than randomness. In fact, the subjectivity and nonrandomness of fuzzy sets are the primary difference between fuzzy set theory and probability theory, which deals with the objective treatment of random phenomena.

Several typical commonly used membership functions are given as follows. These parameterized membership functions play an important role in the study of fuzzy set theory and fuzzy systems.

A triangular membership function is specified by three parameters (a, b, c) as follows,

$$triangle(x; a, b, c) = \max\left(\min\left(\frac{x-a}{b-a}, \frac{c-x}{c-b} \right), 0 \right). \tag{2.5}$$

An example of the triangular membership function defined by $triangle(x; 20, 60, 80)$ is shown in Figure 2.1a.

A trapezoidal membership function is specified by four parameters (a, b, c, d) as follows,

$$trapezoid(x; a, b, c, d) = \max\left(\min\left(\frac{x-a}{b-a}, 1, \frac{d-x}{d-c} \right), 0 \right). \tag{2.6}$$

An example of the trapezoidal membership function defined by $trapezoid(x; 10, 20, 60, 80)$ is illustrated in Figure 2.1b. Obviously, the triangular membership function is a special case of the trapezoidal membership function.

These two forms of membership functions have been widely used in the study of fuzzy sets and fuzzy systems because of their simple expressions and computational efficiency. However, because these membership functions consist of straight line segments, they are not smooth at the turning points between line segments, which might lead to difficulty in some applications, for example, in applications involving derivatives of membership functions. In such cases, smooth membership functions are desirable and the following are typical examples of such membership functions.

A Gaussian membership function is specified by two parameters (σ, c) as follows,

$$Gaussian(x; \sigma, c) = e^{-\left(\frac{x-c}{\sigma}\right)^2}, \tag{2.7}$$

where c and σ represent the center and width of the membership function, respectively. An example of the Gaussian membership function defined by $Gaussian(x; 20, 50)$ is illustrated in Figure 2.1c.

A generalized bell-shaped membership function is specified by three parameters (a, b, c) as follows,

$$bell(x; a, b, c) = \frac{1}{1 + \left|\frac{x-c}{a}\right|^{2b}}, \tag{2.8}$$

where c and a represent the center and width, respectively, of the membership function, and the parameter b is usually positive. An example of the bell-shaped membership function defined by $bell(x; 20, 4, 50)$ is illustrated in Figure 2.1d. A desired generalized bell-shaped membership function can be obtained by a proper selection of the parameter set (a, b, c). Specifically, we can adjust c and a to vary the center and width of the membership function, and then use b to control the slopes at the crossover points. Figure 2.2 illustrates the physical meanings of each parameter in a bell-shaped membership function.

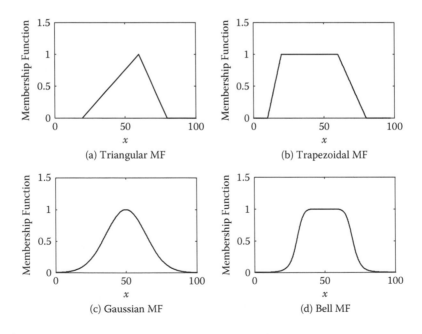

FIGURE 2.1   Examples of various classes of membership functions.

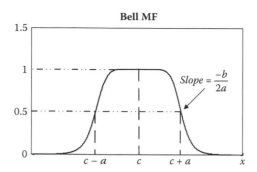

FIGURE 2.2 Physical meaning of parameters in a generalized bell-shaped function.

It should be noted that the list of membership functions described above is by no means exhaustive. Many other membership functions can be created for specific applications.

In addition to membership functions, the following related concepts are also useful and needed in the study of fuzzy sets and fuzzy systems.

Definition 2.2 (Support of a Fuzzy Set)

The *support of a fuzzy set A* in the universe of discourse X is the crisp set defined as follows,

$$S_A := \{x \in X \mid \mu_A(x) > 0\}. \tag{2.9}$$

Example 2.3

Consider the support of the fuzzy set A in Example 2.1. Because $A = 0.1/3 + 0.5/6 + 1.0/8 + 0.5/10 + 0.1/13$, it can be easily observed that the support of the fuzzy set A is $S_A = \{3, 6, 8, 10, 13\}$.

Definition 2.3 (Fuzzy Singleton)

The *fuzzy singleton* is a fuzzy set whose support is a single point in X.

Example 2.4

Reconsider the universe of discourse in Example 2.1: $X = \{\text{positive integers}\}$. Let $B = 1/8$. Then the fuzzy set B is a fuzzy singleton with its support $S_B = \{8\}$, a single point in X.

Definition 2.4 (Center of a Fuzzy Set)

The *center of a fuzzy set A* is the mean value of all points $x \in X$ such that $\mu_A(x)$ achieves its maximum value if the mean value is finite. If the mean value of a fuzzy set is positive (or negative) infinity, then the center is defined as the smallest (largest) among all points that achieve the maximum membership value.

Example 2.5

Reconsider the fuzzy set A defined in Example 2.1. Because $A = 0.1/3 + 0.5/6 + 1.0/8 + 0.5/10 + 0.1/13$, the center of the set A is 8.

Definition 2.5 (Normality of a Fuzzy Set)

A fuzzy set A is *normal* if the largest membership function value attained by any element $x \in X$ equals 1.

It can be observed that all the fuzzy sets in Figure 2.1 are normal fuzzy sets.

Definition 2.6 (α-Level Set and Strong α-Level Set)

An α-*cut or α-level set* of a fuzzy set A is a crisp set that contains all the elements in X that have membership function values in A greater than or equal to α; that is,

$$A_\alpha = \{x \in X \mid \mu_A(x) \geq \alpha\}. \tag{2.10}$$

A strong α-cut or strong α-level set is defined as

$$A_\alpha = \{x \in X \mid \mu_A(x) > \alpha\}. \tag{2.11}$$

Example 2.6

Let $\alpha = 0.5$. Reconsider the fuzzy set A defined in Example 2.1. Because $A = 0.1/3 + 0.5/6 + 1.0/8 + 0.5/10 + 0.1/13$, the α-level set $A_\alpha = \{6, 8, 10\}$, and the strong α-level set $A_\alpha = \{8\}$.

Definition 2.7 (Convexity of a Fuzzy Set)

A fuzzy set A in X is *convex* if and only if

$$\mu_A(\lambda x_1 + (1-\lambda)x_2) \geq \min(\mu_A(x_1), \mu_A(x_2)) \tag{2.12}$$

for all $x_1, x_2 \in X$ and all $\lambda \in [0, 1]$.

The convexity of a fuzzy set can also be defined based on its α-cut set; that is, a fuzzy set A in X is convex if and only if its α-cut set is a convex set for any α in the interval $(0, 1)$.

Definition 2.8 (An Empty Fuzzy Set)

A fuzzy set is *empty* if and only if its membership function is identically zero on X.

Definition 2.9 (Containment of a Fuzzy Set)

A fuzzy set A is *contained* in a fuzzy set B (or, equivalently, A is a subset of B, or A is smaller than or equal to B) if and only if $\mu_A(x) \leq \mu_B(x)$ for all x, denoted as

$$A \subseteq B. \tag{2.13}$$

Example 2.7

Reconsider Example 2.1. Let $A = 0.1/3 + 0.5/6 + 1.0/8 + 0.5/10 + 0.1/13$, and $B = 0.2/3 + 0.6/6 + 1.0/8 + 0.5/10 + 0.1/13$. Then it can be observed that the fuzzy set A is contained in the fuzzy set B; that is, $A \subseteq B$.

Similar to ordinary sets, the operations of complement, union, and intersection can also be defined for fuzzy sets.

Definition 2.10 (Complement of a Fuzzy Set)

The *complement* of a fuzzy set A is denoted by $\bar{A}$, whose membership function is defined as

$$\mu_{\bar{A}}(x) = 1 - \mu_A(x). \tag{2.14}$$

Definition 2.11 (Union of Fuzzy Sets)

The *union* of two fuzzy sets A and B is a fuzzy set C, written as $C = A \cup B$, whose membership function is defined as

$$\mu_C(x) = \max(\mu_A(x), \mu_B(x)) \tag{2.15}$$

or, in abbreviated form,

$$\mu_C(x) = \mu_A(x) \vee \mu_B(x). \tag{2.16}$$

Definition 2.12 (Intersection of Fuzzy Sets)

The *intersection* of two fuzzy sets A and B is a fuzzy set C, written as $C = A \cap B$, whose membership function is defined as

$$\mu_C(x) = \min(\mu_A(x), \mu_B(x)) \tag{2.17}$$

or, in abbreviated form,

$$\mu_C(x) = \mu_A(x) \wedge \mu_B(x). \tag{2.18}$$

Remark 2.3

As pointed out by Zadeh (1965), a more intuitive and appealing definition of the union of fuzzy sets A and B is the smallest fuzzy set containing both A and B. Similarly, the intersection of fuzzy sets A and B is the largest fuzzy set that is contained in both A and B.

Basic operations of fuzzy sets defined in Definitions 2.10–2.12 are illustrated in Figure 2.3 with an example of two fuzzy sets: (a) two fuzzy sets A and B, (b) the complement of A, (c) the union of A and B, and (d) the intersection of A and B.

Definitions 2.11 and 2.12 describe one possible choice of operators for fuzzy OR (union) and fuzzy AND (intersection), respectively. However, the max and min operations do incur some difficulties in the analysis of fuzzy inference systems in some cases. The following definitions are more popular alternatives for fuzzy AND and fuzzy OR operation, respectively,

$$\mu_{A \cap B}(x) = \mu_A(x)\mu_B(x), \tag{2.19}$$

$$\mu_{A \cup B}(x) = \mu_A(x) + \mu_B(x) - \mu_A(x)\mu_B(x). \tag{2.20}$$

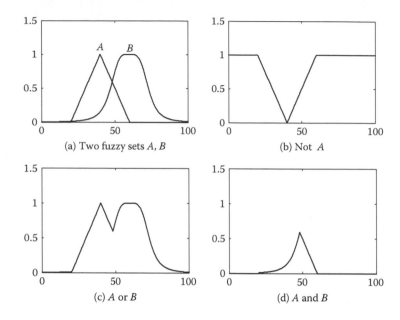

(a) Two fuzzy sets A, B

(b) Not A

(c) A or B

(d) A and B

FIGURE 2.3   Operations on fuzzy sets.

In addition, some other operators have been proposed in the literature under the names of S-norm (or T-conorm) and T-norm operators (Dubois and Prade, 1980) for union and intersection operations of fuzzy sets, respectively. Table 2.1 of S-norm and T-norm operators was listed in Cheng (1998).

TABLE 2.1

Common S-Norm and T-Norm Operators[a]

S-Norm	T-Norm
Maximum: $S_{\max}(a, b) = \max\{a, b\}$	Minimum: $T_{\min}(a, b) = \min\{a, b\}$
Algebraic sum: $S_{as}(a, b) = a + b - ab$	Algebraic product: $T_{ap}(a, b) = ab$
Bounded sum:	Bounded product:
$S_{bs}(a, b) = \min\{1, a + b\}$	$T_{bp}(a, b) = \max\{0, a + b - 1\}$
Drastic sum:	Drastic product:
$S_{ds}(a,b) = \begin{cases} a & \text{if} \quad b = 0 \\ b & \text{if} \quad a = 0 \\ 1 & \text{if} \quad a, b > 0 \end{cases}$	$T_{dp}(a,b) = \begin{cases} a & \text{if} \quad b = 1 \\ b & \text{if} \quad a = 1 \\ 1 & \text{if} \quad a, b < 1 \end{cases}$

[a] $a, b \in [0, 1]$ denote membership functions.

2.3 FUZZY RELATIONS AND FUZZY IF–THEN RULES

Fuzzy relations are one of the most important concepts in fuzzy set theory. They represent a degree of presence or absence of association or interaction between the elements of two or more fuzzy sets. A classical relation among a group of classical sets $X_1, X_2, ..., X_n$ is defined as a subset of the Cartesian product $X_1 \times X_2 \times \cdots \times X_n$. In the context of fuzzy sets, an n-ary fuzzy relation is a fuzzy set R in the Cartesian product space $X_1 \times X_2 \times \cdots \times X_n$.

Definition 2.13 (Fuzzy Relation)

Let $X_1 \times X_2 \times \cdots \times X_n$ be the universe of discourse. An n-ary *fuzzy relation* R in the Cartesian product space $X_1 \times X_2 \times \cdots \times X_n$ is defined as

$$R(x) - \{(x, \mu_R(x)) \mid x = (x_1, ..., x_n) \in X_1 \times X_2 \times \cdots \times X_n\}, \quad (2.21)$$

where $\mu_R : X_1 \times X_2 \times \cdots \times X_n \to [0, 1]$.

Example 2.8

Let $X_1 = X_2 = \Re$ be the set of real numbers. A fuzzy relation of "x_1 is much smaller than x_2" can be defined by the following membership function,

$$\mu_{Much\_smaller} = \frac{1}{1 + e^{-100(x_2 - x_1)}}. \quad (2.22)$$

Similarly, a fuzzy relation of "x_1 is about twice x_2" can be defined by the membership function as follows,

$$\mu_{About\_twice} = e^{-10(x_1 - 2x_2)^2}. \quad (2.23)$$

A fuzzy relation enables one to express ambiguous and qualitative relationships and to link two fuzzy sets in a prespecified manner. Furthermore, fuzzy relations in different product spaces can be combined with each other through a *composition* operation. A general composition operation is defined as follows.

Definition 2.14 (Sup-Star Composition of Fuzzy Relations)

Let $R_1(x, y)$, $(x, y) \in X \times Y$ and $R_2(y, z)$, $(y, z) \in Y \times Z$ be two fuzzy relations. The composition of the fuzzy relations, denoted by $R_1 \circ R_2$, is defined as a fuzzy relation in $X \times Z$ whose membership function is given by

$$\mu_{R_1 \circ R_2}(x, z) = \sup_{y \in Y}(\mu_{R_1}(x, y) * \mu_{R_2}(y, z)), \quad x \in X, y \in Y, z \in Z, \quad (2.24)$$

where $(x, z) \in X \times Z$, $\circ$ is the *sup-star composition* operator and * is any T-norm operator.

The following are two of the most commonly used compositions for fuzzy relations.

Definition 2.15 (Sup-Min Composition)

Let $R_1(x, y)$, $(x, y) \in X \times Y$ and $R_2(y, z)$, $(y, z) \in Y \times Z$ be two fuzzy relations. The *sup-min composition* of R_1 and R_2 is a fuzzy set defined by the membership function

$$\mu_{R_1 \circ R_2}(x, z) = \sup_{y \in Y} \min(\mu_{R_1}(x, y), \mu_{R_2}(y, z)), \quad x \in X, y \in Y, z \in Z, \quad (2.25)$$

where $(x, z) \in X \times Z$.

Definition 2.16 (Sup-Product Composition)

Let $R_1(x, y)$, $(x, y) \in X \times Y$ and $R_2(y, z)$, $(y, z) \in Y \times Z$ be two fuzzy relations. The *sup-product composition* of R_1 and R_2 is a fuzzy set defined by the membership function

$$\mu_{R_1 \circ R_2}(x, z) = \sup_{y \in Y} (\mu_{R_1}(x, y) \mu_{R_2}(y, z)), \quad x \in X, y \in Y, z \in Z, \quad (2.26)$$

where $(x, z) \in X \times Z$.

After introduction of fuzzy relations, fuzzy rules can be described as follows. A fuzzy IF–THEN rule, also known as a *fuzzy rule*, *fuzzy implication*, or *fuzzy conditional statement*, has the following form,

$$\text{IF } x \text{ is } A \text{ THEN } y \text{ is } B, \quad (2.27)$$

where A and B are fuzzy sets often of linguistic values on universes of discourse X and Y, respectively. The IF-part "x is A" is normally called the *antecedent* or *premise* and the THEN-part "y is B" is normally called the *consequence* or *conclusion*. Examples of fuzzy IF–THEN rules are abundant in our daily linguistic expressions. For instance, if pressure is high then volume is small, or if temperature is high then power assumption is large.

The fuzzy rule (2.27) can be abbreviated as $A \to B$, which is also called the *fuzzy implication*. Basically, the expression describes a relation between two variables x and y. This fuzzy implication can be defined as a fuzzy relation R on the product space $X \times Y$. The membership function $\mu_R(x, y) \in [0, 1]$ describes the degree of truth of the implication relation between x and y. Alternatively, the fuzzy relation R can be viewed as the fuzzy set with universe $X \times Y$ and a two-dimensional membership function $\mu_R(x, y)$.

Generally, there are two ways to interpret the fuzzy rule $A \to B$. One way to interpret the implication $A \to B$ is that A is coupled with B, and in this case,

$$R = A \to B = \int_{X \times Y} \mu_A(x) * \mu_B(y) / (x, y), \quad (2.28)$$

where $*$ is any T-norm operation. The other interpretation of implication $A \to B$ is that A entails B, and in this case it can be written as

$$R = A \to B = \bar{A} \cup B. \quad (2.29)$$

2.4   FUZZY REASONING

Fuzzy reasoning (also known as *approximate reasoning*) is an inference procedure used to derive conclusions from a set of fuzzy IF–THEN rules and one or more premises. It is a generalization of classical reasoning with typical traditional inference rules such as modus ponens, modus tollens, and hypothetical syllogism. For the rule of modus ponens, one can infer the truth of a proposition B from the truth of A and the implication $A \rightarrow B$. This procedure can be illustrated as follows.

Premise 1 (fact):	x is A,
Premise 2 (rule):	IF x is A THEN y is B, (2.30)
Consequence (conclusion):	y is B.

However, in most cases of human reasoning, modus ponens is often employed in an approximate manner. This can be written as

Premise 1 (fact):	x is A',
Premise 2 (rule):	IF x is A THEN y is B, (2.31)
Consequence (conclusion):	y is B',

where A' is close to A and B' is close to B. When A, B, A' and B' are fuzzy sets of appropriate universes, the above inference procedure reveals the fundamental principles in fuzzy logic or fuzzy reasoning. This procedure is called *generalized modus ponens* because it has modus ponens as its special case.

In order to define the fuzzy reasoning precisely, the following inference composition rule concept is needed to compute the membership function of the resulting fuzzy set B'.

Definition 2.17 (Compositional Rule of Inference)

Let A, A', and B be fuzzy sets of universes X, X, and Y, respectively. Assume that the fuzzy implication $A \rightarrow B$ is expressed as a fuzzy relation R on $X \times Y$. Then the fuzzy set B' is inferred with its membership function as

$$\mu_{B'}(y) = \sup_{x \in X} \left(\mu_{A'}(x) * \mu_{A \rightarrow B}(x, y) \right), \tag{2.32}$$

where $*$ is any T-norm operator. The compositional rule of inference is also called the sup-star composition.

It is noted that the above fuzzy reasoning can also be expressed as

$$B' = A' \circ R = A' \circ (A \rightarrow B), \tag{2.33}$$

where $\circ$ is the sup-star composition operator.

The equation (2.32) is a general expression for fuzzy reasoning. If one uses min and max as the operators for fuzzy AND and OR, respectively, the fuzzy set B' is induced with its membership function given as

$$\mu_{B'}(y) = \max_{x \in X} \min(\mu_{A'}(x), \mu_R(x, y)). \tag{2.34}$$

Now the inference procedure of the generalized modus ponens can be used to derive conclusions, provided that the fuzzy implication $A \rightarrow B$ is defined as an appropriate binary fuzzy relation. The following two cases are considered.

Case 1: Single Fuzzy Rule
For a single rule with a single antecedent, the following formula can be easily obtained,

$$
\begin{aligned}
\mu_{B'}(y) &= \max_{x \in X} \min(\mu_{A'}(x), \mu_R(x, y)) \\
&= \vee_x [\mu_{A'}(x) \wedge \mu_{A \rightarrow B}(x, y)] \\
&= [\vee_x (\mu_{A'}(x) \wedge \mu_A(x))] \wedge \mu_B(y) \\
&= w \wedge \mu_B(y),
\end{aligned} \tag{2.35}
$$

where w is the degree of match between A and A'.

In the inference procedure, one first finds the degree of match w as the maximum of $\mu_{A'}(x) \wedge \mu_A(x)$; then the membership function of the resulting B' is equal to the membership function of B clipped by w, illustrated in Figure 2.4.

A single fuzzy rule with two antecedents is expressed as

Premise 1 (fact):	x is A' and y is B',
Premise 2 (rule):	IF x is A and y is B THEN z is C, (2.36)
Consequence (conclusion):	z is C'.

The fuzzy rule in Premise 2 in (2.36) can be transformed into a ternary fuzzy relation $R(A \times B \rightarrow C)$ which is specified by the following membership function,

$$\mu_R(x, y, z) = \mu_{(A \times B) \times C}(x, y, z) = \mu_A(x) \wedge \mu_B(y) \wedge \mu_C(z).$$

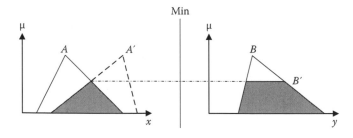

FIGURE 2.4 Fuzzy reasoning for a single rule with a single antecedent.

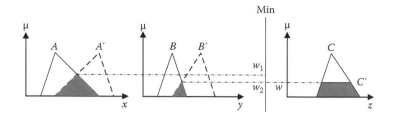

FIGURE 2.5 Approximate reasoning for two antecedents.

The resulting C' is expressed as

$$C' = (A' \times B') \circ (A \times B \rightarrow C),$$

and its membership function is obtained as

$$\mu_{C'}(z) = \vee_{x,y} [\mu_{A'}(x) \wedge \mu_{B'}(y)] \wedge [\mu_A(x) \wedge \mu_B(y) \wedge \mu_C(z)]$$

$$= \vee_{x,y} [\mu_{A'}(x) \wedge \mu_{B'}(y) \wedge \mu_A(x) \wedge \mu_B(y)] \wedge \mu_C(z)$$

$$= \{\vee_x [\mu_{A'}(x) \wedge \mu_A(x)]\} \wedge \{\vee_y [\mu_{B'}(y) \wedge \mu_B(y)]\} \wedge \mu_C(z) \qquad (2.37)$$

$$= w_1 \wedge w_2 \wedge \mu_C(z),$$

where w_1 is the degree of match between A and A', w_2 is the degree of match between B and B', and $w_1 \wedge w_2$ is called the firing strength or degree of fulfillment of this fuzzy rule. The membership function of the resulting C' is equal to the membership function of C clipped by $w_1 \wedge w_2$, illustrated in Figure 2.5. The generalization to more than two antecedents is straightforward.

Case 2: Multiple Fuzzy Rules

The interpretation of multiple rules is usually taken as the union of the fuzzy relations corresponding to the fuzzy rules. In general, the above fuzzy reasoning mechanism can be extended to multiple rules with multiple-antecedent single-consequence.

For instance, given the following facts and rules,

Premise 1 (fact):	x is A' and y is B',
Premise 2 (rule 1):	IF x is A_1 and y is B_1 THEN z is C_1, $\quad$ (2.38)
Premise 3 (rule 2):	IF x is A_2 and y is B_2 THEN z is C_2,
Consequence (conclusion):	z is C'.

One can employ the fuzzy reasoning, shown in Figure 2.6, as an inference procedure to derive the resulting output fuzzy set C'. To verify this inference procedure,

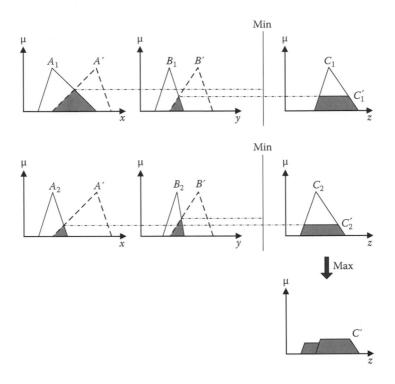

FIGURE 2.6   Fuzzy reasoning for multiple rules with multiple antecedents.

let $R_1 = A_1 \times B_1 \rightarrow C_1$ and $R_2 = A_2 \times B_2 \rightarrow C_2$. Because the max–min composition operator $\circ$ is distributive over the $\cup$ operator, it follows that

$$C' = (A' \times B') \circ (R_1 \cup R_2)$$

$$= [(A' \times B') \circ R_1] \cup [(A' \times B') \circ R_2] \qquad (2.39)$$

$$= C_1' \cup C_2',$$

where C_1' and C_2' are the inferred fuzzy sets for Rules 1 and 2, respectively.

Furthermore, the interpretation of multiple rules with multiple-antecedent multiple-consequence can be treated as a group of multiple rules with multiple-antecedent single-consequence.

2.5   FUZZY MODELS AND FUZZY SYSTEMS

Fuzzy models or fuzzy systems are rule based or knowledge based models or systems originating from the concepts of fuzzy sets, fuzzy IF–THEN rules, and fuzzy reasoning. A typical fuzzy system, especially in applications of fuzzy logic control, consists of four components: knowledge base or fuzzy rule base, inference engine, fuzzification interface, and defuzzification interface (Lee, 1990a). Figure 2.7 shows the block diagram of the structure of a fuzzy system.

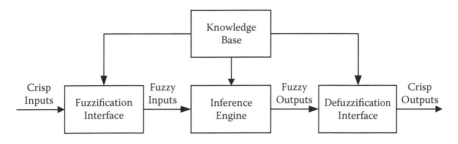

FIGURE 2.7 Basic structure of fuzzy systems.

The knowledge base or fuzzy rule base contains a set of fuzzy IF–THEN rules and a database. The database is the declarative part of the knowledge base that describes the definition of the objects (facts, terms) and the definition of the membership functions used in the fuzzy rules. The fuzzy rules are the procedural part of the knowledge base that contains information on how these objects can be used to infer conclusions.

The inference engine, which is the central component of a fuzzy system, is a reasoning mechanism in the form of fuzzy reasoning described in Section 2.4, and performs the inference procedure upon the fuzzy rules and given conditions to derive conclusions.

The fuzzification interface (or fuzzifier) is a mechanism to transform a real-valued variable into a fuzzy set. It is well known that a crisp value of an input can be transformed into a fuzzy set in the form of a fuzzy singleton that has zero membership function value everywhere except at the particular point where the membership function value is unity. The fuzzification interface employing the technique of fuzzy singleton has been widely adopted in fuzzy control applications. In fact, the observed inputs in typical fuzzy control applications are always of crisp values. The fuzzification interface scales these crisp values, transforms the range of values of the input variables to their corresponding universes of discourse, and then converts the scaled input values into suitable linguistic values that can be defined as a fuzzy set (fuzzy singleton). This method of fuzzification is efficient and simple, and easily implemented. This typical process of fuzzification, often used in fuzzy control applications, is usually called the singleton fuzzification and the corresponding interface is called the singleton fuzzifier.

On the other hand, the defuzzification interface (or defuzzifier) is a mechanism to transform a fuzzy set over an output universe of discourse to a real-valued variable. The objective of defuzzification is to extract a crisp value that best interprets a fuzzy set. The most frequently used defuzzifiers include the center of gravity (or the center of area), center average, and maximum defuzzifiers, among many others.

The center of gravity is defined as

$$z^* = \frac{\int_Z \mu_{C'}(z) z \, dz}{\int_Z \mu_{C'}(z) \, dz}, \tag{2.40}$$

where $\mu_{C'}(z)$ is the aggregated output membership function. The advantage of the center of gravity defuzzifier is its intuitive plausibility, whereas the disadvantage is its high computation cost. To reduce the computational burden of this method, other defuzzification strategies have been developed for specific applications, such as bisector of area, mean of maximum, largest of maximum, and so on. Some other defuzzification methods can be found in Pfluger, Yen, and Langari (1992) and Yager and Filev (1993).

With crisp inputs and outputs, it has been shown that a fuzzy system is in fact a nonlinear mapping from its input space to its output space. More specifically, a fuzzy system describes a transformation from a set of linguistic fuzzy IF–THEN rules, which are often described in natural language, into a mathematical nonlinear mapping, which is often described by a precise mathematical function. It is this attractive and useful feature of fuzzy systems that leads to their wide application in a variety of disciplines, in particular, in fuzzy control applications.

The following three types of fuzzy systems have been widely employed in fuzzy control applications. It should be noted that the main differences among these three fuzzy systems lie in the consequences of their fuzzy rules.

2.5.1 MAMDANI FUZZY SYSTEMS

The Mamdani fuzzy system (Mamdani, 1974, 1976; Mamdani and Assilian, 1975) was proposed as the first attempt to control a steam engine and boiler. A typical fuzzy rule in a Mamdani fuzzy system has the form

$$\text{IF} \quad x_1 \text{ is } A_1, \ldots, \text{ and } x_k \text{ is } A_k,$$

$$\text{THEN} \quad y \text{ is } B, \tag{2.41}$$

where $x = (x_1, \ldots, x_k)$ and y are linguistic variables, $A_1, \ldots, A_k$ are fuzzy sets in the antecedent, and B is a fuzzy set in the consequence, respectively.

A typical fuzzy reasoning of the Mamdani fuzzy system is based on the max–min inference method. Figure 2.8 is an illustration of such fuzzy inference for the Mamdani fuzzy system with two IF–THEN rules.

If different T-norm operators for fuzzy AND and S-norm operators for fuzzy OR, respectively, are adopted for fuzzy reasoning, other types of Mamdani fuzzy systems can be obtained.

2.5.2 TAKAGI–SUGENO FUZZY SYSTEMS

An alternative type of fuzzy systems, known as Takagi–Sugeno (T–S) fuzzy systems, was proposed in Takagi and Sugeno (1985), in an effort to develop a systematic approach to approximating a nonlinear function. A typical fuzzy rule in a T–S fuzzy system has the form

$$\text{IF} \quad x_1 \text{ is } A_1, \ldots \text{ and } x_k \text{ is } A_k,$$

$$\text{THEN} \quad y = f(x_1, \ldots, x_k), \tag{2.42}$$

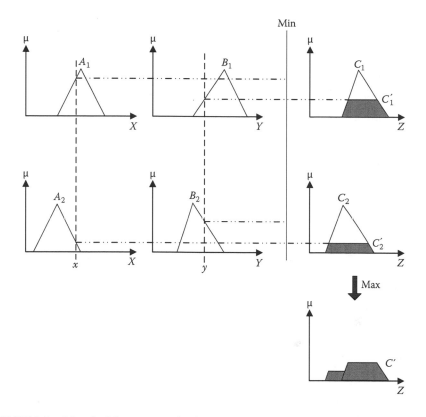

FIGURE 2.8 Mamdani fuzzy reasoning based on max–min inference method.

where $x = (x_1, ..., x_k)$ and y are linguistic variables, $A_1, ..., A_k$ are fuzzy sets in the antecedent, and $y = f(x_1, ..., x_k)$ is a polynomial in the input variable x, but can be any function as long as it can appropriately describe the output of the system within the region specified by the antecedent of the rule. When $f(x_1, ..., x_k)$ is a first-order polynomial, the resulting fuzzy model is called the first-order T–S fuzzy system, which was originally proposed in Takagi and Sugeno (1985) and Sugeno and Kang (1988). If f is a constant, one then has the zero-order T–S fuzzy system, which can be viewed as a special case of the Mamdani fuzzy system, where each rule's consequence is specified by a fuzzy singleton.

The output of a T–S fuzzy system is obtained by the weighted average of the crisp outputs of fuzzy rules. This can avoid the time-consuming procedure of defuzzification. Figure 2.9 gives an example of graphical interpretation of fuzzy reasoning for a T–S fuzzy system with two rules.

2.5.3 FUZZY DYNAMIC SYSTEMS

The original Takagi–Sugeno fuzzy model is not a true dynamic model and thus cannot be adequately utilized for control design because the original model was developed mainly for the purpose of function approximation. Inspired by the original

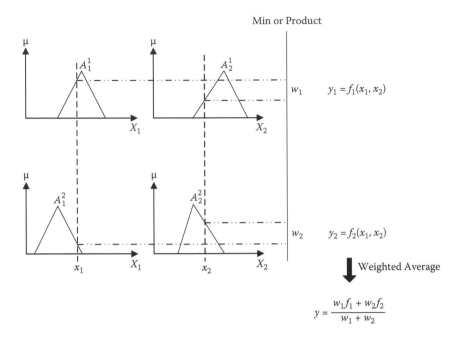

FIGURE 2.9   Takagi and Sugeno's fuzzy inference system.

Takagi–Sugeno fuzzy model, Cao, Rees, and Feng (1997a), Cao and Feng (1995), and Cao and Rees (1995a,b) proposed a new type of fuzzy dynamic model for modeling and control of complex nonlinear dynamic systems. This model can be recognized as an extension of the Takagi–Sugeno fuzzy model in the sense that the T–S model is more function approximation-oriented whereas the fuzzy dynamic model is more systems- and control-oriented. However, fuzzy dynamic systems have frequently been referred to as Takagi–Sugeno fuzzy systems recently in the literature, and thus they are used interchangeably in the rest of this book. The fuzzy dynamic model or T–S fuzzy model consists of a family of local linear dynamic models smoothly connected through fuzzy membership functions. The fuzzy rules of the fuzzy dynamic model have the form

$$R^l: \quad \text{IF} \quad z_1 \text{ is } F_1^l \text{ and } \ldots z_v \text{ is } F_v^l$$

$$\text{THEN} \quad x(t+1) = A_l x(t) + B_l u(t) + a_l \tag{2.43}$$

$$y(t) = C_l x(t)$$

$$l \in L := \{1, 2, \ldots, m\},$$

where R^l denotes the lth fuzzy inference rule, m the number of inference rules, $F_j^l (j = 1, 2, \ldots, v)$ the fuzzy sets, $x(t) \in \mathfrak{R}^n$ the state vector, $u(t) \in \mathfrak{R}^g$ the input vector, $y(t) \in \mathfrak{R}^p$ the output vector, and (A_l, B_l, a_l, C_l) the matrices of the lth local model, and $z(t) := [z_1, z_2, \cdots, z_v]$ the premise variables, which are some measurable variables of the system, for example, the state variables or the output variables.

The final state $x(t)$ of the system is inferred by taking the weighted average of all local models, and the final output $y(t)$ of the system is inferred by taking the weighted average of the output $y(t)$s of all the local models. By using a standard fuzzy inference method, that is, using a singleton fuzzifier, product fuzzy inference, and center-average defuzzifier, the fuzzy dynamic model in (2.43) can be rewritten as (Cao, Rees, and Feng, 1997a; Cao and Feng, 1995)

$$x(t+1) = A(\mu)x(t) + B(\mu)u(t) + a(\mu)$$

$$y(t) = C(\mu)x(t),$$

(2.44)

where

$$A(\mu) = \sum_{l=1}^{m} \mu_l A_l, \; B(\mu) = \sum_{l=1}^{m} \mu_l B_l, \; a(\mu) = \sum_{l=1}^{m} \mu_l a_l, \; C(\mu) = \sum_{l=1}^{m} \mu_l C_l, \quad (2.45)$$

$\mu_l(z)$ is the normalized membership function satisfying

$$\mu_l = \frac{\xi_l(z)}{\sum_{i=1}^{m} \xi_i(z)}, \; \xi_l(z) = \prod_{i=1}^{v} F_i^l(z_i), \; \mu_l \geq 0, \; \sum_{l=1}^{m} \mu_l = 1, \quad (2.46)$$

and $F_i^l(z_i)$ is the grade of membership of z_i in the fuzzy set F_i^l.

In the rest of this book, our focus is on exploiting the fuzzy dynamic model or the so-called T–S fuzzy model (2.43) to develop model based approaches to stability analysis and controller synthesis of fuzzy control systems.

2.6 CONCLUSIONS

In this chapter, we briefly introduce the basic concepts and terminology of fuzzy set theory and fuzzy systems, including fuzzy sets, fuzzy relations, fuzzy rules, fuzzy reasoning, fuzzifiers, defuzzifiers, and fuzzy models. These basic concepts provide the fundamental basis for our study in this book, in particular, on stability analysis and controller synthesis of model-based fuzzy control systems.

3 T–S Fuzzy Modeling and Identification

3.1 INTRODUCTION

Takagi–Sugeno fuzzy models or the so-called fuzzy dynamic models have been widely used to represent complex nonlinear systems. These models are particularly useful when the mechanism of a complex nonlinear system is not clear or its dynamics are not easy to establish. In this chapter, we discuss the detailed description of T–S fuzzy models, their approximation properties as nonlinear functions, and their identification procedures from given nonlinear models or from data generated from the underlying systems.

The rest of the chapter is organized as follows. Section 3.2 presents a basic definition of T–S fuzzy models or T–S fuzzy systems and their important properties. Their universal function approximating property is discussed in Section 3.3. T–S fuzzy model identification from nonlinear models and from data is addressed in Sections 3.4 and 3.5, respectively. Approximation error analysis of T–S fuzzy systems to represent nonlinear systems is carried out in Section 3.6, followed by conclusions in Section 3.7.

3.2 T–S FUZZY MODELS

T–S fuzzy models consist of both fuzzy inference rules and local analytic linear dynamic models as follows,

$$R^l: \quad \text{IF} \quad z_1 \text{ is } F_1^l \text{ and } \ldots z_v \text{ is } F_v^l$$

$$\text{THEN} \quad x(t+1) = A_l x(t) + B_l u(t) + a_l \tag{3.1}$$

$$y(t) = C_l x(t)$$

$$l \in L := \{1, 2, \ldots, m\},$$

where R^l denotes the lth fuzzy inference rule, m the number of inference rules, $F_j^l (j = 1, 2, \ldots, v)$ the fuzzy sets, $x(t) \in \Re^n$ the state vector, $u(t) \in \Re^g$ the input vector, $y(t) \in \Re^p$ the output vector, and (A_l, B_l, a_l, C_l) the matrices of the lth local model, and $z(t) := [z_1, z_2, \ldots, z_v]$ the premise variables, which are some measurable variables of the system, for example, the output variables, the state variables or some of them. It is also assumed without loss of generality that the origin is the equilibrium of the T–S fuzzy system (3.1).

By using a standard fuzzy inference method, that is, using a singleton fuzzifier, product fuzzy inference, and center-average defuzzifier, the T–S fuzzy model in (3.1) can be rewritten as

$$x(t+1) = A(\mu)x(t) + B(\mu)u(t) + a(\mu)$$

$$y(t) = C(\mu)x(t),$$

(3.2)

where

$$A(\mu) = \sum_{l=1}^{m} \mu_l A_l, \; B(\mu) = \sum_{l=1}^{m} \mu_l B_l, \; a(\mu) = \sum_{l=1}^{m} \mu_l a_l, \; C(\mu) = \sum_{l=1}^{m} \mu_l C_l, \quad (3.3)$$

$\mu_l(z)$ is the normalized membership function satisfying

$$\mu_l = \frac{\xi_l(z)}{\sum_{i=1}^{m} \xi_i(z)}, \quad \xi_l(z) = \prod_{i=1}^{v} F_i^l(z_i), \quad \mu_l \geq 0, \quad \sum_{l=1}^{m} \mu_l = 1, \quad (3.4)$$

and $F_i^l(z_i)$ is the grade of membership of z_i in the fuzzy set F_i^l.

The model (3.2) is nonlinear in nature because the membership functions are in general nonlinear functions of the premise variables that contain some or all of the state variables. The T–S fuzzy model (3.2) is in fact the state space fuzzy model. Similarly, the input–output fuzzy model can also be defined as follows (Cao, Rees, and Feng, 1997a),

R^l: IF z_1 is F_1^l and ... z_v is F_v^l

THEN $y(t+1) = G_l(q^{-1})y(t) + H_l(q^{-1})u(t) + D_l$ (3.5)

$$G_l(q^{-1}) = G_1^l + G_2^l q^{-1} + \cdots + G_n^l q^{-n+1}, \qquad G_k^l = [g_{kij}^l]_{p \times p}$$

$$H_l(q^{-1}) = H_0^l + H_1^l q^{-1} + \cdots + H_{n-1}^l q^{-n+1}, \qquad H_k^l = [h_{kij}^l]_{p \times g}$$

$l \in L := \{1, 2, \ldots, m\},$

where, similar to the state space model in (3.1), R^l denotes the lth inference rule, m the number of inference rules, $u(t) \in \Re^g$ the input vector, $y(t) \in \Re^p$ the output vector, (G_l, H_l, D_l) the polynomial matrices of the lth local model of the fuzzy system (3.5), q^{-1} the shift operator defined by $q^{-1}y(t) = y(t-1)$, and $z(t) := [z_1, z_2, \ldots, z_v]$ the premise variables, which are some measurable variables of the system.

Similarly, using a center-average defuzzifier, product inference, and singleton fuzzifier, the dynamic fuzzy model (3.5) can be described by the following global model,

$$y(t+1) = G(q^{-1}, \mu(z))y(t) + H(q^{-1}, \mu(z))u(t) + D(\mu(z)), \quad (3.6)$$

where

$$G(q^{-1}, \mu(z)) = \sum_{l=1}^{m} \mu_l G_l(q^{-1}), \quad H(q^{-1}, \mu(z)) = \sum_{l=1}^{m} \mu_l(z) H_l(q^{-1}) \qquad (3.7)$$

or

$$G(q^{-1}, \mu(z)) = \sum_{j=1}^{n} G_j(\mu)q^{-j}, \quad H(q^{-1}, \mu(z)) = \sum_{j=0}^{n-1} H_j(\mu)q^{-j} \qquad (3.8)$$

with

$$G_j(\mu) = \sum_{l=1}^{m} \mu_l G_j^l, \quad H_j(\mu) = \sum_{l=1}^{m} \mu_l H_j^l, \quad D(\mu(z)) = \sum_{l=1}^{m} \mu_l(z) D_l.$$

The polynomial form of the T–S fuzzy model can be transformed into the state space form in (3.1) or (3.2), or vice versa. For example, the state space expression in the observable companion form can be obtained from (3.5) or (3.6) as follows,

$$x(t+1) = A(\mu)x(t) + B(\mu)u(t) + a(\mu)$$
$$y(t) = C(\mu)x(t), \qquad (3.9)$$

where

$$A(\mu(z)) = \sum_{l=1}^{m} \mu_l(z) A_l, \quad B(\mu) = \sum_{l=1}^{m} \mu_l(z) B_l$$

$$C(\mu(z)) = \sum_{l=1}^{m} \mu_l(z) C_l, \quad a(\mu(z)) = \sum_{l=1}^{m} \mu_l(z) a_l \qquad (3.10)$$

with

$$A_l = \begin{bmatrix} 0 & \cdots & 0 & G_n^l \\ I & \cdots & 0 & G_{n-1}^l \\ \vdots & \vdots & \vdots & \vdots \\ 0 & \cdots & I & G_1^l \end{bmatrix}, \quad B_l = \begin{bmatrix} H_n^l \\ H_{n-1}^l \\ \vdots \\ H_1^l \end{bmatrix}, \quad a_l = \begin{bmatrix} 0 \\ 0 \\ \vdots \\ D_l \end{bmatrix}, \quad C_l = [0 \quad 0 \quad \cdots \quad I].$$

Remark 3.1

T–S fuzzy models include two kinds of knowledge: qualitative, represented by fuzzy IF–THEN rules, and quantitative, represented by local linear models. T–S fuzzy models have a compatible structure with a two-level control system with the lower level providing basic feedback control and the higher level providing supervisory control or scheduling. By using T–S fuzzy models one can formulate these two kinds of knowledge or two-level systems into a unified mathematical framework. This framework provides a possibility for developing a systematic analysis and synthesis methodology for complex nonlinear control systems.

Remark 3.2

T–S fuzzy models are, to a certain extent, similar to the concept of typical piecewise linear approximation methods in nonlinear control. Control of a nonlinear system by piecewise linearization is approached by linearizing the system around a number of nominal operating points, and then applying linear feedback control methods to each local linear model (Kordon et al., 1999). However, analysis of the resulting closed-loop control system is in general difficult and the stability or performance of the closed-loop control system can hardly be guaranteed due to the approxima-tion. On the other hand, a T–S fuzzy model consists of a set of local linear models smoothly connected by membership functions yielding the global model of the sys-tem. Thus the T–S fuzzy model provides a way of designing controllers based on local linear models and analyzing stability or performance of the system based on its global nonlinear model. Moreover, this also provides a framework to consolidate the general industrial practice of nonlinear control system designs such as gain schedul-ing control.

3.3   UNIVERSAL FUNCTION APPROXIMATORS

Fuzzy systems are fundamentally equivalent to mathematical nonlinear mappings. In other words, fuzzy systems can be used to represent nonlinear functions math-ematically. In fact, it has been shown that many fuzzy systems are universal function approximators. In this section, it is shown that T–S fuzzy systems are universal func-tion approximators in the sense that they can be used to approximate any smooth nonlinear functions under certain conditions.

Consider a general nonlinear discrete time system described by a state space model of the form

$$x(t+1) = f(x(t), u(t)), \tag{3.11}$$

where $x(t) \in \mathfrak{R}^n$ is the state vector and $u(t) \in \mathfrak{R}^g$ the input vector of the system, respectively. The function $f(x(t), u(t))$ satisfies the following assumption.

Assumption 3.1

There exists an equilibrium $x_0 = 0 \in \mathfrak{R}^n$ such that $f(0, 0) = 0$ and $f \in C^2$; that is, the function f has a second-order continuous derivative with respect to x and u.

Before presenting the universal function approximation result, one needs to intro-duce the following definitions. Let the inferred fuzzy set $S_l = F_1^l \times F_2^l \times \cdots \times F_v^l$, and assume that $z(t) = x(t)$; that is, the membership functions depend on the system state variables, and consider the following class of membership functions.

Definition 3.1

Suppose the region corresponding to the fuzzy set S_l is divided into three subregions

$$S_l = S_l^0 \cup \partial S_l \cup S_l^\infty. \tag{3.12}$$

The membership function $\mu_l = \mu_l(x(t))$ satisfying the following conditions in the three subregions is called a trapezoid-shape-like membership function (TSLMF).

1.

$$\sum_{l=1}^{m} \mu_l(x) = 1. \tag{3.13a}$$

2. There exists a set of $\bar{x}^l$, $l = 1, 2,...,m$ called the centers of $S_l, l = 1, 2,...,m$ such that

$$\mu_l(\bar{x}^l) = 1, \bar{x}^l \in S_l^0, \qquad l = 1, 2,...,m. \tag{3.13b}$$

3. For a small enough $\varepsilon_\mu > 0$,

$$\mu_l(x) \geq 1 - \varepsilon_\mu, \ \textit{if } x \in S_l^0,$$

$$\varepsilon_\mu < \mu_l(x) < 1 - \varepsilon_\mu, \ \textit{if } x \in \partial S_l, \qquad l = 1, 2,...,m \tag{3.13c}$$

$$\mu_l(x) \leq \varepsilon_\mu, \ \textit{if } x \in S_l^\infty.$$

4. In the subregion S_i^0, $i \neq l$, the rate of $\mu_l \to 0$ is at least

$$O(\|x - \bar{x}^i\|^2), \quad i \neq l, \ i = 1, 2,...,m, \tag{3.13d}$$

where $\varepsilon = O(\|x\|^p)$ implies that

$$\lim_{\|x\| \to 0} \frac{\varepsilon}{\|x\|^p} = c,$$

with c being a constant.

The subregion S_l^0 is called the dominant or active region, the subregion ∂S_l is called the transition region, and the subregion S_l^∞ is called the inactive region.

Given any small enough $\varepsilon_\mu > 0$ one can always choose the parameters in a TSLMF such that (3.13c) holds. ε_μ depends on the decay factors in a TSLMF. Many kinds of membership functions can be classified as TSLMFs. Typical examples are trapezoidal membership functions and triangle membership functions. For instance, the trapezoidal function illustrated in Figure 3.1 is a TSLMF.

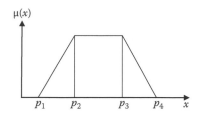

FIGURE 3.1   Trapezoidal function.

In fact, it can be found that

$$S_l^0 = [p_2, p_3], \quad \partial S_l = [p_1, p_2] \cup [p_3, p_4], \quad S_l^\infty = [-\infty, p_1] \cup [p_4, \infty],$$

$$\varepsilon_\mu = 0, \quad \overline{x}^l = (p_3 + p_2)/2.$$

The following membership functions are also TSLMFs.

$$\mu_1(x) = (1 - 1/(1 + \exp(-3(x - \pi/2)))) \cdot (1/(1 + \exp(-3(x + \pi/2))))$$

$$\mu_2(x) = 1 - \mu_1(x).$$

Using TSLMFs one can obtain the corresponding partition of the state space. The following partition is considered.

Definition 3.2
The state space partition is called a well-behaved partition (WBP) if it satisfies the following conditions.

1. Only one of the fuzzy sets S_l, $l = 1, 2, \ldots, m$ contains the origin. Without loss of generality, it is assumed that the origin $x = 0 \in S_1^0$ and $\mu_1(0) = 1$.
2. $X = \overline{S}_1 \cup \overline{S}_2 \cup \ldots \cup \overline{S}_m$, $\overline{S}_i = S_i^0 \cup \partial \overline{S}_i$, $\partial \overline{S}_i = \{x \mid \mu_i(x) \geq \mu_i(x), i = 1, 2, \ldots, m, i \neq l, x \notin S_l^0\}$.
3. $\overline{S}_l$, $l = 1, 2, \ldots, m$ are closed convex sets.

Remark 3.3
It can be seen that if the membership functions of the fuzzy sets satisfy (3.13a), Condition 2 in Definition 3.2 holds. If the membership functions satisfy

$$\mu_l(\lambda x_1 + (1 - \lambda)x_2) \geq \mu_l(x_1) \wedge \mu_l(x_2), \quad 0 \leq \lambda \leq 1, \quad x_1, x_2 \in X,$$

where $\wedge$ denotes the minimum operator, then the following "α–*level set*" A_α of the membership function

$$A_\alpha := \{x \mid \mu(x) \geq \alpha\}$$

is a closed convex set; thus, Condition 3 in Definition 3.2 holds for this kind of membership function. A large number of commonly used membership functions, for example, TSLMFs, satisfy the above two conditions, and thus it is not difficult to obtain a well-behaved partition of the state space.

Let Σ_n be the set of all systems of the form (3.11) satisfying Assumption 3.1. Let Σ_{fm} be the set of all T–S fuzzy models of the form (3.1) or (3.2) with the membership functions as TSLMFs. It has been shown (Cao, Rees, and Feng, 1997a, 2001a) that T–S fuzzy models are universal function approximators in the sense that given any $f(x, u) \in \Sigma_n$ there exists a fuzzy model $\hat{f}(x, u) = A(\mu)x + B(\mu)u + a(\mu) \in \Sigma_{fm}$ that will approximate $f(x, u)$ to any degree of accuracy in any convex compact region. More precisely, let X and U be compact sets in $\mathfrak{R}^n$ and $\mathfrak{R}^g$, respectively, and

$$d_\infty(f_1(x, u) \quad f_2(x, u)) = \sup_{x \in X, u \in U} (\|f_1(x, u) - f_2(x, u)\|) \tag{3.14}$$

be the sup-metric; then $(X \times U, d_\infty)$ is a metric space. The following theorem shows that (Σ_{fm}, d_∞) is dense in $(C^2 (X \times U), d_\infty)$.

Theorem 3.1 (Cao, Rees, and Feng, 1997a)

For any given $f(x, u) \in \Sigma_n$ on the compact set $X \times U \subset \mathfrak{R}^n \times \mathfrak{R}^g$ and arbitrary $\varepsilon > 0$, there exists an $\hat{f}(x, u) \in \Sigma_{fm}$ such that

$$d_\infty(f(x, u) - \hat{f}(x, u)) = \sup_{x \in X, u \in U} (\| f(x, u) - \hat{f}(x, u) \|) < \varepsilon. \tag{3.15}$$

Proof: For a given $f(x, u) \in \Sigma_n$ on the compact set $X \times U \subset \mathfrak{R}^n \times \mathfrak{R}^g$ and arbitrary $\varepsilon > 0$, because f has the second-order continuous derivative with respect to x and u on the compact set $X \times U \subset \mathfrak{R}^n \times \mathfrak{R}^g$, one can obtain the first-order Taylor series expansion of $f(x, u)$ at any point in $X \times U$. Let $f_l(x, u)$ be the lth local model in (3.2) which is equal to the linear first-order Taylor series expansion of $f(x, u)$ at the point $\bar{x}^l \in S_l^0$. Because $f(x, u)$ has continuous second-order derivatives one has $\|D_x^2 f(x, u)\| < M_l$, where $\| \cdot \|$ denotes the induced operator norm and M_l is a positive constant. Let $\bar{M}_1 = \max_l (M_l, l = 1, 2, \ldots, m)$. Then for any $(x, u) \in X \times U$ and m one has

$$f(x, u) - \hat{f}(x, u) = \sum_{l=1}^{m} (f(x, u) - f_l(x, u))\mu_l(x). \tag{3.16}$$

By Taylor's theorem, it follows that

$$\|f(x, u) - f_l(x, u)\| \le \frac{\bar{M}_1}{2} \|x - \bar{x}^l\|^2, \qquad x, \bar{x}^l \in S_l^0 \cup \partial S_l \tag{3.17}$$

$$l = 1, 2, \ldots, m.$$

Because X is compact, there always exists a large enough m and a finite set $\{\bar{x}^1, \bar{x}^2, \ldots, \bar{x}^m\} \subset X$ such that for any $(x, u) \in X \times U$, if $x \in S_l^0 \cup \partial S_l$ one has

$$\|x - \bar{x}^l\|^2 < \frac{\varepsilon}{\bar{M}_1} = \eta_l,$$

$$l = 1, 2, \ldots, m.$$

(3.18)

Thus, it follows from (3.17) and (3.18) that

$$d_\infty(f(x, u) - f_l(x, u)) \le \frac{\varepsilon}{2}, \qquad x, \bar{x}^l \in S_l^0 \cup \partial S_l.$$

(3.19)

If $x \in S_l^0$, from (3.19) and (3.13b) one has

$$d_\infty(f(x, u) - \hat{f}(x, u)) \le d_\infty\Big[(f(x, u) - f_l(x, u))$$

$$+ \sum_{i=1, i \neq l}^{m} (f(x, u) - f_i(x, u) - f(x, u) + f_l(x, u))\mu_i(x)\Big]$$

$$\le \frac{\varepsilon}{2} + \sum_{i=1, i \neq l}^{m} d_\infty(f_l(x, u) - f_i(x, u))\mu_i(x).$$

(3.20)

Because $f_l(x, u), l = 1, 2, \ldots, m$ are continuous functions on the compact set there exists a finite positive constant $\bar{M}_2$ such that

$$\sup_{x \in X, u \in U} \|f_l(x, u)\| \le \bar{M}_2, \quad l = 1, 2, \ldots, m.$$

(3.21)

It then follows from (3.20) and (3.21) that

$$\sum_{i=1, i \neq l}^{m} d_\infty(f_l(x, u) - f_i(x, u))\mu_i(x) \le 2\bar{M}_2 \Delta\mu_l(t) \le 2\bar{M}_2 \varepsilon_\mu \le \frac{\varepsilon}{2},$$

(3.22)

where the parameters in the TSLMF are chosen so that ε_μ in (3.13) satisfies

$$\varepsilon_\mu \le \frac{\varepsilon}{4\bar{M}_2} \text{ and } \mu_l(\bar{x}^l) = 1, l = 1, 2, \ldots, m.$$

From (3.20) and (3.22) it follows that if $x \in S_l^0$,

$$d_\infty(f(x, u) - \hat{f}(x, u)) \le \varepsilon.$$

(3.23)

In addition, it follows from (3.16) that if $x \in \partial \overline{S}_l$,

$$d_\infty(f(x,u)-\hat{f}(x,u)) \leq \sum_{i_l=1}^{m_l} d_\infty\left(f(x,u)-f_{i_l}(x,u)\right)\mu_{i_l}(x) + \sum_{\bar{i}_l=1}^{\bar{m}_l} \mu_{\bar{i}_l}(x)\|f(x,u)-f_{\bar{i}_l}(x,u)\|$$

$$\leq \frac{\varepsilon}{2}\sum_{i_l=1}^{m_l}\mu_{i_l}(x) + 2\overline{M}_2\varepsilon_\mu = \varepsilon, \tag{3.24}$$

where m_l is the number of the local regions that satisfy $\partial \overline{S}_l \cap \partial S_{i_l} \neq \{0\}$ and $\bar{m}_l$ is the number of the local regions that satisfy $\partial \overline{S}_l \cap \partial S_{\bar{i}_l} = \{0\}$.

Because $X \subset S_1^0 \cup \partial \overline{S}_1 \cup \cdots \cup S_m^0 \cup \partial \overline{S}_m$ (the state space partition is a WBP) and for any $x \in X$ (3.23) and (3.24) hold, thus, using this large enough m and the membership functions TSLMFs, one has

$$d_\infty(f(x,u)-\hat{f}(x,u)) \leq \varepsilon, \tag{3.25}$$

and the proof is thus completed. ❏

Remark 3.4

There are many results on the universal function approximation property of various kinds of fuzzy systems, for example, the fuzzy systems with fuzzy basis functions based on the well-known Stone–Weierstrass theorem (Wang and Mendel, 1992; Zeng and Singh, 1994, 1995, 1996). It has also been recently proved that the more commonly used T–S fuzzy model in (3.1) with $a_l \equiv 0$ is also a universal function approximator (Tanaka and Wang, 2001).

Remark 3.5

It should also be noted that the result in Theorem 3.1 only concerns the approximation between two *static* nonlinear functions, that is, $f(x,u)$ and $\hat{f}(x,u)$. However, the error between the state variables of two dynamic systems described in (3.11) and (3.2), which are defined by $f(x,u)$ and $\hat{f}(x,u)$, respectively, might grow in general as time passes. Therefore, much care has to be taken in dealing with the approximation between two dynamic systems instead of two static functions.

Remark 3.6

Simplified T–S fuzzy models without affine terms, that is, the model (3.1) with $a_l \equiv 0$, have been more commonly used, especially in fuzzy controller design, although it is shown that T–S fuzzy models with affine terms have much improved function approximation capabilities (Fantuzzi and Rovatti, 1996).

3.4   T–S FUZZY MODEL IDENTIFICATION FROM NONLINEAR MODELS

Given a nonlinear system model, one can obtain its approximate T–S fuzzy model by linearizing the given nonlinear model around a number of operating points of interest. Suppose a nonlinear system model is given in (3.11), rewritten as

$$x(t+1) = f(x(t), u(t)). \tag{3.26}$$

Our objective is to find its approximate T–S fuzzy model in the form of (3.1). It is supposed that the number of the operating points of interest is given a priori. Then the first task is to determine the local linear models of the T–S fuzzy system. Suppose the operation points of interest are given by $\{(x^1, u^1), (x^2, u^2), \dots, (x^m, u^m)\}$, where the first set is the equilibrium point of the system; that is, $f(x^1, u^1) = 0$. Without loss of generality, we can assume that $x^1 = 0$, $u^1 = 0$. Then by applying the Taylor series expansion of $f(x, u)$ around $x^1 = 0$, $u^1 = 0$, one has

$$
\begin{aligned}
f(x, u) &= \left.\frac{\partial f}{\partial x}\right|_{x=x^1, u=u^1} (x - x^1) + \left.\frac{\partial f}{\partial u}\right|_{x=x^1, u=u^1} (u - u^1) + \varepsilon_1(x, u) \\
&= \left.\frac{\partial f}{\partial x}\right|_{x=0, u=0} x + \left.\frac{\partial f}{\partial u}\right|_{x=0, u=0} u + \varepsilon_1(x, u),
\end{aligned}
\tag{3.27}
$$

where $\varepsilon_1(x, u)$, and the other similar terms in the subsequent expressions, represent the higher-order approximation error. In this case, the local model of the T–S fuzzy model, which approximates the nonlinear system model at the equilibrium, can be expressed as

$$x(k+1) = A_1 x + B_1 u, \tag{3.28}$$

where

$$A_1 = \left.\frac{\partial f}{\partial x}\right|_{x=0, u=0} \qquad B_1 = \left.\frac{\partial f}{\partial u}\right|_{x=0, u=0}. \tag{3.29}$$

Note that the affine term in this case is equal to zero. Similarly, for operating points other than the equilibrium point, one can apply the Taylor series expansion of $f(x, u)$ around (x^l, u^l), $l = 2, 3, \dots, m$, and obtain

$$
\begin{aligned}
f(x, u) &= \left.\frac{\partial f}{\partial x}\right|_{x=x^l, u=u^l} (x - x^l) + \left.\frac{\partial f}{\partial u}\right|_{x=x^l, u=u^l} (u - u^l) + \varepsilon_l(x, u) \\
&= \left.\frac{\partial f}{\partial x}\right|_{x=x^l, u=u^l} x + \left.\frac{\partial f}{\partial u}\right|_{x=x^l, u=u^l} u + \left(-\left.\frac{\partial f}{\partial x}\right|_{x=x^l, u=u^l} x_l - \left.\frac{\partial f}{\partial u}\right|_{x=x^l, u=u^l} u_l \right) + \varepsilon_l(x, u).
\end{aligned}
\tag{3.30}
$$

Then the local model of the T–S fuzzy model that approximates the nonlinear system model at the operating point (x^l, u^l), $l = 2, 3, \ldots, m$ can be expressed as

$$x(k + 1) = A_l x + B_l u + a_l, \tag{3.31}$$

where

$$A_l = \left.\frac{\partial f}{\partial x}\right|_{x=x^l, u=u^l}, \ B_l = \left.\frac{\partial f}{\partial u}\right|_{x=x^l, u=u^l}, \ a_l = -\left.\frac{\partial f}{\partial x}\right|_{x=x^l, u=u^l} x^l - \left.\frac{\partial f}{\partial u}\right|_{x=x^l, u=u^l} u^l. \tag{3.32}$$

When the operating point is not the equilibrium $(0, 0)$, the linearized model has a nonzero affine term a_l whose expression is given in (3.32).

The next task is to determine the fuzzy membership functions for fuzzy sets around those operating points or in the so-called local regions. The ideal choice is to select the membership functions $\mu_l(x, u)$, $l = 1, 2, \ldots, m$, such that the following modeling error is minimized.

$$E = \left\| \sum_{l=1}^{m} \mu_l(x, u)(A_l x + B_l u + a_l) - f(x, u) \right\|. \tag{3.33}$$

In general, this is a nonlinear optimization problem that is difficult to solve. However, in many applications, some simple and typical membership functions, such as triangular, trapezoid, and Gaussian functions, can be utilized. One of the key parameters, that is, the centers of these membership functions can be determined by the operating points (x^l, u^l), $l = 1, 2, \ldots, m$, and the other parameters such as the width and decay rate can be suitably chosen by the designer.

Example 3.1

Consider a model of the computer-simulated truck–trailer formulated in Tanaka and Wang (2001) as follows,

$$x_1(t + 1) = (1 - v \cdot T/L)x_1(t) + v \cdot T/l \cdot u(t)$$

$$x_2(t + 1) = x_2(t) + v \cdot T/L \cdot x_1(t) \tag{3.34}$$

$$x_3(t + 1) = x_3(t) + v \cdot T \cdot \sin(x_2(t) + v \cdot T/2L \cdot x_1(t)),$$

where x_1 denotes the angle difference between the truck and the trailer, x_2 the angle of the trailer, x_3 the vertical position of the rear end of the trailer, and $l = 2.8$, $L = 5.5$, $v = -1.0$, $\bar{t} = 2.0$, $d = 0.01l/\pi$.

Note that the nonlinear function $f(x, u)$ in this case is linear in u and thus one easily has $B = [vT/l \ \ 0 \ \ 0]^T$. Then the linearization procedure can be carried out against the state variable $x = [x_1 \ \ x_2 \ \ x_3]^T$ only. Suppose the operating points are chosen as $x^1 = [0 \ \ 0 \ \ 0]^T$, $x^2 = [0 \ \ \pi/3 \ \ 0]^T$, $x^3 = [\pi/3 \ \ 0 \ \ 0]^T$, $x^4 = [\pi/3 \ \ \pi/3 \ \ 0]^T$.

Then one has

$$
\frac{\partial f}{\partial x} =
\begin{bmatrix}
1 - vT/L & 0 & 0 \\
vT/L & 1 & 0 \\
v^2T^2 \cos(x_2 + vT/(2L)x_1)/(2L) & vT \cos(x_2 + vT/(2L)x_1) & 1
\end{bmatrix}
$$

$$
=
\begin{bmatrix}
1.3636 & 0 & 0 \\
-0.3636 & 1 & 0 \\
0.3636 \cos(x_2 - 0.1818x_1) & -2\cos(x_2 - 0.1818x_1) & 1
\end{bmatrix}.
\tag{3.35}
$$

For the four operating points, one can get the corresponding matrices and affine terms, respectively, as follows.

$$
A_1 =
\begin{bmatrix}
1.3636 & 0 & 0 \\
-0.3636 & 1 & 0 \\
0.3636 & -2 & 1
\end{bmatrix}, \quad
B_1 =
\begin{bmatrix}
-0.7143 \\
0 \\
0
\end{bmatrix}, \quad
a_1 =
\begin{bmatrix}
0 \\
0 \\
0
\end{bmatrix},
$$

$$
A_2 =
\begin{bmatrix}
1.3636 & 0 & 0 \\
-0.3636 & 1 & 0 \\
0.1818 & -1 & 1
\end{bmatrix}, \quad
B_2 =
\begin{bmatrix}
-0.7143 \\
0 \\
0
\end{bmatrix}, \quad
a_2 =
\begin{bmatrix}
0 \\
-\pi/3 \\
\pi/3
\end{bmatrix},
\tag{3.36}
$$

$$
A_3 =
\begin{bmatrix}
1.3636 & 0 & 0 \\
-0.3636 & 1 & 0 \\
0.3571 & -1.9639 & 1
\end{bmatrix}, \quad
B_3 =
\begin{bmatrix}
-0.7143 \\
0 \\
0
\end{bmatrix}, \quad
a_3 =
\begin{bmatrix}
-0.4545\pi \\
0.1212\pi \\
-0.1190\pi
\end{bmatrix},
$$

$$
A_4 =
\begin{bmatrix}
1.3636 & 0 & 0 \\
-0.3636 & 1 & 0 \\
0.2381 & -1.3097 & 1
\end{bmatrix}, \quad
B_4 =
\begin{bmatrix}
-0.7143 \\
0 \\
0
\end{bmatrix}, \quad
a_4 =
\begin{bmatrix}
-0.4545\pi \\
-0.2121\pi \\
0.3572\pi
\end{bmatrix}.
$$

Choose the following Gaussian-type functions,

$$
h_1 = \exp\left(-\frac{x_1^2 + x_2^2 + x_3^2}{\sigma_1^2}\right)
$$

$$
h_2 = \exp\left(-\frac{x_1^2 + (x_2 - \pi/3)^2 + x_3^2}{\sigma_2^2}\right)
\tag{3.37}
$$

$$
h_3 = \exp\left(-\frac{(x_1 - \pi/3)^2 + x_2^2 + x_3^2}{\sigma_3^2}\right)
$$

$$
h_4 = \exp\left(-\frac{(x_1 - \pi/3)^2 + (x_2 - \pi/3)^2 + x_3^2}{\sigma_4^2}\right),
$$

where σ_1, σ_2, σ_3, and σ_4 are the widths of the corresponding functions, respectively. Then one can obtain the normalized membership functions for each local model as

$$\mu_1(x) = \frac{h_1}{h_1 + h_2 + h_3 + h_4}$$

$$\mu_2(x) = \frac{h_2}{h_1 + h_2 + h_3 + h_4}$$

$$\mu_3(x) = \frac{h_3}{h_1 + h_2 + h_3 + h_4}$$ (3.38)

$$\mu_4(x) = \frac{h_4}{h_1 + h_2 + h_3 + h_4}.$$

Note that σ_1, σ_2, σ_3, and σ_4 can be tuned based on modeling error or knowledge.

If a linear instead of affine T–S fuzzy model is to be constructed, one simple approach is just to ignore the affine terms in the linearization procedure described in this section, although the approximating error would be increased in this case. Note that another approach to obtaining the approximate linear T–S fuzzy model rather than the affine T–S fuzzy model was suggested in Teixeira and Zak (1999) by linearization and optimization directly.

3.5   T–S FUZZY MODEL IDENTIFICATION FROM DATA

When the nonlinear system model is available, the construction approach to T–S fuzzy models described in the last section can be applied. However, in many engineering applications, the mathematical models of complex engineering systems are very difficult, even impossible, to obtain. In this case, we may have to resort to a data approach for identifying the T–S fuzzy model of a complex system. The input–output T–S fuzzy model (3.5) or (3.7) in the global form is employed for this purpose in this section.

The objective of the T–S fuzzy model identification is to identify both the membership functions and the local model parameters in (3.5) or (3.7). Many approaches to identifying the membership functions have been developed. One approach is described in the following subsection.

3.5.1   IDENTIFICATION OF MEMBERSHIP FUNCTIONS

The key idea for identification of membership functions is to use fuzzy clustering to get the number of rules and to determine the characteristic parameters of the membership functions. In the subsequent discussion the following membership functions, which are TSLMFs, are considered.

$$\mu_l(z, \bar{z}_l, \sigma_l) = \left(\sum_{j=1}^{m} \frac{\| z - \bar{z}_l \|^{\sigma_l}}{\| z - \bar{z}_j \|^{\sigma_l}} \right)^{-1}.$$ (3.39)

Identification of the membership functions includes the partition of the premise variable space of the fuzzy model (3.5) into a set of subregions. The number of subregions corresponds to the number of rules, the centers of the subregions correspond to the centers of the membership functions, and the degree of the overlap among the subregions corresponds to the decay factors of the membership functions.

Using fuzzy clustering algorithms, one can obtain the number of rules and the centers and decay factors of the membership functions at the same time. The following fuzzy dynamic clustering algorithm was proposed in Cao, Rees, and Feng (1997a), and is suitable for taking into account dynamic behavior of the local dynamic models.

At first suppose the number of rules is fixed; that is, the number m is fixed. Define a performance criterion function as follows,

$$J(\mu, \overline{z}, \alpha) = w_1 \left(\sum_{t=1}^{N} \sum_{l=1}^{m} \mu_l(t)^\omega \parallel z(t) - \overline{z}_l \parallel^2 \right) + w_2 \left(\sum_{t=1}^{N} \sum_{l=1}^{m} \mu_l(t)^\omega \parallel e_l(t) \parallel^2 \right) \quad (3.40)$$

subject to

$$\sum_{l=1}^{m} \mu_l(t) = 1, \quad \forall t,$$

$$e_l(t) = y(t) - \varphi(t-1)^T \alpha^l, \quad (3.41)$$

$$\varphi(t-1)^T = [(1,1,\ldots,1), y(t-1)^T, \ldots, y(t-n)^T, u(t-1)^T, \ldots, u(t-n)^T],$$

$$\alpha^l = \begin{bmatrix} \alpha_1^l & \alpha_2^l & \cdots & \alpha_p^l \end{bmatrix}, \quad \alpha_j^l = \begin{bmatrix} \alpha_{j1}^l & \cdots & \alpha_{jn}^l \end{bmatrix}$$

$$\overline{n} = p + p \times n + g \times n,$$

where N is the sampling points of the time, m is the number of rules, $\overline{z} = [\overline{z}_1 \quad \overline{z}_2 \quad \cdots \quad \overline{z}_m]$ is an m-tuple of mean prototypes, $\parallel z(t) - \overline{z}_l \parallel$ is the distance of the feature point $z(t)$ to the mean prototype $\overline{z}_l$, $\hat{y}(t) = \varphi(t-1)^T \alpha^l$, $l = 1, 2, \ldots, m$ are the m predicting equations of the local linear models called the equation prototypes, ω is used to control the shape of the membership functions, and w_1 and w_2 are the weighting factors.

Remark 3.7

The first term in (3.40) requires the distance from the feature vectors to the prototypes to be as small as possible, which corresponds to the fuzzy input space partition. The second term in (3.40) requires this input space partition to be an optimal partition such that the estimation errors between the output of the fuzzy system and the local linear models are minimized. In this sense the fuzzy clustering algorithm based on the criterion function (3.40) can be considered as a kind of dynamic behavior clustering algorithm.

Minimizing (3.40) with respect to $\mu, \overline{z}, \alpha$, respectively, one has the following result.

Theorem 3.2

Given the weighting factors w_1, w_2, and ω, the necessary conditions for minimizing $J(\mu, \overline{z}, \alpha)$ are given as

$$\overline{z}_l = \frac{\sum_{t=1}^{N} \mu_l(t)^{\omega} z(t)}{\sum_{t=1}^{N} \mu_l(t)^{\omega}}, \quad l = 1, 2, \ldots, m \tag{3.42a}$$

$$\alpha^l = \left[\Phi^T D_l \Phi \right]^{-1} \Phi^T D_l Y, \quad l - 1, 2, \ldots, m \tag{3.42b}$$

$$\mu_l(t) = \left\{ \sum_{j=1}^{m} \frac{\left(w_1 \| z(t) - \overline{z}_l \|^2 + w_2 \| e_l(t) \|^2 \right)^{\sigma}}{\left(w_1 \| z(t) - \overline{z}_j \|^2 + w_2 \| e_j(t) \|^2 \right)^{\sigma}} \right\}^{-1}, \quad \sigma = 1/(\omega - 1) \tag{3.42c}$$

$$l = 1, 2, \ldots, m,$$

where

$$\Phi = [\varphi(1) \ \varphi(2) \ \ldots \ \varphi(N)]^T, \ Y = [y(1) \ y(2) \ \ldots \ y(N)]^T, \ D_l = diag[\mu_l(t)]_{N \times N}.$$

Proof: Consider the following Lagrangian:

$$\overline{J}(\mu, \overline{z}, \alpha, \lambda) = w_1 \left(\sum_{t=1}^{N} \sum_{l=1}^{m} \mu_l(t)^{\omega} \left\| z(t) - \overline{z}_l \right\|^2 \right) + w_2 \left(\sum_{t=1}^{N} \sum_{l=1}^{m} \mu_l(t)^{\omega} \| e_l(t) \|^2 \right)$$

$$+ \sum_{t=1}^{N} \left[\lambda\omega \left(1 - \sum_{l=1}^{m} \mu_l(t) \right) \right]. \tag{3.43}$$

Because $\mu_l(t)$, $l = 1, 2, \ldots, m$ are independent of each other, minimizing $\overline{J}(\mu, \overline{z}, \alpha, \lambda)$ with respect to $\mu_l(t)$ is equivalent to minimizing the following individual objective function with respect to each $\mu_l(t)$.

$$\overline{J}_1(\mu) = w_1 \mu_l(t)^{\omega} \left\| z(t) - \overline{z}_l \right\|^2 + w_2 \mu_l(t)^{\omega} \| e_l(t) \|^2 + \lambda\omega \left(1 - \sum_{l=1}^{m} \mu_l(t) \right). \tag{3.44}$$

Differentiating (3.44) with respect to $\mu_l(t)$ and setting it to 0, one obtains (3.42c). Note that minimizing $\overline{J}(\mu, \overline{z}, \alpha, \lambda)$ with respect to $\overline{z}$ is equivalent to minimizing the following objective function with respect to $\overline{z}$.

$$\overline{J}_2(\overline{z}) = \sum_{t=1}^{N} \sum_{l=1}^{m} \mu_l(t)^{\omega} \left\| z(t) - \overline{z}_l \right\|^2. \tag{3.45}$$

Differentiating (3.45) with respect to $\overline{z}$ and setting it to 0, one obtains (3.42a). Minimizing $\overline{J}(\mu,\overline{z},\alpha,\lambda)$ with respect to α is equivalent to minimizing the following objective function with respect to α,

$$\overline{J}_3(\alpha) = \sum_{t=1}^{N} \sum_{l=1}^{m} \mu_l(t)^\omega \parallel e_l(t) \parallel^2. \tag{3.46}$$

Differentiating (3.46) with respect to α and setting it to 0, one has (3.42b). The proof is thus completed. $\square$

Based on Theorem 3.2, the following algorithm can be proposed.

Fuzzy Input Space Clustering Algorithm (FISCA)

Step 1. Choose the weighting factors w_1, w_2, and ω, and pick a termination threshold $\varepsilon > 0$ and a set of initial membership functions $\mu_l^{(0)}$, $l = 1, 2, \dots m$ such that $\Sigma_{l=1}^{m}\mu_l^{(0)} = 1$.

Step 2. Update $\mu^{(k)} \rightarrow \mu^{(k+1)}$ according to (3.42) if $I_t = 0$,
Otherwise,   choose $\mu_i^{(k+1)}(t) = 0, \forall i \in \tilde{I}_t$, and $\mu_i^{(k+1)}(t)$, $i \in I_t$ such   that $\Sigma_{i \in I_t} \mu_i^{(k+1)}(t) = 1$,
where $I_t := \{i \mid 1 \leq i \leq m;\ w_1\|z(t)-\overline{z}_i\|^2 + w_2\|e_i(t)\|^2 = 0\}$, $\tilde{I}_t := \{1, 2, \dots,$
$m\} - I_t$.

Step 3. If $\parallel \mu^{(k+1)} \rightarrow \mu^{(k)} \parallel \leq \varepsilon$, then stop; otherwise go to Step 2.

In this algorithm, w_1, w_2, and ω, are adjustable parameters. w_1 and w_2 are the weighting factors in (3.40). If w_1 is larger than w_2 it implies that more attention is paid to the mean clustering accuracy; otherwise it implies that more attention is paid to the equation clustering accuracy. ω is a shape parameter of the membership functions and determines the decay factor.

It should be noted that the optimal partition of the input space has been obtained after the FISCA. However, the solutions given in (3.42c) are not suitable candidates in general for the membership functions due to the local model prediction error terms. In fact, the estimation error terms in (3.42c) should be removed from the membership functions because the partition information should only be represented by the centers of the membership functions. Thus, the membership functions can be chosen as

$$\mu_l(z) = \left(\sum_{j=1}^{m} \frac{\parallel z - \overline{z}_l \parallel^{2\sigma}}{\parallel z - \overline{z}_j \parallel^{2\sigma}} \right)^{-1}, \quad l = 1, 2, \dots, m. \tag{3.47}$$

Then the following criterion can be used to determine the number of rules.

$$h_z(m) = w_1 \left[\sum_{t=1}^{N} \sum_{l=1}^{m} \mu_l(t)^\omega \parallel z(t) - \overline{z}_l\|^2 - \|\overline{z}_l - \overline{z}\|^2 \right]$$

$$+ w_2 \left[\sum_{t=1}^{N} \sum_{l=1}^{m} \mu_l(t)^\omega (\|e_l(t)\|^2 - \|\overline{e}\|^2) \right], \tag{3.48}$$

where

$$z = \frac{1}{N} \sum_{t=1}^{N} z(t), \quad \bar{e} = \frac{1}{N} \sum_{t=1}^{N} \| y(t) - \varphi(t-1)^T \alpha \|, \quad \alpha = [\Phi^T \Phi]^{\#} \Phi^T Y,$$

and $[\Phi(\mu)^T \Phi(\mu)]^{\#}$ represents the pseudo inverse of $[\Phi(\mu)^T \Phi(\mu)]$ if it is not invertible.

The number of rules m is determined so that $h_z(m)$ reaches a minimum as m increases. It can be seen that the first term in (3.48) has two parts: one is the variance of the data in a mean cluster, and the other is the variance of the mean clusters themselves. The second term in (3.48) also has two parts: one is the variance of the data in the equation cluster and the other is the variance of the equation clusters themselves. Therefore the optimal clustering is to minimize the variance in each cluster and to maximize the variance among the clusters. Using the criterion $h_z(m)$ and the FISCA, one can obtain the following identification algorithm for the membership functions.

Membership Function Identification Algorithm
Step 1. Pick a termination threshold $\bar{\varepsilon} > 0$ and set $m = 1$.
Step 2. Run the FISCA and obtain a set of membership functions.
Step 3. If $\| h_z(m+1) - h_z(m) \| < \bar{\varepsilon}$, then stop; otherwise, $m = m + 1$, go to Step 2.

After membership functions are determined, the next objective is to identify the parameters of the local affine models, which are described in the next subsection.

3.5.2 IDENTIFICATION OF LOCAL MODELS

The multi-input multi-output fuzzy model (3.5) or (3.7) can be represented by p multiple-input single-output systems. Without loss of generality only the multiple-input single-output system is considered and $y(t)$ is still used to represent one of the output components in (3.7). In this case, one of the output components in (3.7) can be rewritten as

$$y(t) = \sum_{i=1}^{M} v_i(t, \mu) \theta_i, \tag{3.49}$$

where $v_i(t, \mu)$ is the product of the membership function μ and a certain term involving outputs or inputs in (3.7); for example, one of $v_i(t, \mu)$ is $\mu_i(z(t-1)) y(t-1)$; θ_i is one of the parameters of the local models in (3.7) consisting of the coefficients of the polynomials G_l and H_l, $l = 1, 2, \ldots, m$; and M is the number of all terms in (3.7) that represent the output component. Let $\theta = (\theta_1, \theta_2, \ldots, \theta_M)^T$, and

$$\hat{y}(t) = v^T(t, \mu) \hat{\theta}, \tag{3.50}$$

where $\hat{y}(t)$ and $\hat{\theta}$ denote the prediction of the output $y(t)$ and the estimation of the parameter θ, respectively.

The task of parameter estimation is to determine $\hat{\theta}$ to minimize the following criterion,

$$J(\hat{\theta}) = \| Y - \Phi(\mu)\hat{\theta} \|^2, \tag{3.51}$$

where

$$Y = (y(1), y(2), \ldots, y(N))^T \tag{3.52}$$

$$\Phi(\mu) = \begin{bmatrix} v_1(1,\mu) & v_2(1,\mu) & \cdots & v_M(1,\mu) \\ v_1(2,\mu) & v_2(2,\mu) & \cdots & v_M(2,\mu) \\ \vdots & \vdots & \vdots & \vdots \\ v_1(N,\mu) & v_2(N,\mu) & \cdots & v_M(N,\mu) \end{bmatrix}. \tag{3.53}$$

The minimum point of (3.51) can be described by

$$\hat{\theta} = [\Phi(\mu)^T \Phi(\mu)]^{\#} \Phi(\mu)^T Y, \tag{3.54}$$

where $[\Phi(\mu)^T \Phi(\mu)]^{\#}$ represents the pseudo inverse of $[\Phi(\mu)^T \Phi(\mu)]$ if it is not invertible.

It can be seen that the model (3.49) is a linear model with respect to θ and the parameter estimation algorithm only involves the least-squares method. The convergence of the identification algorithm is thus ensured as a conventional least-squares algorithm. After the parameter θ is identified, then the parameters of the polynomials G_l and H_l, $l = 1, 2, \ldots, m$ can be determined easily based on the definition of θ. However, it should be noted that unless persistent excitation conditions are satisfied the convergence of the estimated parameters to the true system parameters would not be guaranteed, although the convergence to some constant parameters is ensured.

3.6   APPROXIMATION ERROR ANALYSIS

It has been shown in Section 3.3 that the T–S fuzzy models can be used to approximate any smooth nonlinear functions to any desired accuracy in any convex compact region of interest. To further analyze the approximation error, one has the following fact on the so-called trapezoid-shape-like membership functions (TSLMFs).

Fact 3.1

Let $\Delta\mu_l(x) = \sum_{i=1, i \neq l}^{m} \mu_i(x)$. It follows from the properties of the TSLMF that there exists a set of sufficiently small constants $\varepsilon_{\mu l}$, $l = 1, 2, \ldots, m$, such that

$$\Delta\mu_l(x) \leq \varepsilon_\mu, \quad \forall\, x \in S_l^0 \tag{3.55a}$$

$$\Delta\mu_l(x) = O\left(\|x - \bar{x}^l\|^2\right) \leq \varepsilon_{\mu l} \|x - \bar{x}^l\|, \quad \forall\, x \in S_l^0 \tag{3.55b}$$

$$l = 1, 2, \ldots, m.$$

Then the following corollary arises directly from (3.55) and the proof of Theorem 3.1.

Corollary 3.1

The approximate T–S fuzzy model $\hat{f}(x,u)$ in Theorem 3.1 described in (3.1) or (3.2) has the following properties, for a given $\varepsilon > 0$ and $l = 1, 2, \ldots, m$.

1. There exists a positive constant ε_f such that

$$\|f(x,u) - \hat{f}(x,u)\| \le \varepsilon_f \|x - \bar{x}^l\| \le \varepsilon, \quad \forall\, x \in \bar{S}_l, \tag{3.56a}$$

$$\| f(x,u) - f_l(x,u) \| \le \varepsilon_f \| x - \bar{x}^l \| \le \varepsilon \quad \forall \quad x \in \bar{S}_l. \tag{3.56b}$$

2. There exists a set of positive constants $\varepsilon_{fl}, l = 1, 2, \ldots, m$ such that

$$\|\Delta f_l(x)\| \le \sum_{i=1, i \ne l}^{m} \mu_i(x) \|f_i(x,u) - f_l(x,u)\| \le \varepsilon_{fl} \|x - \bar{x}^l\| \le \varepsilon, \quad \forall\, x \in \bar{S}_l \tag{3.57}$$

where

$$\Delta f_l(x) := \sum_{i=1, i \ne l}^{m} \mu_i(x)(f_l(x,u) - f_i(x,u)).$$

Furthermore, if the partition of the state space is WBP as defined in Section 3.3, then only one local region, say S_1, contains the origin, and the local regions S_i, $i = 2, 3, \ldots, m$ can be chosen such that $\|x - \bar{x}^l\| \le \|x\|$, $\forall l$. Thus, one has the following properties of the T–S fuzzy model under such situations.

$$\|f(x,u) - \hat{f}(x,u)\| \le \varepsilon_f \|x\| \tag{3.58a}$$

$$\|f(x,u) - f_l(x,u)\| \le \varepsilon_{fl} \|x\|, \quad \forall\, x \in \bar{S}_l \tag{3.58b}$$

$$\|\Delta f_l(x)\| \le \varepsilon_{fl} \|x\|, \quad \forall\, x \in \bar{S}_l \tag{3.58c}$$

$$l = 1, 2, \ldots, m.$$

These results further characterize the properties of the approximating error between the T–S fuzzy models and their underlying nonlinear systems, and thus provide valuable information on robust stabilization of general nonlinear systems via T–S fuzzy model based approaches.

3.7 CONCLUSIONS

This chapter has presented a general T–S fuzzy system model that consists of a number of local linear or affine models and a set of smooth nonlinear membership functions. The universal function approximation property of the T–S fuzzy model has been established. Two approaches to identification of the T–S fuzzy model have been presented: one based on a linearization technique when the nonlinear system model is available, and the other data-driven when the nonlinear system model is not available. Some more detailed characterizations of the approximation error between the T–S fuzzy model and the underlying nonlinear system have also been given. These characterizations are useful when the stabilization of general nonlinear systems via model based fuzzy control approaches is considered.

4 Stability Analysis of T–S Fuzzy Systems

4.1 INTRODUCTION

Stability is one of the most important and fundamental issues for control systems. One critical problem with conventional fuzzy logic control systems is the difficulty encountered in stability analysis of such systems and in systematic design of control laws. It is believed the problem arises from the model-free concept of conventional fuzzy control systems although this concept is very useful in many control engineering applications, especially when models of the systems to be controlled are not available. With the introduction of T–S fuzzy models, stability analysis of fuzzy control systems becomes much easier, and more important, it can be handled in the same way as conventional model based control systems. In particular, stability analysis of T–S fuzzy systems has been pursued mainly based on Lyapunov stability theory with a variety of Lyapunov functions. Three of the most popular are the so-called *common* (or *global*) quadratic, *piecewise* quadratic, and *fuzzy* (or *nonquadratic*) Lyapunov functions. This chapter is devoted to stability analysis of T–S fuzzy systems based on these Lyapunov functions.

The rest of the chapter is organized as follows. Stability analyses of T–S fuzzy systems based on common, piecewise, and fuzzy quadratic Lyapunov functions are presented in Sections 4.2, 4.3, and 4.4, respectively. Comparison of a variety of stability results is carried out via numerical examples in Section 4.5, and some remarks conclude the chapter in Section 4.6.

4.2 STABILITY ANALYSIS BASED ON COMMON QUADRATIC LYAPUNOV FUNCTIONS

Consider the T–S fuzzy model in (3.1) with $u \equiv 0$ and $a_l \equiv 0$ as follows.

$$R^l: \quad \text{IF} \quad z_1 \text{ is } F_1^l \text{ and } \dots z_v \text{ is } F_v^l$$

$$\text{THEN} \quad x(t+1) = A_l x(t), \tag{4.1}$$

$$l \in L := \{1, 2, \dots, m\},$$

which can also be described by

$$x(t+1) = \sum_{l=1}^{m} \mu_l(z) A_l x(t). \tag{4.2}$$

One of the first results on stability analysis based on common quadratic Lyapunov functions is suggested in Tanaka and Sugeno (1992), and since then numerous modifications and improved methods have been proposed. Before presenting the stability results, we first introduce a useful lemma.

Lemma 4.1 (Tanaka and Sugeno, 1992)

If P is a positive definite matrix and matrices A_l, A_j are of appropriate dimensions such that

$$A_l^T P A_l - P < 0 \text{ and } A_j^T P A_j - P < 0,$$

then

$$A_i^T P A_j + A_j^T P A_i - 2P < 0. \tag{4.3}$$

Proof: One can easily verify that

$$A_i^T P A_j + A_j^T P A_i - 2P = -(A_i - A_j)^T P(A_i - A_j) + \left(A_i^T P A_i - P\right) + \left(A_j^T P A_j - P\right)$$

$$\leq \left(A_i^T P A_i - P\right) + \left(A_j^T P A_j - P\right).$$

Then the claimed result follows directly. ❑

By defining a Lyapunov function candidate as

$$V(x) = x^T P x, \tag{4.4}$$

where the matrix P is positive definite, the following result can be readily obtained (Tanaka and Sugeno, 1992).

Theorem 4.1

The T–S fuzzy system (4.1), or equivalently (4.2), is globally exponentially stable

1. If there exists a positive definite matrix P such that the following linear matrix inequalities (LMIs) are satisfied,

$$A_l^T P A_l - P < 0, \quad l \in L; \tag{4.5}$$

   or equivalently,

2. If there exists a positive definite matrix X such that the following linear matrix inequalities (LMIs) are satisfied,

$$\begin{bmatrix} -X & X A_l^T \\ A_l X & -X \end{bmatrix} < 0, \quad l \in L. \tag{4.6}$$

Proof: Choose a Lyapunov function candidate as in (4.4). There exist positive constants α, β such that

$$\alpha \, \| x \|^2 \leq V(t) \leq \beta \, \| x \|^2 . \tag{4.7}$$

It follows from Lemma 4.1 that if the condition in (4.5) is satisfied, one has

$$A_i^T P A_j + A_j^T P A_i - 2P < 0, \forall l, j \in L, \tag{4.8}$$

which in turn implies that there exists a constant $\rho > 0$ such that

$$A_i^T P A_j + A_j^T P A_i - 2P + \rho I < 0. \tag{4.9}$$

Then along all trajectories of the system in (4.1) with $x(t) \neq 0$, one has

$$\Delta V(t) = V(x(t+1)) - V(x(t))$$

$$= \left[\sum_{l=1}^{m} \mu_l A_l x(t) \right]^T P \left[\sum_{l=1}^{m} \mu_l A_l x(t) \right] - x(t)^T P x(t)$$

$$= \sum_{l=1}^{m} \sum_{j=1}^{m} \mu_l \mu_j x(t)^T A_l^T P A_j x(t) - x(t)^T P x(t)$$

$$= \sum_{l=1}^{m} \sum_{j=1}^{m} \mu_l \mu_j x(t)^T \left(A_l^T P A_j - P \right) x(t)$$

$$= \frac{1}{2} \sum_{l=1}^{m} \sum_{j=1}^{m} \mu_l \mu_j x(t)^T \left(A_l^T P A_j + A_j^T P A_l - 2P \right) x(t)$$

$$\leq -\rho \| x(t) \|^2 . \tag{4.10}$$

Therefore, the claimed exponential stability result in Part 1 is established based on standard Lyapunov stability theory. The equivalent condition, 2, follows directly from the Schur complement Lemma A.2 together with $X = P^{-1}$, and thus the proof is completed. ❑

Remark 4.1

Note that the form of LMIs in (4.6) is more suitable to controller synthesis as can be observed in the next few chapters. Also note that (4.6) implies that its feasible solution of X is positive definite. The conditions (4.5) or (4.6) are linear matrix inequalities in the variable P or X, respectively. The feasibilities of these LMIs, as well as other LMIs in the rest of this and subsequent sections and chapters, are easily tested by commercially available software packages such as LMI Toolbox (Gahinet et al., 1995).

Remark 4.2

It should also be noted that the stability conditions in Theorem 4.1 are independent of membership functions of the T–S fuzzy systems and this will inevitably result in some degree of conservatism because the useful information of membership functions is not utilized.

Example 4.1

Consider a simple T–S fuzzy system in the form of (4.1) with local matrices

$$A_1 = \begin{bmatrix} 0.95 & 0.01 \\ -0.05 & 0.95 \end{bmatrix}, \quad A_2 = \begin{bmatrix} 0.95 & 0.05 \\ -0.05 & 0.95 \end{bmatrix},$$

and membership functions as shown in Figure 4.1.

By applying the MATLAB® LMI toolbox to solve (4.5) with the above matrices, one can easily find that there exists a positive definite matrix,

$$P = \begin{bmatrix} 1.3652 & 0.0332 \\ 0.0332 & 1.3568 \end{bmatrix},$$

satisfying (4.5) and thus it can be concluded from Theorem 4.1 that the T–S fuzzy system is globally exponentially stable. Simulation results with four different initial conditions are plotted in Figure 4.2 where convergence to the equilibrium, that is, the origin, can be observed.

However, stability analysis approaches based on common quadratic Lyapunov functions tend to be conservative. This can be illustrated by the following example.

Example 4.2

Consider a T–S fuzzy system as follows,

$$R^l: \quad \text{IF} \qquad x_1 \text{ is } F^l$$

$$\text{THEN} \quad x(t+1) = A_l x(t) \tag{4.11}$$

$$l \in L := \{1, \dots, 7\},$$

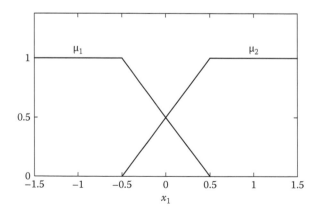

FIGURE 4.1 Membership functions in Example 4.1.

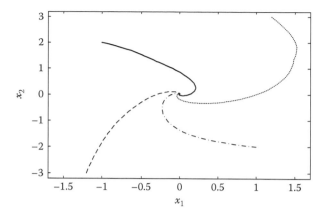

FIGURE 4.2 Trajectories from four initial conditions.

where F^l represents a set of fuzzy labels {negative very large, negative large, negative small, around zero, positive small, positive large, positive very large} and the corresponding membership functions are defined in Figure 4.3.

The system matrices are given as

$$A_1 = \begin{bmatrix} 1 & 0.5 \\ -0.3 & 0.8 \end{bmatrix}, A_2 = \begin{bmatrix} 1 & 0.4875 \\ -0.275 & 0.8 \end{bmatrix}, A_3 = \begin{bmatrix} 1 & 0.475 \\ -0.25 & 0.8 \end{bmatrix}, A_4 = \begin{bmatrix} 1 & 0.45 \\ -0.2 & 0.8 \end{bmatrix},$$

$$A_5 = \begin{bmatrix} 1 & 0.425 \\ -0.15 & 0.8 \end{bmatrix}, A_6 = \begin{bmatrix} 1 & 0.4125 \\ -0.125 & 0.8 \end{bmatrix}, A_7 = \begin{bmatrix} 1 & 0.4 \\ -0.1 & 0.8 \end{bmatrix}.$$

By using the MATLAB LMI toolbox, one can easily verify that there exists no positive definite matrix P for the fuzzy system to show its stability. In other

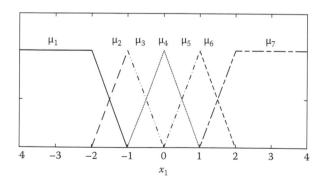

FIGURE 4.3 Membership functions in Example 4.2.

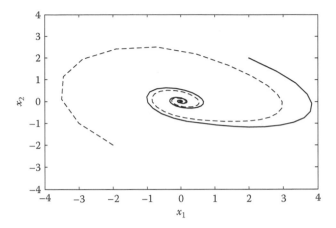

FIGURE 4.4   Trajectories from two initial conditions.

words, the fuzzy system does not admit a global quadratic Laypunov function. However, the simulation results shown in Figure 4.4 indicate that the system is stable.

This example confirms that the stability analysis approaches based on common (global) quadratic Lyapunov functions tend to be conservative, and even worse, such functions might not exist for many highly complex nonlinear systems as demonstrated in this example and many others (Feng, 2004a; Johansson, Rantzer, and Arzen, 1999). This is one of the main limitations of this kind of approach.

4.3   STABILITY ANALYSIS BASED ON PIECEWISE QUADRATIC LYAPUNOV FUNCTIONS

Because common quadratic Lyapunov functions tend to be conservative in stability analysis of T–S fuzzy systems, it is thus desirable to develop less conservative stability results for these systems. Piecewise quadratic Lyapunov functions are one of the options available for this purpose. In order to facilitate development of approaches based on piecewise quadratic Lyapunov functions, one needs to partition the premise variable space of a fuzzy system, or the state space in the case of $z(t) = x(t)$. The following partition is referred to as the first kind in the sequel (Cao, Rees, and Feng, 1996a,b, 1997b).

Define m regions in the premise variable space as follows.

$$S_l = \{z \mid \mu_l(z) > \mu_i(z), \ i = 1, 2, \ldots, m, \ i \neq l\}, \ l \in L. \tag{4.12}$$

Then the global model of the T–S fuzzy system (4.2) can be expressed in each local region as

$$x(t+1) = (A_l + \Delta A_l(\mu)) x(t), \ z(t) \in S_l, \ l \in L, \tag{4.13}$$

where

$$\Delta A_l(\mu) = \sum_{i=1,i\neq l}^{m} \mu_i \Delta A_{li}, \quad \Delta A_{li} = A_i - A_l.$$

Note that the number of regions in this kind of partition is the same as the number of fuzzy rules or the number of local linear models, and that the fuzzy model (4.13) is different from the local model in the T–S fuzzy model (4.1) because it considers all interactions among the local models of (4.1) in terms of uncertainty ΔA_l and is, in fact, the global fuzzy model (4.2) expressed in the local region S_l.

For the purpose of stability analysis and stabilization, we introduce the following upper bounds for the uncertainty term of the fuzzy model (4.13),

$$[\Delta A_l(\mu)]^T [\Delta A_l(\mu)] \leq E_{lA}^T E_{lA}, \; l \in L. \tag{4.14}$$

Note that there are many ways to obtain these upper bounds; interested readers can refer to Cao, Rees, and Feng (1996a,b, 1997b) for details. For subsequent use, we need the following lemma.

Lemma 4.2

Let A and E be matrices of appropriate dimensions, and P be a symmetric matrix satisfying

$$\frac{1}{\varepsilon} I - P > 0, \quad \varepsilon > 0;$$

then

$$A^T PE + E^T PA + E^T PE \leq A^T P \left(\frac{1}{\varepsilon} I - P \right)^{-1} PA + \frac{1}{\varepsilon} E^T E.$$

Proof: Let

$$Y = \left(\frac{1}{\varepsilon} I - P \right)^{-1/2} PA - \left(\frac{1}{\varepsilon} I - P \right)^{1/2} E.$$

Then the claimed result follows directly from expanding $Y^T Y \geq 0$; that is,

$$Y^T Y = \left[\left(\frac{1}{\varepsilon} I - P \right)^{-1/2} PA - \left(\frac{1}{\varepsilon} I - P \right)^{1/2} E \right]^T \left[\left(\frac{1}{\varepsilon} I - P \right)^{-1/2} PA - \left(\frac{1}{\varepsilon} I - P \right)^{1/2} E \right]$$

$$= A^T P \left(\frac{1}{\varepsilon} I - P \right)^{-1} PA - A^T PE - E^T PA + E^T \left(\frac{1}{\varepsilon} I - P \right) E \geq 0,$$

and thus the proof is completed. ❑

In addition, define a set Ω that represents all possible system transitions among regions; that is,

$$\Omega := \{(l, j) \mid z(t) \in S_l, z(t+1) \in S_j, \forall l, j \in L, l \neq j\}. \tag{4.15}$$

Remark 4.3

The set Ω can be determined by reachability analysis, which is described by Algorithm 2 in Ferrari-Trecate et al. (2002). The basic idea is to compute the evolution from the initial set in order to explore all possible state trajectories. After running the algorithm, all possible trajectories can be identified and thus the set Ω can be determined. If it is possible for the transitions to happen between all regions, then $\Omega = L \times L := \{(l, j) \mid l, j \in L, j \neq l\}$.

One then is ready to present a stability result based on the following piecewise quadratic Lyapunov function candidate.

$$V(x) = x^T P_l x, \quad z \in S_l, \quad l \in L. \tag{4.16}$$

Theorem 4.2 (Feng, 2004a)

The T–S fuzzy system (4.1) (or equivalently (4.13)) is globally exponentially stable

1. If there exists a set of positive definite matrices $P_l, l \in L$ such that the following LMIs are satisfied,

$$\begin{bmatrix} A_l^T P_l A_l - P_l + E_{lA}^T E_{lA} & A_l^T P_l \\ P_l A_l & -(I - P_l) \end{bmatrix} < 0, \quad l \in L \tag{4.17}$$

$$\begin{bmatrix} A_l^T P_j A_l - P_l + E_{lA}^T E_{lA} & A_l^T P_j \\ P_j A_l & -(I - P_j) \end{bmatrix} < 0, \quad l, j \in \Omega; \tag{4.18}$$

or equivalently,

2. If there exists a set of positive definite matrices $X_l, l \in L$ such that the following LMIs are satisfied,

$$\begin{bmatrix} -X_l & X_l A_l^T & X_l E_{lA}^T \\ A_l X_l & -(X_l - I) & 0 \\ E_{lA} X_l & 0 & -I \end{bmatrix} < 0, \, l \in L, \tag{4.19}$$

$$\begin{bmatrix} -X_l & X_l A_l^T & X_l E_{lA}^T \\ A_l X_l & -(X_j - I) & 0 \\ E_{lA} X_l & 0 & -I \end{bmatrix} < 0, \, l, j \in \Omega. \tag{4.20}$$

Proof: Consider the piecewise quadratic Lyapunov function candidate defined in (4.16). It follows directly that there exist positive constants α, β such that

$$\alpha \, \|x\|^2 \le V(t) \le \beta \, \|x\|^2. \tag{4.21}$$

Using Lemma 4.2, one has

$$(A_l + \Delta A_l)^T P_j (A_l + \Delta A_l) - P_l$$

$$= A_l^T P_j A_l - P_l + A_l^T P_j \Delta A_l + \Delta A_l^T P_j A_l + \Delta A_l^T P_j \Delta A_l$$

$$\le A_l^T P_j A_l - P_l + A_l^T P_j \left(\frac{1}{\varepsilon} I - P_j \right)^{-1} P_j A_l + \frac{1}{\varepsilon} \Delta A_l^T \Delta A_l$$

$$\le A_l^T P_j A_l - P_l + A_l^T P_j \left(\frac{1}{\varepsilon} I - P_j \right)^{-1} P_j A_l + \frac{1}{\varepsilon} E_{lA}^T E_{lA}, \tag{4.22}$$

where $j = l$ when x stays in the region S_l, $j \ne l$ when x transits from the region S_l to S_j.

On the other hand, using the Schur complement Lemma A.2, it follows that (4.17) and (4.18) imply

$$A_l^T P_j A_l - P_l + A_l^T P_j (I - P_j)^{-1} P_j A_l + E_{lA}^T E_{lA} < 0, \tag{4.23}$$

which in turn implies

$$\frac{1}{\varepsilon} A_l^T P_j A_l - \frac{1}{\varepsilon} P_l + \frac{1}{\varepsilon} A_l^T P_j (I - P_j)^{-1} P_j A_l + \frac{1}{\varepsilon} E_{lA}^T E_{lA} < 0, \forall \varepsilon > 0. \tag{4.24}$$

Rewriting (4.24) as

$$A_l^T \frac{1}{\varepsilon} P_j A_l - \frac{1}{\varepsilon} P_l + A_l^T \frac{1}{\varepsilon} P_j \left(\frac{1}{\varepsilon} I - \frac{1}{\varepsilon} P_j \right)^{-1} \frac{1}{\varepsilon} P_j A_l + \frac{1}{\varepsilon} E_{lA}^T E_{lA} < 0, \tag{4.25}$$

and denoting $(1/\varepsilon) P_l$ as P_l and $(1/\varepsilon) P_j$ as P_j lead to

$$A_l^T P_j A_l - P_l + A_l^T P_j \left(\frac{1}{\varepsilon} I - P_j \right)^{-1} P_j A_l + \frac{1}{\varepsilon} E_{lA}^T E_{lA} < 0. \tag{4.26}$$

Thus it follows from (4.22) and (4.26) that

$$(A_l + \Delta A_l)^T P_j (A_l + \Delta A_l) - P_l < 0, \tag{4.27}$$

which in turn implies that there exists a constant $\rho > 0$ such that

$$(A_l + \Delta A_l)^T P_j (A_l + \Delta A_l) - P_l + \rho I < 0. \tag{4.28}$$

Then along trajectories of the system (4.13) with $x(t) \neq 0$, one has

$$
\begin{aligned}
\Delta V(t) &= V(x(t+1)) - V(x(t)) \\
&= [(A_l + \Delta A_l)x(t)]^T P_j[(A_l + \Delta A_l)x(t)] - x(t)^T P_l x(t) \\
&= x(t)^T \left[(A_l + \Delta A_l)^T P_j (A_l + \Delta A_l) - P_l \right] x(t) \\
&\leq x(t)^T (-\rho I)x(t) \\
&= -\rho \| x \|^2 .
\end{aligned}
\tag{4.29}
$$

Therefore, the desired result in Condition 1 follows directly from (4.21) and (4.29) based on standard Lyapunov theory.

Similar to the proof of Theorem 4.1, the equivalence between conditions in Part 1 and Part 2 follows directly from the Schur complements together with $X = P^{-1}$, and thus the proof is completed. ❑

The conditions (4.17) and (4.18), or equivalently, (4.19) and (4.20), are linear matrix inequalities in the variables $P_l, l \in L$ or equivalently, $X_l (= P_l^{-1})$, $l \in L$. A solution to these inequalities ensures $V(x) = x^T P_l x$, $z(t) \in S_l$ or equivalently, $V(x) = x^T X_l^{-1} x$, $z(t) \in S_l$, to be a Lyapunov function for the system. The LMIs in (4.17) or (4.19) guarantee that the function decreases along all system trajectories within each region. The LMIs in (4.18) or (4.20) guarantee that the function decreases when the system trajectories transit from one region to another.

Note that the uncertainty terms are introduced in (4.13) for the first kind of partition approach described in (4.12), and this would normally lead to the increase of conservatism of the corresponding stability analysis results because the worst case of uncertainties is considered as shown in (4.14). Some search approaches for approximate upper bounds instead of the worst-case bounds for these uncertainties are presented in Cao, Rees, and Feng (1996a,b, 1997b).

Example 4.3

To illustrate the application of Theorem 4.2, we reconsider Example 4.2. By applying Theorem 4.2, one finds the solutions to those LMIs as follows.

$$
P_1 = \begin{bmatrix} 0.0191 & 0.0064 \\ 0.0064 & 0.0334 \end{bmatrix}, P_2 = \begin{bmatrix} 0.0188 & 0.0067 \\ 0.0067 & 0.0337 \end{bmatrix}, P_3 = \begin{bmatrix} 0.0185 & 0.0071 \\ 0.0071 & 0.0348 \end{bmatrix},
$$

$$
P_4 = \begin{bmatrix} 0.0181 & 0.0079 \\ 0.0079 & 0.0361 \end{bmatrix}, P_5 = \begin{bmatrix} 0.0178 & 0.0087 \\ 0.0087 & 0.0381 \end{bmatrix}, P_6 = \begin{bmatrix} 0.0172 & 0.0091 \\ 0.0091 & 0.0409 \end{bmatrix},
$$

$$
P_7 = \begin{bmatrix} 0.0170 & 0.0095 \\ 0.0095 & 0.0434 \end{bmatrix},
$$

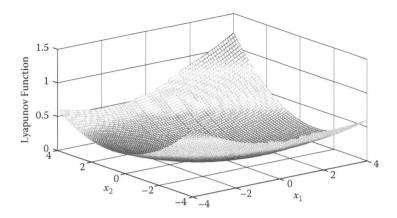

FIGURE 4.5 Surface of Lyapunov function in Example 4.3.

and it is thus verified that the system is globally exponentially stable. The surface of the corresponding Lyapunov function is shown in Figure 4.5. Note that the uncertainty bounds are chosen as the maximum differences between the system matrices of adjacent regions.

Another approach to space partition, which is referred to be the second kind in the sequel, was suggested by Johansson, Rantzer, and Arzen (1999). The partition is based on the natural induction of the fuzzy system (4.1) into a number of polyhedral regions $\{S_l\}_{l \in \bar{L}} \subseteq \Re^v$ of the premise variable space. These consist of crisp (operating) and fuzzy (interpolation) regions. The crisp region is defined as the region where $\mu_l(z) = 1$ for some $l \in \bar{L}$; all other membership functions evaluate to zero. The system dynamic of a crisp region is governed by one of the local models of the fuzzy system (4.1). On the other hand, the fuzzy region is defined as the region where $0 < \mu_l(z) < 1$ and the system dynamic is governed by a convex combination of several local linear models. In the extreme case where all the regions of a T–S fuzzy system are crisp, that is, $\mu_l(z) = 1$ for some l and all other membership functions are equal to zero, then the global fuzzy model (4.2) becomes a piecewise linear system,

$$x(t+1) = A_l x(t), \quad z(t) \in S_l, \quad l \in \bar{L}. \tag{4.30}$$

Note that the number of regions in the set $\bar{L}$ for the second kind of space partition is different from the number of the local linear models in general which is the number of regions in the first kind of space partition approach.

With such a partition, we can rewrite the global fuzzy model (4.2) in each region as a convex combination of linear models,

$$x(t+1) = \sum_{k \in \aleph(l)} \mu_k(z)\{A_k x(t)\}, z(t) \in S_l, \quad l \in \bar{L} \tag{4.31}$$

with $0 \le \mu_k(z) \le 1$, $\Sigma_{k \in \aleph(l)} \mu_k(z) = 1$. For each region S_l, the set $\aleph(l)$ contains the indices for the system matrices used in the interpolation within that region. For a crisp region, $\aleph(l)$ contains a single element.

In comparison with the global fuzzy model (4.2), the fuzzy model (4.31) is described in each local region. This is similar to (4.13) for the first kind of space partition approach.

Similar to (4.15), we define a set $\bar{\Omega}$ that represents all possible transitions among regions of the system (4.31); that is,

$$\bar{\Omega} := \{(l, j) \mid z(t) \in S_l, z(t+1) \in S_j, \forall l, j \in \bar{L}, l \ne j\}. \tag{4.32}$$

Then one is ready to present a stability result based on a similar piecewise quadratic Lyapunov function candidate defined in (4.16).

Theorem 4.3 (Wang and Feng, 2004)

The T–S fuzzy system (4.1), or equivalently (4.31), is globally exponentially stable

1. If there exists a set of positive definite matrices $P_l, l \in \bar{L}$ such that the following LMIs are satisfied,

$$A_k^T P_l A_k - P_l < 0, \ l \in \bar{L}, \ k \in \aleph(l), \tag{4.33}$$

$$A_k^T P_j A_k - P_l < 0, \ (l, j) \in \bar{\Omega}, \ k \in \aleph(l); \tag{4.34}$$

or equivalently,
2. If there exists a set of positive definite matrices X_l, $l \in \bar{L}$ such that the following LMIs are satisfied,

$$\begin{bmatrix} -X_l & X_l A_k^T \\ A_k X_l & -X_l \end{bmatrix} < 0, \ l \in \bar{L}, \ k \in \aleph(l), \tag{4.35}$$

$$\begin{bmatrix} -X_l & X_l A_k^T \\ A_k X_l & -X_j \end{bmatrix} < 0, \ (l, j) \in \bar{\Omega}, \ k \in \aleph(l). \tag{4.36}$$

Proof: The proof is similar to that of Theorem 4.2 and thus omitted. ❑

When the positive definite matrices in (4.33) and (4.34) (or equivalently (4.35) and (4.36)) are chosen as a common one, that is, $P_1 = P_2 = \cdots = P_m = P$ (or equivalently, $X_1 = X_2 = \cdots X_m = X = P^{-1}$), then the result of Theorem 4.3 reduces to that of Theorem 4.1. It thus can be easily seen that common quadratic Lyapunov functions are a special case of the more general piecewise quadratic Lyapunov functions, and

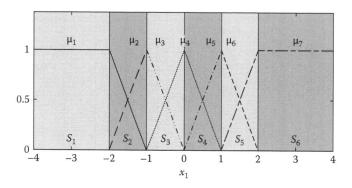

FIGURE 4.6 Space partitions.

the latter is less conservative. However, the computation cost of the latter would be higher in general.

Example 4.4

Reconsider the T–S fuzzy system defined in Example 4.2. The state space partition with the second kind of partition approach is shown in Figure 4.6. With the second kind of partition, one has

$$x(t+1) = A_1 x(t), \qquad\qquad x(t) \in S_1$$

$$x(t+1) = \mu_1(x)A_1 x(t) + \mu_2(x)A_2(t), \qquad x(t) \in S_2$$

$$x(t+1) = \mu_3(x)A_3 x(t) + \mu_4(x)A_4(t), \qquad x(t) \in S_3$$

$$x(t+1) = \mu_4(x)A_4 x(t) + \mu_5(x)A_5 x(t), \qquad x(t) \in S_4$$

$$x(t+1) = \mu_6(x)A_6 x(t) + \mu_7(x)A_7(t), \qquad x(t) \in S_5$$

$$x(t+1) = A_7 x(t), \qquad\qquad x(t) \in S_6.$$

By applying Theorem 4.3, one finds the solutions to those LMIs as follows,

$$P_1 = \begin{bmatrix} 36.0140 & 12.2633 \\ 12.2633 & 63.3344 \end{bmatrix},\ P_2 = \begin{bmatrix} 35.9967 & 13.0170 \\ 13.0170 & 65.1237 \end{bmatrix},$$

$$P_3 = \begin{bmatrix} 34.6821 & 14.7888 \\ 14.7888 & 66.3461 \end{bmatrix},\ P_4 = \begin{bmatrix} 32.9278 & 16.7999 \\ 16.7999 & 68.8253 \end{bmatrix},$$

$$P_5 = \begin{bmatrix} 31.8061 & 19.3030 \\ 19.3030 & 70.1430 \end{bmatrix},\ P_6 = \begin{bmatrix} 30.4213 & 20.4640 \\ 20.4640 & 69.1775 \end{bmatrix},$$

and it is thus verified that the system is globally exponentially stable. The surface of the corresponding Lyapunov function is shown in Figure 4.7.

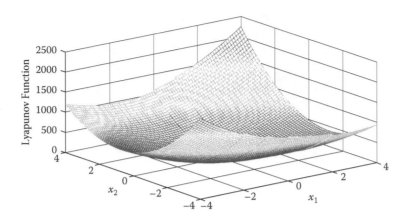

FIGURE 4.7 Surface of Lyapunov function in Example 4.4.

4.4 STABILITY ANALYSIS BASED ON FUZZY QUADRATIC LYAPUNOV FUNCTIONS

In addition to common and piecewise quadratic Lyapunov functions discussed in the previous two sections, the so-called nonquadratic or fuzzy quadratic Lyapunov functions can also be used to determine stability of T–S fuzzy systems (Choi and Park, 2003; Guerra and Vermeiren, 2004; Zhou et al., 2005). The fuzzy Lyapunov function is defined as

$$V(x) = \sum_{l=1}^{m} \mu_l(z) x^T P_l x. \tag{4.37}$$

A useful lemma is introduced.

Lemma 4.3

If P_l, P_j are positive definite square matrices and A, B are matrices of appropriate dimensions such that

$$A^T P_j A - P_l < 0 \text{ and } B^T P_j B - P_l < 0,$$

then

$$A^T P_j B + B^T P_j A - 2P_l < 0. \tag{4.38}$$

Proof: One can easily verify that

$$A^T P_j B + B^T P_j A - 2P_l = -(A - B)^T P_j (A - B) + \left(A^T P_j A - P_l \right) + \left(B^T P_j B - P_l \right)$$

$$\leq \left(A^T P_j A - P_l \right) + \left(B^T P_j B - P_l \right).$$

Thus the claimed result follows directly. ❑

Then one has the following stability result.

Theorem 4.4 (Zhou et al., 2005)

The T–S fuzzy system (4.1), or equivalently (4.2), is globally exponentially stable

1. If there exists a set of positive definite matrices $P_l, l \in L$ such that the following LMIs are satisfied,

$$A_l^T P_j A_l - P_l < 0, \quad j,l \in L, \tag{4.39}$$

   or equivalently,

2. If there exists a set of positive definite matrices $X_l, l \in L$ such that the following LMIs are satisfied,

$$\begin{bmatrix} -X_l & X_l A_l^T \\ A_l X_l & -X_j \end{bmatrix} < 0, \quad j,l \in L. \tag{4.40}$$

Proof: Choose a Lyapunov function candidate as in (4.37); one has that there exist positive constants α, β such that

$$\alpha \| x \|^2 \le V(t) \le \beta \| x \|^2. \tag{4.41}$$

It follows from Lemma 4.3 that if the condition in (4.39) is satisfied, one has

$$A_l^T P_j A_j + A_j^T P_j A_l - 2P_l < 0, \forall l, j \in L, \tag{4.42}$$

which in turn implies that there exists a constant $\rho > 0$ such that

$$A_l^T P_j A_j + A_j^T P_j A_l - 2P_l + \rho I < 0. \tag{4.43}$$

Then along all trajectories of the system in (4.1) with $x(t) \ne 0$, one has

$$\Delta V(t) = V(x(t+1)) - V(x(t))$$

$$= \left[\sum_{i=1}^{m} \mu_i(z(t)) A_i x(t) \right]^T \sum_{j=1}^{m} \mu_j(z(t+1)) P_j \left[\sum_{l=1}^{m} \mu_l(z(t)) A_l x(t) \right]$$

$$- x(t)^T \sum_{l=1}^{m} \mu_l(z(t)) P_l x(t)$$

$$= \sum_{i=1}^{m} \sum_{l=1}^{m} \sum_{j=1}^{m} \mu_i(z(t)) \mu_l(z(t)) \mu_j(z(t+1)) \left[x(t)^T A_i^T P_j A_l x(t) - x(t)^T P_l x(t) \right]$$

$$= \sum_{i=1}^{m} \sum_{l=1}^{m} \sum_{j=1}^{m} \mu_i(z(t)) \mu_l(z(t)) \mu_j(z(t+1)) x(t)^T \left(A_i^T P_j A_l - P_l \right) x(t)$$

$$= \frac{1}{2} \sum_{i=1}^{m} \sum_{l=1}^{m} \sum_{j=1}^{m} \mu_i(z(t)) \mu_l(z(t)) \mu_j(z(t+1)) x(t)^T \left(A_l^T P_j A_i + A_i^T P_j A_l - 2P_l \right) x(t)$$

$$\le -\rho \| x(t) \|^2. \tag{4.44}$$

Therefore, the claimed exponential stability result in Part 1 is established based on standard Lyapunov stability theory. Similar to the stability results in the last two sections, the equivalent Condition 2 follows directly from the Schur complements together with $X_l = P_l^{-1}$, and thus the proof is completed. ❑

Note that (4.40) implies that its feasible solution of X_l is positive definite. Also note that when the positive definite matrices in (4.39) (or equivalently (4.40)) are chosen as a common one, that is, $P_1 = \cdots = P_m = P$ (or equivalently, $X_1 = \cdots = X_m = X = P^{-1}$), then the result of Theorem 4.4 reduces to that of Theorem 4.1. It thus can be easily seen that common quadratic Lyapunov functions are a special case of the more general fuzzy Lyapunov functions, and the latter are less conservative. However, similar to piecewise quadratic Lyapunov functions, the computation cost of the latter would be much higher.

Remark 4.4
In the case of continuous-time T–S fuzzy systems, stability analysis via fuzzy Lyapunov functions is much more involved or difficult than that for their discrete-time counterparts due to the fact that the membership functions are in general functions of the system states and the time derivative of fuzzy Lyapunov functions involves derivatives of the membership functions, and thus the derivatives of the system states. As a result, the derivative of the fuzzy Lyapunov function becomes much more complex; in fact it becomes a nonlinear function in terms of the system matrices thus leading to difficulty in stability analysis.

Example 4.5
Reconsider the T–S fuzzy system defined in Example 4.2. By applying Theorem 4.4, one obtains the solutions to those LMIs as follows,

$$P_1 = \begin{bmatrix} 9.2116 & 3.3799 \\ 3.3799 & 17.3330 \end{bmatrix}, \; P_2 = \begin{bmatrix} 9.0738 & 3.5579 \\ 3.5579 & 17.4344 \end{bmatrix},$$

$$P_3 = \begin{bmatrix} 9.0060 & 3.7291 \\ 3.7291 & 17.4134 \end{bmatrix}, \; P_4 = \begin{bmatrix} 8.9435 & 4.0682 \\ 4.0682 & 17.2726 \end{bmatrix},$$

$$P_5 = \begin{bmatrix} 8.9279 & 4.4118 \\ 4.4118 & 17.1085 \end{bmatrix}, \; P_5 = \begin{bmatrix} 8.9265 & 4.5887 \\ 4.5887 & 17.0471 \end{bmatrix},$$

$$P_7 = \begin{bmatrix} 8.9191 & 4.7735 \\ 4.7735 & 17.0290 \end{bmatrix},$$

and it is thus verified that the system is globally exponentially stable. The surface of the corresponding Lyapunov function is shown in Figure 4.8.

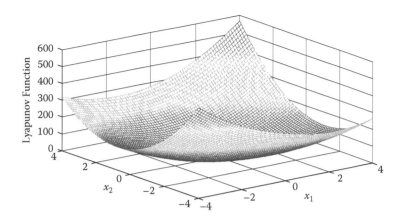

FIGURE 4.8 Surface of Lyapunov function in Example 4.5.

4.5 STABILITY ANALYSIS OF T–S FUZZY AFFINE SYSTEMS BASED ON PIECEWISE QUADRATIC LYAPUNOV FUNCTIONS

Stability analysis of T–S fuzzy models with affine terms, that is, $a_l \neq 0$ for some l in (3.1), is much more involved. The analysis based on common quadratic Lyapunov functions is suggested in Kim and Kim (2001, 2002), and the analysis based on piecewise quadratic Lyapunov functions is suggested in Feng (2004a), Johansson, Rantzer, and Arzen (1999), and Wang and Feng (2004). Here for illustration a result based on the second kind of space partition and piecewise quadratic Lyapunov functions is presented.

In this case, define $L_0 \subseteq \bar{L}$ as the set of indices for the regions that contain the origin and $L_1 \subseteq \bar{L}$ the set of indices for the regions that do not contain the origin. For convenient notation, also define

$$\bar{A}_k = \begin{bmatrix} A_k & a_k \\ 0 & 1 \end{bmatrix}, \bar{x} = \begin{bmatrix} x \\ 1 \end{bmatrix}, \tag{4.45}$$

where $a_k = 0$ for all $k \in \aleph(l)$ with $l \in L_0$. Then using this notation, similar to (4.31), the fuzzy system (3.1) with $u \equiv 0$ can be expressed as

$$\bar{x}(t+1) = \sum_{k \in \aleph(l)} \mu_k(z)\bar{A}_k\bar{x}(t), \quad z(t) \in S_l, l \in \bar{L}. \tag{4.46}$$

The following Lyapunov function candidate is used:

$$V(x) = \begin{cases} x^T P_l x, & z \in S_l, \ l \in L_0 \\ \bar{x}^T \bar{P}_l \bar{x}, & z \in S_l, \ l \in L_1. \end{cases} \tag{4.47}$$

This function combines the power of quadratic Lyapunov functions near the equilibrium point with the flexibility of piecewise affine functions in the large.

Because the matrix P_l or $\bar{P}_l$ is only used to describe the Lyapunov function in the local region S_l, the S-procedure (Boyd et al., 1994) can be used to reduce conservatism of the stability results. To this end, the characteristics or information of the local regions can be utilized. As partitions of premise variable space induced from fuzzy membership functions are polyhedra, the matrix $\bar{E}_l$, $l \in \bar{L}$ can be constructed for each region such that

$$\bar{E}_l \bar{x} \geq 0, \qquad z \in S_l, \, l \in \bar{L},$$

where $\bar{E}_l = [E_l \quad e_l], \, l \in \bar{L}$ with E_l being an $n \times n$ matrix, e_l being an $n \times 1$ vector, and moreover $e_l = 0_{n \times 1}$ for $l \in L_0$. It should be noted that the above vector inequality is defined as elementwise; that is, each entry of the vector is nonnegative. A systematic procedure for constructing these matrices $\bar{E}_l, \, l \in L$ for a given T–S fuzzy system can be found in Johansson, Rantzer, and Arzen (1999). The procedure is based directly on the information in the fuzzy rule base.

Then one is ready to present the following stability result (Feng, 2004a; Wang and Feng, 2004).

Theorem 4.5

The T–S fuzzy system (3.1) with $u \equiv 0$, or equivalently (4.46) is globally exponentially stable, if there exist symmetric matrices P_l, $l \in L_0$, $\bar{P}_l$, $l \in L_1$ and symmetric matrices U_l, Q_{lk}, Q_{ljk} such that U_l, Q_{lk}, Q_{ljk} have nonnegative entries, and the following LMIs are satisfied,

$$0 < P_l - E_l^T U_l E_l, \, l \in L_0, \tag{4.48}$$

$$A_k^T P_l A_k - P_l + E_l^T Q_{lk} E_l < 0, \, l \in L_0, \, k \in \aleph(l), \tag{4.49}$$

$$0 < \bar{P}_l - \bar{E}_l^T U_l \bar{E}_l, \, l \in L_1, \tag{4.50}$$

$$\bar{A}_k^T \bar{P}_l \bar{A}_k - \bar{P}_l + \bar{E}_l^T Q_{lk} \bar{E}_l < 0, \, l \in L_1, \, k \in \aleph(l), \tag{4.51}$$

$$A_k^T P_j A_k - P_l + E_l^T Q_{ljk} E_l < 0, \, (l, j) \in \bar{\Omega}, \, l, j \in L_0, \, k \in \aleph(l), \tag{4.52}$$

$$\bar{A}_k^T \bar{P}_j \bar{A}_k - \bar{P}_l + \bar{E}_l^T Q_{ljk} \bar{E}_l < 0, \, (l, j) \in \bar{\Omega}, \, l, j \in L_1, \, k \in \aleph(l), \tag{4.53}$$

$$\bar{A}_k^T \hat{P}_j \bar{A}_k - \bar{P}_l + \bar{E}_l^T Q_{ljk} \bar{E}_l < 0, \, (l, j) \in \bar{\Omega}, \, l \in L_1, \, j \in L_0, \, k \in \aleph(l), \tag{4.54}$$

$$A_k^T \tilde{P}_j A_k - P_l + E_l^T Q_{ljk} E_l < 0, \, (l, j) \in \bar{\Omega}, \, l \in L_0, \, j \in L_1, \, k \in \aleph(l), \tag{4.55}$$

where

$$\hat{P}_j = [I_{n\times n} \quad 0_{n\times 1}]^T P_j [I_{n\times n} \quad 0_{n\times 1}] \text{ for } (4.54),$$

$$\tilde{P}_j = [I_{n\times n} \quad 0_{n\times 1}]\overline{P}_j [I_{n\times n} \quad 0_{n\times 1}]^T \text{ for } (4.55).$$

Proof: Consider the Lyapunov function candidate (4.47), or in a more compact form,

$$V(t) = \overline{x}(t)^T \overline{P}_i \overline{x}(t), \, x \in S_i, \, i \in L. \tag{4.56}$$

It is obvious from (4.56) that there exists a constant $\gamma > 0$ such that for $x \in S_i$, $i \in L_0$,

$$V(t) \le \gamma \|x\|^2, \tag{4.57}$$

and for $x \in S_i$, $i \in L_1$,

$$V(t) \le \gamma \|\overline{x}\|^2 = \gamma(\|x\|^2 + 1) \le \gamma\left(\|x\|^2 + \frac{\|x\|^2}{c}\right) = \frac{\gamma(1+c)}{c}\|x\|^2, \tag{4.58}$$

where $c := \min_{x \in S_i, i \in L_1} \|x\|^2 > 0$ because $x \ne 0$ for $x \in S_i$, $i \in L_1$.

Combining (4.57) and (4.58) leads to the existence of a constant $\beta > 0$ such that $V(t) \le \beta \|x\|^2$. Moreover, (4.48) and (4.50) imply, respectively, that there exists a constant $\alpha > 0$, such that,

$$\alpha \|x\|^2 \le x^T (P_i - E_i^T U_i E_i)x \le x^T P_i x, \text{ for } x \in S_i, i \in L_0, \tag{4.59}$$

$$\alpha \|x\|^2 \le \alpha \|\overline{x}\|^2 \le \overline{x}^T (\overline{P}_i - \overline{E}_i^T U_i \overline{E}_i)\overline{x} \le \overline{x}^T \overline{P}_i \overline{x}, \text{ for } x \in S_i, i \in L_1. \tag{4.60}$$

That is,

$$\alpha \|x\|^2 \le V(t), \text{ for } x \in S_i, i \in L. \tag{4.61}$$

Thus one has

$$\alpha \|x\|^2 \le V(t) \le \beta \|x\|^2. \tag{4.62}$$

In addition, it follows from (4.49), (4.51)–(4.53) that there exists a constant $\rho > 0$ such that

$$\overline{A}_k^T \overline{P}_j \overline{A}_k - \overline{P}_i + \rho I < 0, \tag{4.63}$$

where $j = i$ when $x(t)$ stays in the subspace S_i, $j \ne i$ when $x(t)$ transits from the subspace S_i to S_j. Defining $n_i = \text{card}(\aleph(i))$, then along trajectories of the system with $x(t) \neq 0$, one has

$$\Delta V(t) = V(t+1) - V(t)$$

$$= \left(\sum_{k=1}^{n_i} \mu_k(x) \cdot \overline{A}_k \overline{x}(t) \right)^T \overline{P}_j \left(\sum_{k=1}^{n_i} \mu_k(x) \cdot \overline{A}_k \overline{x}(t) \right) - \overline{x}^T(t) \overline{P}_i \overline{x}(t)$$

$$= \sum_{k=1}^{n_i} \mu_k^2(x) \cdot \overline{x}^T(t) \left\{ \overline{A}_k^T \overline{P}_j \overline{A}_k - \overline{P}_i \right\} \overline{x}(t)$$

$$+ \sum_{k<m}^{n_i} \mu_k(x) \mu_m(x) \cdot \overline{x}^T(t) \left\{ \overline{A}_k^T \overline{P}_j \overline{A}_m + \overline{A}_m^T \overline{P}_j \overline{A}_k - 2\overline{P}_i \right\} \overline{x}(t), \qquad (4.64)$$

where $\mu_k(x) \geq 0$, and $\Sigma_{k=1}^{n_i} \mu_k(x) > 0$.
It then follows from (4.63), (4.64), and Lemma 4.3 that

$$\Delta V(t) \leq -\rho \parallel x(t) \parallel^2 . \qquad (4.65)$$

Therefore, the desired result follows directly from (4.62) and (4.65) based on standard Lyapunov theory, and the proof is thus completed. ❑

The above conditions are linear matrix inequalities in the variables $P_l, \overline{P}_l, U_l, Q_{lk}$, and Q_{ljk}. A solution to these inequalities ensures $V(x)$ defined in (4.47) is a Lyapunov function for the T–S fuzzy system. The LMIs in (4.48) and (4.50) guarantee that the function is positive definite for each region, those in (4.49) and (4.51) guarantee that the function decreases within each region, and those in (4.52–(4.55) guarantee that the function decreases when the system transits from one region to another. In addition, $E_l^T U_l E_l$, $\overline{E}_l^T U_l \overline{E}_l$, $E_l^T Q_{lk} E_l$, $\overline{E}_l^T Q_{lk} \overline{E}_l$, $E_l^T Q_{ljk} E_l$, and $\overline{E}_l^T Q_{ljk} \overline{E}_l$ in these LMIs are the terms of the S-procedure used to reduce conservatism of the stability result. It should also be noted that the matrices P_l, $l \in L_0$, $\overline{P}_l$, $l \in L_1$ are not required to be positive definite.

Example 4.6
Consider the T–S fuzzy affine system in the form of (3.1) with system matrices as

$$A_1 = \begin{bmatrix} 0.9 & -0.1 \\ 0.1 & 1 \end{bmatrix}, \ a_1 = \begin{bmatrix} 0 \\ -0.02 \end{bmatrix},$$

$$A_2 = \begin{bmatrix} 0.8 & -0.6 \\ 0.0034 & 0.082 \end{bmatrix}, \ a_2 = \begin{bmatrix} 0 \\ 0 \end{bmatrix},$$

$$A_3 = \begin{bmatrix} 0.9 & -0.1 \\ 0.1 & 1 \end{bmatrix}, \ a_3 = \begin{bmatrix} 0 \\ 0.02 \end{bmatrix},$$

and the membership functions described in Figure 4.9.

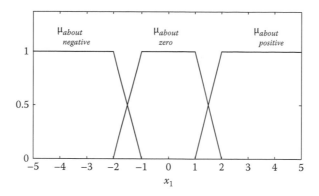

FIGURE 4.9   Membership functions in Example 4.6.

The matrices characterizing the local regions can be found and are given by

$$\bar{E}_1 = \begin{bmatrix} -1 & 0 & -2 \\ 1 & 0 & 5 \end{bmatrix}, \bar{E}_2 = \begin{bmatrix} -1 & 0 & -1 \\ 1 & 0 & 2 \end{bmatrix}, E_{3a} = \begin{bmatrix} -1 & 0 \\ 0 & 0 \end{bmatrix},$$

$$E_{3b} = \begin{bmatrix} 1 & 0 \\ 0 & 0 \end{bmatrix}, \bar{E}_4 = \begin{bmatrix} -1 & 0 & 2 \\ 1 & 0 & -1 \end{bmatrix}, \bar{E}_5 = \begin{bmatrix} -1 & 0 & 5 \\ 1 & 0 & -2 \end{bmatrix}.$$

There does not exist a common positive definite matrix P for the system. However, by using Theorem 4.5, one obtains the following feasible solutions to those LMIs,

$$\bar{P}_1 = \begin{bmatrix} 15.3838 & 3.8354 & -9.6328 \\ 3.8354 & 21.9349 & 9.4560 \\ -9.6328 & 9.4560 & -6.3677 \end{bmatrix}, \bar{P}_2 = \begin{bmatrix} 11.9567 & 3.2102 & -11.9983 \\ 3.2102 & 22.0191 & 9.0108 \\ -11.9983 & 9.0108 & -0.8551 \end{bmatrix},$$

$$P_{3a} = P_{3b} = \begin{bmatrix} 14.0127 & -5.4555 \\ -5.4555 & 11.1096 \end{bmatrix},$$

$$\bar{P}_4 = \begin{bmatrix} 11.9567 & 3.2102 & 11.9983 \\ 3.2102 & 22.0191 & -9.0108 \\ 11.9983 & -9.0108 & -0.8551 \end{bmatrix}, \bar{P}_5 = \begin{bmatrix} 15.3838 & 3.8354 & 9.6328 \\ 3.8354 & 21.9349 & -9.4560 \\ 9.6328 & -9.4560 & -6.3677 \end{bmatrix},$$

and thus one can verify that the fuzzy system is globally exponentially stable. Note that LMI solutions $\bar{P}_1$, $\bar{P}_2$, $\bar{P}_4$, and $\bar{P}_5$ are indefinite. This is due to the S-procedure that can reduce the conservatism of the stability results. The three-dimensional level curve for the piecewise quadratic Lyapunov function is shown in Figure 4.10.

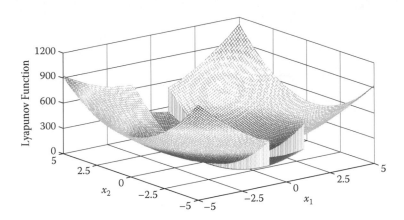

FIGURE 4.10 Surface of Lyapunov function in Example 4.6.

4.6 COMPARISON OF STABILITY RESULTS VIA NUMERICAL EXAMPLES

In this section, the stability results discussed in the previous sections for T–S fuzzy systems based on common, piecewise, and fuzzy quadratic Lyapunov functions, respectively, are compared via numerical examples. For this purpose, the following modified stability results in terms of exponential decay rate parameter λ are adopted where the parameter λ is used for the performance indicator of comparison; the proofs of those modified stability results are straightforward and thus omitted.

Theorem 4.6
If there exist two positive constants η, ξ, and a positive definite matrix P such that the following linear matrix inequalities (LMIs) are satisfied,

$$\eta I < P < I, \tag{4.66}$$

$$A_l^T P A_l - P + \xi I < 0, \, l \in L, \tag{4.67}$$

then the T–S fuzzy system (4.1), or equivalently (4.2), is globally exponentially stable with the decay rate λ defined as follows,

$$\|x(t)\| \le \alpha e^{-\lambda t} \|x(0)\|, \tag{4.68}$$

where $\alpha = \sqrt{1/\eta}$ and $\lambda = -\ln(1-\xi)/2$.

Theorem 4.7
The T–S fuzzy system (4.1), or equivalently (4.13), is globally exponentially stable with the decay rate λ as defined in (4.68), if there exist positive constants η, ξ,

ε, and a set of positive definite matrices $P_l, l \in L$ such that the following LMIs are satisfied,

$$\eta I < P_l < I, \ l \in L, \tag{4.69}$$

$$\begin{bmatrix} A_l^T P_l A_l - P_l + \xi I + \dfrac{1}{\varepsilon} E_{lA}^T E_{lA} & A_l^T P_l \\[2ex] P_l A_l & -\left(\dfrac{1}{\varepsilon} I - P_l\right) \end{bmatrix} < 0, \quad l \in L, \tag{4.70}$$

$$\begin{bmatrix} A_l^T P_j A_l - P_l + \xi I + \dfrac{1}{\varepsilon} E_{lA}^T E_{lA} & A_l^T P_j \\[2ex] P_j A_l & -\left(\dfrac{1}{\varepsilon} I - P_j\right) \end{bmatrix} < 0, \quad l, j \in \Omega. \tag{4.71}$$

Theorem 4.8

The T–S fuzzy system (4.1), or equivalently (4.31), is globally exponentially stable with the decay rate λ as defined in (4.68), if there exist positive constants η, ξ, and a set of positive definite matrices P_l, $l \in \bar{L}$ such that the following LMIs are satisfied,

$$\eta I < P_l < I, l \in \bar{L}, \tag{4.72}$$

$$A_k^T P_l A_k - P_l + \xi I < 0, \ l \in \bar{L}, \ k \in \aleph(l), \tag{4.73}$$

$$A_k^T P_j A_k - P_l + \xi I < 0, \ (l, j) \in \bar{\Omega}, \ k \in \aleph(l). \tag{4.74}$$

Theorem 4.9

The T–S fuzzy system (4.1), or equivalently (4.2), is globally exponentially stable with the decay rate λ as defined in (4.68), if there exist positive constants η, ξ, and a set of positive definite matrices P_l, $l \in L$ such that the following LMIs are satisfied,

$$\eta I < P_l < I, \ l \in L, \tag{4.75}$$

$$A_l^T P_j A_l - P_l + \xi I < 0, \ j, l \in L. \tag{4.76}$$

The objective of the comparison is to find the maximum decay rate for each stability result and then compare their performances. For this purpose, the following optimization algorithms can be developed based on Theorems 4.6–4.9, respectively.

TABLE 4.1

Maximum Decay Rates in Example 4.7

Algorithm	4.1	4.2	4.3	4.4
Maximum λ	0.0318	0.0171	0.0318	0.0320

Algorithm 4.1: $\max\limits_{P} \xi$, subject to LMIs (4.66) and (4.67)

Algorithm 4.2: $\max\limits_{P_l, l \in L} \xi$, subject to LMIs (4.69)–(4.71)

Algorithm 4.3: $\max\limits_{P_l, l \in L} \xi$, subject to LMIs (4.72)–(4.74)

Algorithm 4.4: $\max\limits_{P_l, l \in L} \xi$, subject to LMIs (4.75) and (4.76)

Example 4.7

Reconsider the T–S fuzzy system defined in Example 4.1. By applying Algorithms 4.1–4.4, respectively, one can obtain the corresponding maximum decay rates summarized in Table 4.1.

Example 4.8

Reconsider the T–S fuzzy system defined in Example 4.2. It is noted that there is no feasible solution for the LMIs based on common quadratic Lyapunov functions. By applying Algorithms 4.2–4.4, respectively, one can obtain the corresponding maximum decay rates summarized in Table 4.2.

It can be observed from these examples that the stability analysis approaches based on piecewise or fuzzy quadratic Lyapunov functions are in general less conservative than the approaches based on common quadratic Lyapunov functions in the sense of larger exponential decay rates. However, which approach is better is case-dependent. It should also be noted that Algorithm 4.2, which is based on piecewise quadratic Lyapunov functions, might lead to significant conservatism in some cases due to the introduction of uncertainties as demonstrated in Example 4.7.

TABLE 4.2

Maximum Decay Rates in Example 4.8

Algorithm	4.1	4.2	4.3	4.4
Maximum λ	infeasible	0.0022	0.0082	0.0025

4.7   CONCLUSIONS

This chapter has presented a number of stability analysis results for T–S fuzzy systems via Lyapunov function-based approaches. Three classes are considered, that is, common, piecewise, and fuzzy quadratic Lyapunov functions. It has been shown via analysis and numerical examples that the approaches based on piecewise or fuzzy quadratic Lyapunov functions are in general less conservative than those based on common quadratic Lyapunov functions, and thus are more useful in practical applications.

5 Stabilization Controller Synthesis of T–S Fuzzy Systems

5.1 INTRODUCTION

Controller design is one of the most important issues in systems and control. The objective of stabilization controller synthesis is to find a controller for a given system such that the resulting closed-loop control system is globally asymptotically (or exponentially) stable. Controller synthesis of T–S fuzzy systems has been well developed based on Lyapunov function approaches. Similar to the case of stability analysis, three of the most often used Lyapunov functions in controller synthesis of T–S fuzzy systems are common, piecewise, and fuzzy quadratic. This chapter presents a number of results of stabilization controller synthesis of T–S fuzzy systems based on those three Lyapunov functions.

The rest of the chapter is organized as follows. First, the issue of stabilization controller synthesis of T–S fuzzy systems based on common quadratic Lyapunov functions is addressed in Section 5.2. Then, this synthesis based on piecewise quadratic Lyapunov functions is discussed in Section 5.3. Stabilization controller design of T–S fuzzy systems based on fuzzy quadratic Lyapunov functions is presented in Section 5.4, followed by comparisons of those stabilization results via simulation examples in Section 5.5. Finally, some remarks conclude the chapter.

5.2 STABILIZATION BASED ON COMMON QUADRATIC LYAPUNOV FUNCTIONS

Consider the T–S fuzzy system without affine terms in (3.1) rewritten as

$$R^l: \quad \text{IF} \quad z_1 \text{ is } F_1^l, \text{AND}, \ldots z_v \text{ is } F_v^l$$

$$\text{THEN} \quad x(t+1) = A_l x(t) + B_l u(t) \tag{5.1}$$

$$y(t) = C_l x(t)$$

$$l \in L := \{1, 2, \ldots, m\},$$

or equivalently,

$$x(t+1) = A(\mu)x(t) + B(\mu)u(t)$$

$$y(t) = C(\mu)x(t), \tag{5.2}$$

where $A(\mu)$, $B(\mu)$, and $C(\mu)$ are as defined in (3.3).

Two kinds of control schemes are mainly used: one is the smooth fuzzy control scheme defined as

$$C^l: \quad \text{IF} \quad z_1 \text{ is } F_1^l, \text{ AND}, \dots z_v \text{ is } F_v^l$$

$$\text{THEN} \quad u(t) = K_l x(t) \tag{5.3}$$

$$l \in L := \{1, 2, \dots, m\},$$

which can be rewritten as

$$u(t) = \sum_{l=1}^{m} \mu_l(z) K_l x(t), \tag{5.4}$$

and the other is the switching control scheme defined as

$$u(t) = K_l x(t), \, z(t) \in S_l, l \in L, \tag{5.5}$$

where the space partition of the first kind is adopted here. The first control scheme is often called parallel distributed compensation and the second one is called local compensation.

The closed-loop fuzzy control system consisting of the T–S fuzzy system (5.2) and the smooth controller (5.4) can be described as

$$x(t+1) = \sum_{j=1}^{m} \sum_{l=1}^{m} \mu_j \mu_l (A_l + B_l K_j) x(t). \tag{5.6}$$

By defining a Lyapunov function candidate as

$$V(x) = x^T X^{-1} x, \tag{5.7}$$

where the matrix X is positive definite, one can easily obtain the following stabilization result (Tanaka, Ikeda, and Wang, 1998; Tanaka and Wang, 2001).

Theorem 5.1
The closed-loop fuzzy control system (5.6) is globally exponentially stable if there exist a positive definite matrix X and a set of matrices $Q_l, l \in L$ such that the following LMIs are satisfied:

$$\begin{bmatrix} -X & XA_l^T + Q_j^T B_l^T \\ A_l X + B_l Q_j & -X \end{bmatrix} < 0, \quad l, j \in L. \tag{5.8}$$

Moreover, the controller gains are given by

$$K_l = Q_l X^{-1}, \quad l \in L. \tag{5.9}$$

Proof: Consider the Lyapunov function candidate defined in (5.7) and denote

$$A_{clj} = A_l + B_l K_j. \tag{5.10}$$

By directly applying Theorem 4.1, one obtains that if there exists a positive definite matrix X such that the following LMIs are satisfied,

$$\begin{bmatrix} -X & XA_{clj}^T \\ A_{clj}X & -X \end{bmatrix} < 0, \ l, j \in L, \tag{5.11}$$

then the closed-loop control system (5.6) is globally exponentially stable. Substituting (5.10) into (5.11) leads to

$$\begin{bmatrix} -X & XA_l^T + XK_j^T B_l^T \\ A_l X + B_l K_j X & -X \end{bmatrix} < 0, \ l, j \in L. \tag{5.12}$$

By letting $Q_j = K_j X$, (5.12) becomes (5.8). Thus the closed-loop fuzzy control system (5.6) is globally exponentially stable if there exist a positive definite matrix X and a set of matrices $Q_l, l \in L$ such that LMIs in (5.8) are satisfied, and the controller gains can be determined by (5.9). The proof is thus completed. ❏

The stabilization controller synthesis result in Theorem 5.1 can be improved in many ways, either by reducing the number of LMIs in (5.8), by reducing the conservatism of the obtained LMIs, or both. In fact, a number of improved results on stabilization controller synthesis have been obtained (Kim and Lee, 2000; Liu and Zhang, 2003; Ma, Sun, and He, 1998; Tanaka, Ikada, and Wang, et al., 1998; Teixeira, Assuncao, and Avellar 2003; Tuan et al., 2001; Wang and Sun, 2005). One typical result is given below (Kim and Lee, 2000).

Theorem 5.2
The closed-loop fuzzy control system (5.6) is globally exponentially stable if there exist a positive definite matrix X, a set of matrices $Q_l, l \in L$, a set of symmetric matrices $\Phi_l, l \in L$, and a set of matrices $\Phi_{lj} = \Phi_{jl}^T, l, j \in L, l < j$ such that the following LMIs are satisfied:

$$\begin{bmatrix} -X + \Phi_l & XA_l^T + Q_l^T B_l^T \\ A_l X + B_l Q_l & -X \end{bmatrix} < 0, \quad l \in L, \tag{5.13}$$

$$\begin{bmatrix} -X + \Phi_{lj} & XA_l^T + Q_j^T B_l^T \\ A_l X + B_l Q_j & -X \end{bmatrix} + \begin{bmatrix} -X + \Phi_{jl} & XA_j^T + Q_l^T B_j^T \\ A_j X + B_j Q_l & -X \end{bmatrix} < 0, \ l, j \in L, \ l < j, \tag{5.14}$$

$$\Phi := \begin{bmatrix} \Phi_1 & \Phi_{12} & \cdots & \Phi_{1m} \\ \Phi_{21} & \Phi_2 & \cdots & \Phi_{2m} \\ \vdots & \vdots & \ddots & \vdots \\ \Phi_{m1} & \Phi_{m2} & \cdots & \Phi_m \end{bmatrix} > 0. \tag{5.15}$$

Moreover, the controller gains are given by

$$K_l = Q_l X^{-1}, \ l \in L. \tag{5.16}$$

Proof: By applying the Schur complement to (5.13) and (5.14), respectively, one has

$$(A_l X + B_l Q_l)^T X^{-1} (A_l X + B_l Q_l) - X + \Phi_l < 0, \tag{5.17}$$

$$\frac{1}{2}(A_l X + B_l Q_j + A_j X + B_j Q_l)^T X^{-1} (A_l X + B_l Q_l + A_j X + B_j Q_l) - 2X + \Phi_{lj} + \Phi_{jl} < 0. \tag{5.18}$$

Pre- and postmultiplying X^{-1} to (5.17) and (5.18) and noting (5.16), one has, respectively,

$$(A_l + B_l K_l)^T X^{-1} (A_l + B_l K_l) - X^{-1} + X^{-1} \Phi_l X^{-1} < 0, \tag{5.19}$$

$$\frac{1}{2}(A_l + B_l K_j + A_j + B_j K_l)^T X^{-1} (A_l + B_l K_l + A_j + B_j K_l)$$
$$- 2X^{-1} + X^{-1} \Phi_{lj} X^{-1} + X^{-1} \Phi_{jl} X^{-1} < 0. \tag{5.20}$$

Consider the Lyapunov function candidate defined in (5.7) and denote $A_{clj} = A_l + B_l K_j$. Taking its difference along the system trajectories with $x(t) \neq 0$ leads to

$$\Delta V(t) = V(x(t+1)) - V(x(t))$$

$$= \left[\sum_{l=1}^{m} \mu_l A_{clj} x(t) \right]^T X^{-1} \left[\sum_{l=1}^{m} \mu_l A_{clj} x(t) \right] - x(t)^T X^{-1} x(t)$$

$$= \sum_{l=1}^{m} \sum_{j=1}^{m} \mu_l \mu_j x(t)^T A_{clj}^T X^{-1} A_{cjl} x(t) - x(t)^T X^{-1} x(t)$$

$$= \sum_{l=1}^{m} \sum_{j=1}^{m} \mu_l^2 x(t)^T \left(A_{cll}^T X^{-1} A_{cll} - X^{-1} \right) x(t)$$

$$+ \sum_{l=1}^{m} \sum_{l<j}^{m} \mu_l \mu_j x(t)^T \left(A_{clj}^T X^{-1} A_{cjl} + A_{cjl}^T X^{-1} A_{clj} - 2X^{-1} \right) x(t)$$

$$= \sum_{l=1}^{m} \sum_{j=1}^{m} \mu_l^2 x(t)^T \left(A_{cll}^T X^{-1} A_{cll} - X^{-1} \right) x(t)$$

$$+ \sum_{l=1}^{m} \sum_{l<j}^{m} \mu_l \mu_j x(t)^T \left[\frac{1}{2}(A_{clj} + A_{cjl})^T X^{-1} (A_{clj} + A_{cjl}) - 2X^{-1} \right] x(t). \tag{5.21}$$

Then using (5.19) and (5.20) in (5.21) leads to

$$\Delta V(t) \leq -\sum_{l=1}^{m} \sum_{j=1}^{m} \mu_l^2 x(t)^T X^{-1} \Phi_l X^{-1} x(t)$$

$$-\sum_{l=1}^{m} \sum_{l<j}^{m} \mu_l \mu_j x(t)^T \left(X^{-1} \Phi_{lj} X^{-1} + X^{-1} \Phi_{jl} X^{-1} \right) x(t)$$

$$\leq - \begin{bmatrix} \mu_1 x \\ \vdots \\ \mu_m x \end{bmatrix}^T \begin{bmatrix} X^{-1} \Phi_1 X^{-1} & \cdots & X^{-1} \Phi_{1m} X^{-1} \\ \vdots & \ddots & \vdots \\ X^{-1} \Phi_{m1} X^{-1} & \cdots & X^{-1} \Phi_m X^{-1} \end{bmatrix} \begin{bmatrix} \mu_1 x \\ \vdots \\ \mu_m x \end{bmatrix} \qquad (5.22)$$

$$\leq -\bar{\rho} \sum_{l=1}^{m} \mu_l^2 x^T x \text{ (for some } \bar{\rho} > 0)$$

$$\leq -\rho \parallel x(t) \parallel^2.$$

Thus the closed-loop fuzzy control system (5.6) is globally exponentially stable based on the result of Theorem 4.1. Moreover, the controller gains can be determined by (5.13). The proof is thus completed. ❑

On the other hand, with the switching control law (5.5), the closed-loop fuzzy control system can be described as

$$x(t+1) = A_{cl} x(t), z(t) \in S_l, l \in L, \qquad (5.23)$$

where

$$A_{cl} = A_l + \Delta A_l(\mu) + (B_l + \Delta B_l(\mu)) K_l. \qquad (5.24)$$

The system described in (5.23) can be recognized as a piecewise linear system with uncertainties. By defining the following upper bounds for the uncertainties ΔB_l, $l \in L$ as those in (4.14) for ΔA_l,

$$[\Delta B_l(\mu)]^T [\Delta B_l(\mu)] \leq E_{lB}^T E_{lB}, \quad l \in L, \qquad (5.25)$$

one has the following result.

Theorem 5.3

The closed-loop fuzzy control system (5.23) is globally exponentially stable if there exist a positive definite matrix X and a set of matrices Q_l, $l \in L$ such that the following

LMIs are satisfied,

$$
\begin{bmatrix}
-X & XA_l^T + Q_l^T B_l^T & XE_{lA}^T & Q_l^T E_{lB}^T \\
A_l X + B_l Q_l & -(X-I) & 0 & 0 \\
E_{lA} X & 0 & -\dfrac{1}{2}I & 0 \\
E_{lB} Q_l & 0 & 0 & -\dfrac{1}{2}I
\end{bmatrix} < 0, \; l \in L. \qquad (5.26)
$$

Moreover, the controller gains are given by

$$
K_l = Q_l X^{-1}, \quad l \in L. \qquad (5.27)
$$

Proof: Based on the result in Theorem 4.1 and its proof, one learns that the system (5.23) is globally exponentially stable if there exists a positive definite matrix X satisfying the following inequalities,

$$
A_{cl}^T X^{-1} A_{cl} - X^{-1} < 0, \; l \in L. \qquad (5.28)
$$

One then needs to show that the inequalities (5.26) imply (5.28). Using Lemma 4.2, one has

$$
\begin{aligned}
A_{cl}^T X^{-1} A_{cl} - X^{-1} &= [A_l + \Delta A_l + (B_l + \Delta B_l) K_l]^T X^{-1} [A_l + \Delta A_l + (B_l + \Delta B_l) K_l] - X^{-1} \\
&= (A_l + B_l K_l)^T X^{-1} (A_l + B_l K_l) + (A_l + B_l K_l)^T X^{-1} (\Delta A_l + \Delta B_l K_l) \\
&\quad + (\Delta A_l + \Delta B_l K_l)^T X^{-1} (A_l + B_l K_l) \\
&\quad + (\Delta A_l + \Delta B_l K_l)^T X^{-1} (\Delta A_l + \Delta B_l K_l) - X^{-1} \\
&\leq (A_l + B_l K_l)^T X^{-1} (A_l + B_l K_l) \\
&\quad + (A_l + B_l K_l)^T X^{-1} \left(\frac{1}{\varepsilon} I - X^{-1} \right)^{-1} X^{-1} (A_l + B_l K_l) \\
&\quad + \frac{1}{\varepsilon} (\Delta A_l + \Delta B_l K_l)^T (\Delta A_l + \Delta B_l K_l) - X^{-1} \\
&\leq (A_l + B_l K_l)^T \left(X^{-1} + X^{-1} \left(\frac{1}{\varepsilon} I - X^{-1} \right)^{-1} X^{-1} \right) (A_l + B_l K_l) \\
&\quad + \frac{2}{\varepsilon} \Delta A_l^T \Delta A_l + \frac{2}{\varepsilon} K_l^T \Delta B_l^T \Delta B_l K_l - X^{-1} \\
&\leq (A_l + B_l K_l)^T (X - \varepsilon I)^{-1} (A_l + B_l K_l) + \frac{2}{\varepsilon} E_{lA}^T E_{lA} \\
&\quad + \frac{2}{\varepsilon} K_l^T E_{lB}^T E_{lB} K_l - X^{-1}.
\end{aligned} \qquad (5.29)
$$

It then can be easily seen that the following inequality,

$$(A_l + B_l K_l)^T (X - \varepsilon I)^{-1} (A_l + B_l K_l) + \frac{2}{\varepsilon} E_{lA}^T E_{lA} + \frac{2}{\varepsilon} K_l^T E_{lB}^T E_{lB} K_l - X^{-1} < 0, \quad (5.30)$$

is equivalent to

$$X(A_l + B_l K_l)^T (X - \varepsilon I)^{-1} (A_l + B_l K_l) X + \frac{2}{\varepsilon} X E_{lA}^T E_{lA} X + \frac{2}{\varepsilon} X K_l^T E_{lB}^T E_{lB} K_l X - X < 0.$$

$$(5.31)$$

Multiplying $1/\varepsilon$ to (5.31) and letting $X/\varepsilon \to X$ leads to

$$X(A_l + B_l K_l)^T (X - I)^{-1} (A_l + B_l K_l) X + 2 X E_{lA}^T E_{lA} X + 2 X K_l^T E_{lB}^T E_{lB} K_l X - X < 0.$$

$$(5.32)$$

Noting (5.27) and applying the Schur complement formula several times, it is easily shown that inequality (5.32) is in turn equivalent to the linear matrix inequality (5.26). Thus, it has been shown that the inequalities (5.26) imply (5.28). Therefore, it can be concluded that the closed-loop control system is globally exponentially stable and the proof is thus completed. ❑

Remark 5.1
By comparing the results in Theorems 5.1–5.3, one notices that the numbers of LMIs in (5.8), (5.13)–(5.15), and (5.26) are $m \times m$, $(m(m+1)+2)/2$, and m, respectively. It can be easily seen that the computation cost of (5.26) is smaller than that of (5.13)–(5.15), and much smaller than that of (5.8). However, it should also be noted that the price to pay for such computational efficiency in (5.26) is the introduction of uncertainty terms as described in (5.21), which might lead to some conservatism. In addition, the control law used in Theorem 5.3 is a switching control law whereas that for Theorems 5.1 and 5.2 is a smooth one.

Example 5.1
Consider a T–S fuzzy system in the form of (5.1) with matrices,

$$A_1 = \begin{bmatrix} 1 & 0.4 \\ -0.5 & 1 \end{bmatrix}, A_2 = \begin{bmatrix} 1 & 0.5 \\ -0.4 & 1 \end{bmatrix}, B_1 = \begin{bmatrix} 0 \\ 1 \end{bmatrix}, B_2 = \begin{bmatrix} 0 \\ 1 \end{bmatrix},$$

and membership functions as shown in Figure 5.1.

The open-loop system is unstable. By applying Theorem 5.1, one obtains the following solutions:

$$X = \begin{bmatrix} 10.3657 & -5.5820 \\ -5.5820 & 13.4462 \end{bmatrix}, K_1 = [-0.0801 \quad -1.2569], K_2 = [-0.0801 \quad -1.2569].$$

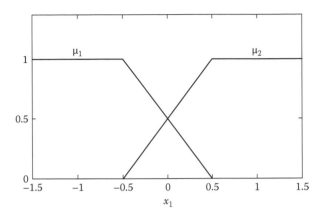

FIGURE 5.1   Membership functions in Example 5.1.

With the obtained controller, the closed-loop control system exhibits stable responses under any initial conditions. One typical simulation result is illustrated in Figure 5.2.

Similar to the case of stability analysis, the stabilization controller synthesis based on common quadratic Lyapunov functions tends to be conservative in general. This can be illustrated by the following example.

Example 5.2

Consider a T–S fuzzy system in the form of (5.1) with matrices

$$A_1 = \begin{bmatrix} 1 & 0.2 \\ -0.5 & 1 \end{bmatrix}, A_2 = \begin{bmatrix} 1 & 0.5 \\ -0.2 & 1 \end{bmatrix}, B_1 = \begin{bmatrix} 0 \\ 1 \end{bmatrix}, B_2 = \begin{bmatrix} 1 \\ 0 \end{bmatrix},$$

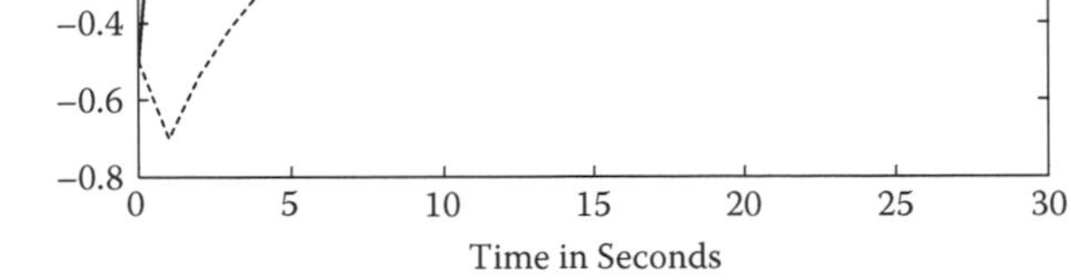

FIGURE 5.2   State responses of the closed-loop system in Example 5.1.

and the membership functions as

$$\mu_1 = \begin{cases} 1 & if\ x_1 \le 0 \\ 0 & otherwise \end{cases},$$

$$\mu_2 = 1 - \mu_1.$$

Noted that the system is in fact a piecewise linear system and the open-loop system is unstable. It is also found that there is no feasible solution to the stabilization results in Theorems 5.1, 5.2, or 5.3.

This clearly indicates that the approaches based on common quadratic Lyapunov functions are in general conservative. Better approaches to controller synthesis of T–S fuzzy systems are desirable. The following two sections are devoted to those approaches based on piecewise and fuzzy quadratic Lyapunov functions, respectively.

5.3 STABILIZATION BASED ON PIECEWISE QUADRATIC LYAPUNOV FUNCTIONS

In this section, several stabilization methods are presented based on piecewise quadratic Lyapunov functions (Cao, Rees, and Feng, 1997a; Chen et al., 2005b; Feng, 2003, 2004a,b).

With space partition of the first kind and the switching controller defined as

$$u(t) = K_l x(t),\ z(t) \in S_l,\ l \in L, \tag{5.33}$$

the closed-loop fuzzy control system can be described by

$$x(t+1) = A_{cl} x(t),\ z(t) \in S_l, \tag{5.34}$$

where $A_{cl} = A_l + \Delta A_l(\mu) + (B_l + \Delta B_l(\mu))K_l$, and the upper bounds for ΔA_l and ΔB_l are given in (4.14) and (5.25), respectively. Define the set Ω as in (4.15) representing all possible system transitions among regions.

By using a piecewise quadratic Lyapunov function candidate of the form

$$V(x) = x^T X_l^{-1} x,\ z(t) \in S_l, \tag{5.35}$$

one has the following stabilization result for the T–S fuzzy system described in (5.1).

Theorem 5.4
The closed-loop fuzzy control system (5.34) is globally exponentially stable if there exist a set of positive definite matrices $X_l,\ l \in L$ and a set of matrices $Q_l,\ l \in L$ such

that the following LMIs are satisfied:

$$
\begin{bmatrix}
-X_l & X_lA_l^T + Q_l^T B_l^T & X_l E_{lA}^T & Q_l^T E_{lB}^T \\
A_l X_l + B_l Q_l & -(X_l - I) & 0 & 0 \\
E_{lA} X_l & 0 & -\dfrac{1}{2}I & 0 \\
E_{lB} Q_l & 0 & 0 & -\dfrac{1}{2}I
\end{bmatrix} < 0,\ l \in L, \qquad (5.36)
$$

$$
\begin{bmatrix}
-X_l & X_lA_l^T + Q_l^T B_l^T & X_l E_{lA}^T & Q_l^T E_{lB}^T \\
A_l X_l + B_l Q_l & -(X_j - I) & 0 & 0 \\
E_{lA} X_l & 0 & -\dfrac{1}{2}I & 0 \\
E_{lB} Q_l & 0 & 0 & -\dfrac{1}{2}I
\end{bmatrix} < 0,\ l, j \in \Omega. \qquad (5.37)
$$

Moreover, the controller gains are given by

$$
K_l = Q_l X_l^{-1}, \quad l \in L. \qquad (5.38)
$$

Proof: Based on the result in Theorem 4.2 and its proof, one learns that the system (5.34) is globally exponentially stable if there exist a set of positive definite matrices X_l, $l \in L$ satisfying the following inequalities,

$$
A_{cl}^T X_l^{-1} A_{cl} - X_l^{-1} < 0,\ l \in L, \qquad (5.39)
$$

$$
A_{cl}^T X_j^{-1} A_{cl} - X_l^{-1} < 0,\ l, j \in \Omega. \qquad (5.40)
$$

We first show that the inequality (5.37) implies (5.40). Using Lemma 4.2, the left-hand side of inequality (5.40) can be expressed as

$$
\begin{aligned}
A_{cl}^T X_j^{-1} A_{cl} - X_l^{-1} &= [A_l + \Delta A_l + (B_l + \Delta B_l)K_l]^T X_j^{-1}[A_l + \Delta A_l + (B_l + \Delta B_l)K_l] - X_l^{-1} \\
&= (A_l + B_l K_l)^T X_j^{-1}(A_l + B_l K_l) + (A_l + B_l K_l)^T X_j^{-1}(\Delta A_l + \Delta B_l K_l) \\
&\quad + (\Delta A_l + \Delta B_l K_l)^T X_j^{-1}(A_l + B_l K_l) \\
&\quad + (\Delta A_l + \Delta B_l K_l)^T X_j^{-1}(\Delta A_l + \Delta B_l K_l) - X_l^{-1} \\
&\leq (A_l + B_l K_l)^T X_j^{-1}(A_l + B_l K_l) \\
&\quad + (A_l + B_l K_l)^T X_j^{-1}\left(\frac{1}{\varepsilon}I - X_j^{-1}\right)^{-1} X_j^{-1}(A_l + B_l K_l) \\
&\quad + \frac{1}{\varepsilon}(\Delta A_l + \Delta B_l K_l)^T(\Delta A_l + \Delta B_l K_l) - X_l^{-1}
\end{aligned}
$$

$$\leq (A_l + B_l K_l)^T \left[X_j^{-1} + X_j^{-1} \left(\frac{1}{\varepsilon} I - X_j^{-1} \right)^{-1} X_j^{-1} \right] (A_l + B_l K_l)$$

$$+ \frac{2}{\varepsilon} \Delta A_l^T \Delta A_l + \frac{2}{\varepsilon} K_l^T \Delta B_l^T \Delta B_l K_l - X_l^{-1}$$

$$\leq (A_l + B_l K_l)^T (X_j - \varepsilon I)^{-1} (A_l + B_l K_l) + \frac{2}{\varepsilon} E_{lA}^T E_{lA}$$

$$+ \frac{2}{\varepsilon} K_l^T E_{lB}^T E_{lB} K_l - X_l^{-1}. \tag{5.41}$$

It then can be easily seen that the following inequality,

$$(A_l + B_l K_l)^T (X_j - \varepsilon I)^{-1} (A_l + B_l K_l) + \frac{2}{\varepsilon} E_{lA}^T E_{lA} + \frac{2}{\varepsilon} K_l^T E_{lB}^T E_{lB} K_l - X_l^{-1} < 0 \tag{5.42}$$

is equivalent to

$$X_l (A_l + B_l K_l)^T (X_j - \varepsilon I)^{-1} (A_l + B_l K_l) X_l$$

$$+ \frac{2}{\varepsilon} X_l E_{lA}^T E_{lA} X_l + \frac{2}{\varepsilon} X_l K_l^T E_{lB}^T E_{lB} K_l X_l - X_l < 0. \tag{5.43}$$

Multiplying $1/\varepsilon$ in (5.43) and letting $X_l/\varepsilon \to X_l$ and $X_j/\varepsilon \to X_j$ leads to

$$X_l (A_l + B_l K_l)^T (X_j - I)^{-1} (A_l + B_l K_l) X_l$$

$$+ 2 X_l E_{lA}^T E_{lA} X_l + 2 X_l K_l^T E_{lB}^T E_{lB} K_l X_l - X_l < 0. \tag{5.44}$$

Let $Q_l = K_l X_l$. Using Schur complement formulas a few times, it is easily shown that the inequality (5.44) is in turn equivalent to the linear matrix inequality (5.37). Thus, we have shown that the inequalities (5.37) imply (5.40). Following the same procedure, it can also be shown that the inequalities (5.36) imply (5.39). Therefore, it can be concluded that the closed-loop control system is globally exponentially stable and the proof is thus completed. ❑

With space partition of the second kind and the switching controller defined as

$$u(t) = K_l x(t), \ z(t) \in S_l, \ l \in \overline{L}, \tag{5.45}$$

the closed–loop fuzzy control system can be described by

$$x(t+1) = \sum_{k \in \aleph(l)} \mu_k (z(t)) A_{clk} x(t), \ z(t) \in S_l, \tag{5.46}$$

where $A_{clk} := A_k + B_k K_l$ and the set $\aleph(l)$ is as defined in (4.31). That is, for each region S_l, the set $\aleph(l)$ contains the indices for the system matrices used in the interpolation within that region. For a crisp region, $\aleph(l)$ contains a single element. By using the same piecewise quadratic Lyapunov function candidate as in (5.35) one has the following stabilization result.

Theorem 5.5 (Wang and Feng, 2004)

The closed-loop fuzzy control system (5.46) is globally exponentially stable if there exist a set of positive definite matrices X_l, $l \in \bar{L}$ and a set of matrices, Q_l, $l \in \bar{L}$ such that the following LMIs are satisfied:

$$\begin{bmatrix} -X_l & X_l A_k^T + Q_l^T B_k^T \\ A_k X_l + B_k Q_l & -X_l \end{bmatrix} < 0, \ l \in \bar{L}, \ k \in \aleph(l), \tag{5.47}$$

$$\begin{bmatrix} -X_l & X_l A_k^T + Q_l^T B_k^T \\ A_k X_l + B_k Q_l & -X_j \end{bmatrix} < 0, \ (l, j) \in \bar{\Omega}, \ k \in \aleph(l). \tag{5.48}$$

Moreover, the controller gains are given by

$$K_l = Q_l X_l^{-1}, \ l \in \bar{L}. \tag{5.49}$$

Proof: Based on the result in Theorem 4.3 and its proof, one learns that the fuzzy control system (5.46) is globally exponentially stable if there exists a set of positive definite matrices X_l, $l \in \bar{L}$, satisfying the following inequalities:

$$A_{clk}^T X_l^{-1} A_{clk} - X_l^{-1} < 0, \ l \in \bar{L}, \ k \in \aleph(l), \tag{5.50}$$

$$A_{clk}^T X_j^{-1} A_{clk} - X_l^{-1} < 0, \ (l, j) \in \bar{\Omega}, \ k \in \aleph(l). \tag{5.51}$$

We first show that (5.48) is equivalent to (5.51). Using Schur's complement lemma, (5.48) is equivalent to

$$-X_l + (A_k X_l + B_k Q_l)^T X_j^{-1} (A_k X_l + B_k Q_l) < 0. \tag{5.52}$$

Because $Q_l = K_l X_l$ and $A_{clk} = A_k + B_k K_l$, (5.52) becomes

$$-X_l + (A_{clk} X_l)^T X_j^{-1} (A_{clk} X_l) < 0. \tag{5.53}$$

Multiplying both sides of (5.53) by X_l^{-1} leads to

$$A_{clk}^T X_j^{-1} A_{clk} - X_l^{-1} < 0,$$

which is exactly (5.51). Thus, we have shown that (5.48) is equivalent to (5.51). Similarly, one can easily show that (5.47) is equivalent to (5.50). Therefore, it can be concluded that the closed-loop fuzzy control system is globally exponentially stable and the proof is thus completed. ❏

The stabilization results in Theorems 5.4 and 5.5 can be further improved in the sense of less conservatism by introducing extra slack variables V_l, $l \in \bar{L}$ in LMIs as in de Oliveira, Bernussou, and Geromel (1999). It is noted that the matrices V_l, $l \in \bar{L}$ are not even required to be symmetric. One improved result for Theorem 5.5 is summarized in the following theorem.

Theorem 5.6

The closed-loop fuzzy control system (5.46) is globally exponentially stable if there exist a set of positive definite matrices X_l, $l \in \bar{L}$ and two sets of matrices Q_l, $l \in \bar{L}$, and V_l, $l \in \bar{L}$ such that the following LMIs are satisfied,

$$\begin{bmatrix} X_l - V_l^T - V_l & V_l^T A_k^T + Q_l^T B_k^T \\ A_k V_l + B_k Q_l & -X_l \end{bmatrix} < 0, \ l \in \bar{L}, \ k \in \aleph(l), \tag{5.54}$$

$$\begin{bmatrix} X_l - V_l^T - V_l & V_l^T A_k^T + Q_l^T B_k^T \\ A_k V_l + B_k Q_l & -X_j \end{bmatrix} < 0, \ (l, j) \in \bar{\Omega}, \ k \in \aleph(l). \tag{5.55}$$

Moreover, the controller gains are given by

$$K_l = Q_l V_l^{-1}, \quad l \in \bar{L}. \tag{5.56}$$

Proof: Based on the result in Theorem 4.3 and its proof, one learns that the fuzzy control system (5.46) is globally exponentially stable if there exist symmetric positive definite matrices X_l, $l \in \bar{L}$ satisfying the inequalities (5.50) and (5.51). We first show (5.55) implies (5.51). By noting (5.56), it follows from (5.55) that

$$\begin{bmatrix} X_l - V_l^T - V_l & V_l^T A_{clk}^T \\ A_{clk} V_l & -X_j \end{bmatrix} < 0,$$

which implies the inequality

$$\begin{bmatrix} -V_l^T X_l^{-1} V_l & V_l^T A_{clk}^T \\ A_{clk} V_l & -X_j \end{bmatrix} < 0, \tag{5.57}$$

because $(X_l - V_l)^T X_l^{-1} (X_l - V_l) = X_l - V_l^T - V_l + V_l^T X_l^{-1} V_l > 0$.
 Via the Schur complement, (5.57) is equivalent to

$$V_l^T A_{clk}^T X_j^{-1} A_{clk} V_l - V_l^T X_l^{-1} V_l < 0, \tag{5.58}$$

which in turn implies that

$$A_{clk}^T X_j^{-1} A_{clk} - X_l^{-1} < 0. \tag{5.59}$$

Thus we have shown that (5.55) implies (5.51). Similarly, one can also show that (5.54) implies (5.50). Therefore, it can be concluded that the closed-loop fuzzy control system is globally exponentially stable and the proof is thus completed. ❏

Remark 5.2
The conservatism and the number of LMIs in Theorems 5.4–5.6 can be further reduced by employing a similar technique to that of Theorem 5.2 (Kim and Lee, 2000).

Remark 5.3
When the positive definite matrices in (5.47) and (5.48) are chosen as a common one, that is, $X_1 = X_2 = \cdots X_m = X$, then the result of Theorem 5.5 reduces to that of Theorem 5.1. It thus can be easily seen that the stabilization result based on piecewise quadratic Lyapunov functions is less conservative than that based on common quadratic Lyapunov functions.

Example 5.3
Reconsider the T–S fuzzy system defined in Example 5.2. By applying Theorem 5.4, one obtains the following solutions:

$$X_1 = \begin{bmatrix} 50.9661 & -22.6562 \\ -22.6562 & 95.7950 \end{bmatrix}, X_2 = \begin{bmatrix} 95.7950 & 22.6562 \\ 22.6562 & 50.9661 \end{bmatrix},$$

$$K_1 = \begin{bmatrix} 0.2277 & -1.0561 \end{bmatrix}, K_2 = \begin{bmatrix} -1.0561 & -0.2277 \end{bmatrix}.$$

It thus follows from Theorem 5.4 that the closed-loop control system is globally exponentially stable. Simulation results with a number of initial conditions are recorded in Figure 5.3. Similarly, by applying Theorems 5.5 and 5.6, the feasible solutions were also obtained, which confirm the stability of the closed-loop control system. The solutions via Theorem 5.5 are given as

$$X_1 = \begin{bmatrix} 2.6118 & -0.7557 \\ -0.7557 & 4.7911 \end{bmatrix}, X_2 = \begin{bmatrix} 4.7911 & 0.7557 \\ 0.7557 & 2.6118 \end{bmatrix},$$

$$K_1 = \begin{bmatrix} 0.4015 & -1.0255 \end{bmatrix}, K_2 = \begin{bmatrix} -1.0255 & -0.4015 \end{bmatrix};$$

and the solutions via Theorem 5.6 are given as

$$X_1 = \begin{bmatrix} 1.9489 & -0.6261 \\ -0.6261 & 3.7949 \end{bmatrix}, X_2 = \begin{bmatrix} 3.7949 & 0.6261 \\ 0.6261 & 1.9489 \end{bmatrix},$$

$$V_1 = \begin{bmatrix} 1.9535 & -0.3270 \\ -0.5499 & 3.4783 \end{bmatrix}, V_2 = \begin{bmatrix} 3.4783 & 0.5499 \\ 0.3270 & 1.9535 \end{bmatrix},$$

$$K_1 = \begin{bmatrix} 0.3832 & -1.0234 \end{bmatrix}, K_2 = \begin{bmatrix} -1.0234 & -0.3832 \end{bmatrix}.$$

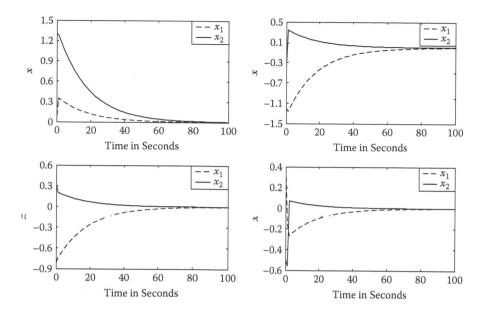

FIGURE 5.3　State responses of the closed-loop system in Example 5.3.

It can be easily observed from this example that the stabilization controller synthesis results based on piecewise quadratic Lyapunov functions are less conservative than those obtained based on common quadratic Lyapunov functions.

5.4　STABILIZATION BASED ON FUZZY QUADRATIC LYAPUNOV FUNCTIONS

Consider a smooth control law as in (5.3), rewritten as follows:

$$u(t) = \sum_{l=1}^{m} \mu_l K_l x(t).$$

Then the closed-loop fuzzy control system consisting of the T–S fuzzy system (5.1) and the smooth controller (5.3) can be described as in (5.6), rewritten as

$$x(t+1) = \sum_{j=1}^{m} \sum_{l=1}^{m} \mu_j \mu_l A_{clj} x(t),$$

where $A_{clj} = A_l + B_l K_j$.

By defining a Lyapunov function candidate as

$$V(x) = \sum_{l=1}^{m} \mu_l(z) x^T X_l^{-1} x, \tag{5.60}$$

where X_l is positive definite, one has the following stabilization result.

Theorem 5.7

The closed-loop fuzzy control system (5.6) is globally exponentially stable if there exist a set of positive definite matrices X_l, $l \in L$, two sets of matrices Q_l, $l \in L$, and V_l, $l \in L$ such that the following LMIs are satisfied:

$$\begin{bmatrix} X_l - V_j^T - V_j & V_j^T A_l^T + Q_j^T B_l^T \\ A_l V_j + B_l Q_j & -X_i \end{bmatrix} < 0, \; i, j, l \in L. \tag{5.61}$$

Moreover, the controller gains are given by

$$K_j = Q_j V_j^{-1}, \quad j \in L. \tag{5.62}$$

Proof: It follows from Theorem 4.4 that the system (5.6) is globally exponentially stable if the following LMIs are satisfied:

$$A_{clj}^T X_i^{-1} A_{clj} - X_l^{-1} < 0, \; i, j, l \in L. \tag{5.63}$$

We show that (5.61) implies (5.63). By noting (5.62), it follows from (5.61) that

$$\begin{bmatrix} X_l - V_j^T - V_j & V_j^T A_{clj}^T \\ A_{clj} V_j & -X_i \end{bmatrix} < 0,$$

which implies the following inequality,

$$\begin{bmatrix} -V_j^T X_l^{-1} V_j & V_j^T A_{clj}^T \\ A_{clj} V_j & -X_i \end{bmatrix} < 0, \tag{5.64}$$

because $(X_l - V_j)^T X_l^{-1}(X_l - V_j) = X_l - V_j^T - V_j + V_j^T X_l^{-1} V_j \geq 0$.

Via the Schur complement, (5.64) is equivalent to

$$V_j^T A_{clj}^T X_i^{-1} A_{clj} V_j - V_j^T X_l^{-1} V_j < 0, \tag{5.65}$$

which in turn implies that

$$A_{clj}^T X_i^{-1} A_{clj} - X_l^{-1} < 0. \tag{5.66}$$

Thus we have shown that (5.61) implies (5.63). Therefore, it can be concluded that the closed-loop fuzzy control system is globally exponentially stable and the proof is thus completed. ❑

Remark 5.4

When the positive definite matrices in (5.61) are chosen as a common one, that is, $X_1 = X_2 = \cdots = X_m = X = V_1 = V_2 = \cdots = V_m$, then the result of Theorem 5.7 reduces to that of Theorem 5.1. It thus can be easily seen that the stabilization result based on fuzzy quadratic Lyapunov functions is less conservative than that based on common quadratic Lyapunov functions.

The number of LMIs and the conservatism in Theorem 5.7 can be reduced by following a similar idea to that of Theorem 5.2 (Kim and Lee, 2000). The result is given below.

Theorem 5.8

The closed-loop fuzzy control system (5.6) is globally exponentially stable if there exist a set of positive definite matrices X_l, $l \in L$, two sets of matrices Q_l, $l \in L$, G_l, $l \in L$, a set of symmetric matrices Φ_l^i, $l, i \in L$, and a set of matrices $\Phi_{lj}^i = (\Phi_{jl}^i)^T$, $l, j, i \in L, l < j$ such that the following LMIs are satisfied,

$$
\begin{bmatrix}
X_j - G_j^T - G_j + \Phi_j^i & G_j^T A_j^T + Q_j^T B_j^T \\
A_j G_j + B_j Q_j & -X_i
\end{bmatrix} < 0, \ i, j \in L,
\tag{5.67}
$$

$$
\begin{bmatrix}
X_l - G_j^T - G_j + \Phi_{lj}^i & G_j^T A_l^T + Q_j^T B_l^T \\
A_l G_j + B_l Q_j & -X_i
\end{bmatrix} +
$$
$$
\begin{bmatrix}
X_j - G_l^T - G_l + \Phi_{jl}^i & G_l^T A_j^T + Q_l^T B_j^T \\
A_j G_l + B_j Q_l & -X_i
\end{bmatrix} < 0 \quad l, j, i \in L, \ l < j,
\tag{5.68}
$$

$$
\Phi^i := \begin{bmatrix}
2\Phi_1^i & \Phi_{12}^i & \cdots & \Phi_{1m}^i \\
\Phi_{21}^i & 2\Phi_2^i & \cdots & \Phi_{2m}^i \\
\vdots & \vdots & \ddots & \vdots \\
\Phi_{m1}^i & \Phi_{m2}^i & \cdots & 2\Phi_m^i
\end{bmatrix} > 0, \ i \in L.
\tag{5.69}
$$

Moreover, the controller gains are given by

$$
K_l = Q_l G_l^{-1}, \ l \in L.
\tag{5.70}
$$

Proof: The proof follows from proofs of Theorem 5.7 and Theorem 5.2, and is thus omitted. ❑

Remark 5.5

By comparing (5.8) and (5.61), one notices that the number of LMIs in (5.61) would be much larger than those in (5.8) especially when the fuzzy rule number m is large. This may lead to difficulty in many practical applications.

Remark 5.6

As indicated in Chapter 4, the main difficulty in using fuzzy Lyapunov functions arises when continuous-time systems are considered. It appears that the difficulty cannot be easily overcome and still presents a considerable challenge for the fuzzy logic control community.

Example 5.4

Reconsider the T–S fuzzy system defined in Example 5.2. By applying Theorem 5.7, one obtains the following solutions:

$$X_1 = \begin{bmatrix} 63.5664 & -46.8256 \\ -46.8256 & 428.7889 \end{bmatrix}, \ X_2 = \begin{bmatrix} 94.7503 & 12.3532 \\ 12.3532 & 367.7674 \end{bmatrix},$$

$$V_1 = \begin{bmatrix} 60.7578 & 3.9556 \\ -33.8931 & 362.2198 \end{bmatrix}, \ V_2 = \begin{bmatrix} 60.7578 & 3.9556 \\ -33.8931 & 362.2198 \end{bmatrix},$$

$$K_1 = \begin{bmatrix} -0.6652 & -0.4839 \end{bmatrix}, \ K_2 = \begin{bmatrix} -0.6652 & -0.4839 \end{bmatrix}.$$

It thus follows from Theorem 5.7 that the closed-loop control system is globally exponentially stable. Simulation results with a number of initial conditions are recorded in Figure 5.4.

Similarly, by applying Theorem 5.8, one obtains

$$X_1 = \begin{bmatrix} 0.0929 & -0.0792 \\ -0.0792 & 0.5979 \end{bmatrix}, \ X_2 = \begin{bmatrix} 0.1768 & 0.0273 \\ 0.0273 & 0.4587 \end{bmatrix},$$

$$G_1 = \begin{bmatrix} 0.1069 & -0.0807 \\ -0.0523 & 0.4780 \end{bmatrix}, \ G_2 = \begin{bmatrix} 0.1936 & 0.0314 \\ -0.0523 & 0.4425 \end{bmatrix},$$

$$K_1 = \begin{bmatrix} -0.5362 & -0.7941 \end{bmatrix}, \ K_2 = \begin{bmatrix} -0.8501 & -0.4820 \end{bmatrix}.$$

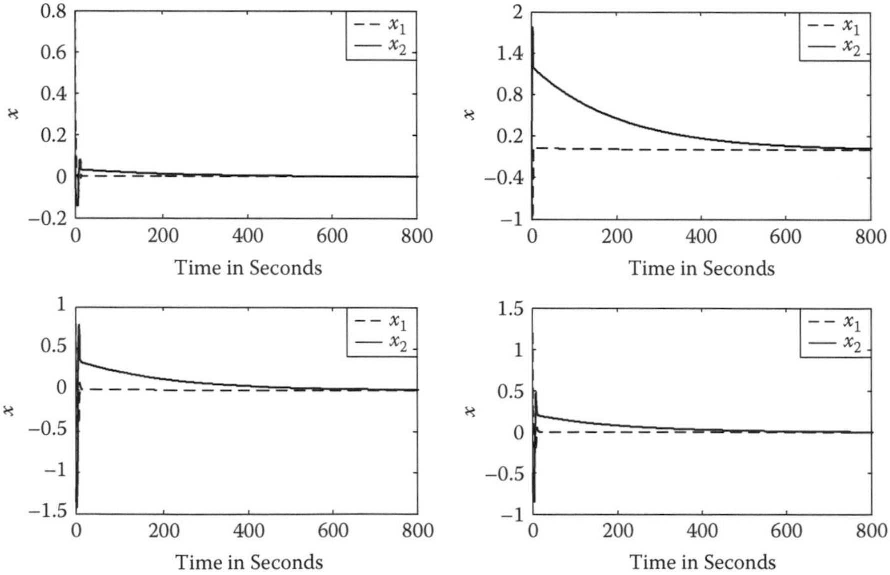

FIGURE 5.4   State responses of the closed-loop system in Example 5.4.

It can be easily observed from this example that the stabilization controller synthesis results based on fuzzy quadratic Lyapunov functions are less conservative than those obtained based on common quadratic Lyapunov functions.

5.5 COMPARISON OF STABILIZATION RESULTS VIA NUMERICAL EXAMPLES

In this section, the stabilization controller synthesis results discussed in the previous sections for T–S fuzzy systems based on common, piecewise, and fuzzy quadratic Lyapunov functions, respectively, are compared via numerical examples. For this purpose, the following modified stabilization results in terms of exponential decay rate parameter λ are adopted where the parameter λ is used for the performance indicator for comparison. The proofs of these results are straightforward and thus omitted.

Theorem 5.9
If there exist two positive constants η, ξ, a positive definite matrix X, and a set of matrices Q_l, $l \in L$ such that the following LMIs are satisfied,

$$I < X < \eta I, \tag{5.71}$$

$$\begin{bmatrix} -X & XA_l^T + Q_j^T B_l^T & X \\ A_l X + B_l Q_j & -X & 0 \\ X & 0 & -\frac{1}{\xi} I \end{bmatrix} < 0, \ l, j \in L, \tag{5.72}$$

then the closed-loop fuzzy control system (5.6) is globally exponentially stable with the decay rate λ defined as follows,

$$\|x(t)\| \le \alpha e^{-\lambda t} \|x(0)\|, \tag{5.73}$$

where $\alpha = \sqrt{\eta}$ and $\lambda = -\ln(1-\xi)/2$.

Moreover, the controller gains are given by

$$K_l = Q_l X^{-1}, \ l \in L. \tag{5.74}$$

Theorem 5.10
The closed-loop fuzzy control system (5.34) is globally exponentially stable with the decay rate λ as defined in (5.73) if there exist positive constants η, ξ, ε, a set of positive definite matrices X_l, $l \in L$, and a set of matrices Q_l, $l \in L$ such that the following LMIs are satisfied:

$$I < X_l < \eta I, \ l \in L \tag{5.75}$$

$$
\begin{bmatrix}
-X_l & X_lA_l^T + Q_l^TB_l^T & X_lE_{lA}^T & Q_l^TE_{lB}^T & X_l \\
A_lX_l + B_lQ_l & -(X_l - \varepsilon I) & 0 & 0 & 0 \\
E_{lA}X_l & 0 & -\dfrac{\varepsilon}{2}I & 0 & 0 \\
E_{lB}Q_l & 0 & 0 & -\dfrac{\varepsilon}{2}I & 0 \\
X_l & 0 & 0 & 0 & -\dfrac{1}{\xi}I
\end{bmatrix} < 0, \ l \in L, \tag{5.76}
$$

$$
\begin{bmatrix}
-X_l & X_lA_l^T + Q_l^TB_l^T & X_lE_{lA}^T & Q_l^TE_{lB}^T & X_l \\
A_lX_l + B_lQ_l & -(X_j - \varepsilon I) & 0 & 0 & 0 \\
E_{lA}X_l & 0 & -\dfrac{\varepsilon}{2}I & 0 & 0 \\
E_{lB}Q_l & 0 & 0 & -\dfrac{\varepsilon}{2}I & 0 \\
X_l & 0 & 0 & 0 & -\dfrac{1}{\xi}I
\end{bmatrix} < 0, \ l, j \in \Omega. \tag{5.77}
$$

Moreover, the controller gains are given by

$$
K_l = Q_lX_l^{-1}, \quad l \in L. \tag{5.78}
$$

Theorem 5.11
The closed-loop fuzzy control system (5.46) is globally exponentially stable with the decay rate λ as defined in (5.73) if there exist two positive constants η, ξ, a set of positive definite matrices X_l, $l \in \bar{L}$, and a set of matrices, Q_l, $l \in \bar{L}$ such that the following LMIs are satisfied:

$$
I < X_l < \eta I, \ l \in \bar{L}, \tag{5.79}
$$

$$
\begin{bmatrix}
-X_l & X_lA_k^T + Q_l^TB_k^T & X_l \\
A_kX_l + B_kQ_l & -X_l & 0 \\
X_l & 0 & -\dfrac{1}{\xi}I
\end{bmatrix} < 0, \ l \in \bar{L}, \ k \in \aleph(l), \tag{5.80}
$$

$$
\begin{bmatrix}
-X_l & X_lA_k^T + Q_l^TB_k^T & X_l \\
A_kX_l + B_kQ_l & -X_j & 0 \\
X_l & 0 & -\dfrac{1}{\xi}I
\end{bmatrix} < 0, \ (l, j) \in \bar{\Omega}, \ k \in \aleph(l). \tag{5.81}
$$

Moreover, the controller gains are given by

$$K_l = Q_l X_l^{-1}, \ l \in \bar{L}. \tag{5.82}$$

Theorem 5.12

The closed-loop fuzzy control system (5.6) is globally exponentially stable if there exist two positive constants η, ξ, a set of positive definite matrices X_l, $l \in L$, and sets of matrices Q_l, $l \in L$, such that the following LMIs are satisfied:

$$I < X_l < \eta I, \ l \in L, \tag{5.83}$$

$$\begin{bmatrix} -X_j & X_j^T A_l^T + Q_j^T B_l^T & X_j \\ A_l X_j + B_l Q_j & -X_i & 0 \\ X_j & 0 & -\dfrac{1}{\xi} I \end{bmatrix} < 0, \ i, j, l \in L. \tag{5.84}$$

Moreover, the controller gains are given by

$$K_l = Q_l X_l^{-1}, \ l \in L. \tag{5.85}$$

To reduce the conservatism of the result in Theorem 5.12, the following theorem (which corresponds to Theorem 5.7) is presented.

Theorem 5.13

The closed-loop fuzzy control system (5.6) is globally exponentially stable if there exist two positive constants η, ξ, a set of positive definite matrices X_l, $l \in L$, two sets of matrices Q_l, $l \in L$, and V_l, $l \in L$ such that the following LMIs are satisfied:

$$I < X_l < \eta I, \ l \in L, \tag{5.86}$$

$$\begin{bmatrix} X_l - V_j^T - V_j & V_j^T A_l^T + Q_j^T B_l^T & V_j^T \\ A_l V_j + B_l Q_j & -X_i & 0 \\ V_j & 0 & -\dfrac{1}{\xi} I \end{bmatrix} < 0, \ i, j, l \in L. \tag{5.87}$$

Moreover, the controller gains are given by

$$K_l = Q_l V_l^{-1}, \ l \in L. \tag{5.88}$$

The objective of comparison is to find the maximum decay rate for each stabilization result and then compare their performances. For this purpose, the following optimization algorithms can be developed based on Theorems 5.9–5.13, respectively.

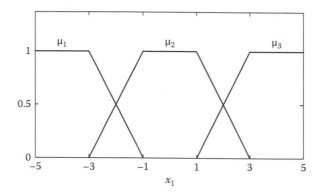

FIGURE 5.5 Membership functions in Example 5.5.

Algorithm 5.1: $\max_{X,Q_l,l\in L} \xi$, subject to LMIs (5.71) and (5.72).

Algorithm 5.2: $\max_{X_l,Q_l,l\in L} \xi$, subject to LMIs (5.75)–(5.77)

Algorithm 5.3: $\max_{X_l,Q_l,l\in \bar{L}} \xi$, subject to LMIs (5.79)–(5.81)

Algorithm 5.4: $\max_{X_l,Q_l,l\in L} \xi$, subject to LMIs (5.83) and (5.84)

Algorithm 5.5: $\max_{X_l,Q_l,V_l,l\in L} \xi$, subject to LMIs (5.86) and (5.87)

Example 5.5

Consider the T–S fuzzy system defined in (5.1) with matrix parameters as follows:

$$A_1 = \begin{bmatrix} 1 & 0.4 \\ -0.5 & 1 \end{bmatrix}, \ A_2 = \begin{bmatrix} 1 & 0.5 \\ -0.4 & 1 \end{bmatrix}, \ A_3 = \begin{bmatrix} 1.2 & 0.6 \\ -0.5 & 1.2 \end{bmatrix},$$

$$B_1 = \begin{bmatrix} 0 \\ 1 \end{bmatrix}, \ B_2 = \begin{bmatrix} 0 \\ 1 \end{bmatrix}, \ B_3 = \begin{bmatrix} 0 \\ 1 \end{bmatrix},$$

and the membership functions described in Figure 5.5. Note that the open-loop system is unstable. By applying Algorithms 5.1–5.5, respectively, one obtains the corresponding maximum decay rates, which are summarized in Table 5.1.

TABLE 5.1

Maximum Decay Rates in Example 5.5

Algorithm	5.1	5.2	5.3	5.4	5.5
Maximum λ	0.0777	0.0357	0.0808	0.0777	0.0792

TABLE 5.2

Maximum Decay Rates in Example 5.6

Algorithm	5.1	5.2	5.3	5.4	5.5
Maximum λ	Infeasible	0.0011	0.0101	Infeasible	0.0016

Example 5.6

Consider the T–S fuzzy system defined in (5.1) with matrix parameters as follows:

$$A_1 = \begin{bmatrix} 1.0625 & -0.2125 \\ 0.3825 & 1.0625 \end{bmatrix}, \ A_2 = \begin{bmatrix} 1.0625 & -0.2125 \\ 0.6375 & 1.0625 \end{bmatrix}, \ A_3 = \begin{bmatrix} 1.0625 & 0.3825 \\ 0.2125 & 1.0625 \end{bmatrix},$$

$$B_1 = \begin{bmatrix} 0 \\ 1 \end{bmatrix}, \ B_2 = \begin{bmatrix} 0 \\ 1 \end{bmatrix}, \ B_2 = \begin{bmatrix} 1 \\ 0 \end{bmatrix},$$

and the same membership functions described in Figure 5.5. Also note that the open-loop system is unstable. By applying Algorithms 5.1–5.5, respectively, one obtains the corresponding maximum decay rates, which are summarized in Table 5.2, where the approach based on the common quadratic Lyapunov function and the approach based on the fuzzy quadratic Lyapunov functions without any improvement (Theorem 5.12, Algorithm 5.4) have no feasible solution.

It can be observed from these examples that the stabilization approaches based on piecewise quadratic Lyapunov functions or fuzzy quadratic Lyapunov functions are in general less conservative than those based on common quadratic Lyapunov functions. It can also be observed that the approach based on fuzzy quadratic Lyapunov functions without any modification (Theorem 5.12) is of similar conservatism to that based on common quadratic Lyapunov functions (Theorem 5.9) in the sense that they lead to the same solution or infeasibility for these two particular examples. However, much care should be taken when the comparison is made for other systems. In particular, it might depend on the particular systems under study, whether the stabilization approaches based on piecewise quadratic Lyapunov functions are less conservative than those based on fuzzy quadratic Lyapunov functions or vice versa. It should also be noted that Algorithm 5.2, which is based on piecewise quadratic Lyapunov functions, might lead to significant conservatism in some cases due to the introduction of uncertainties as demonstrated in Example 5.5.

5.6 CONCLUSIONS

This chapter has presented a number of approaches to stabilization controller synthesis of T–S fuzzy systems based on three commonly used Lyapunov functions. Both smooth and switching fuzzy controllers have been considered. It is shown that stabilization controller gains can be obtained by solving linear matrix inequalities. It is also shown via analysis and numerical examples that the stabilization control approaches based on piecewise quadratic Lyapunov functions or fuzzy quadratic Lyapunov functions are in general less conservative than their counterparts based on common quadratic Lyapunov functions.

6 Robust H_∞ Controller Synthesis of T–S Fuzzy Systems

6.1 INTRODUCTION

Stabilization controller synthesis of T–S fuzzy systems has been studied based on common, piecewise, and fuzzy quadratic Lyapunov functions, respectively, in the last chapter, where T–S fuzzy systems are supposed to be ideal in the sense that there are no modeling uncertainties such as unmodeled dynamics or disturbances. However, those uncertainties are unavoidable for most systems in practice. Therefore, robustness issues of control systems under such circumstances have to be addressed. This chapter presents a number of results on robust H_∞ controller synthesis of T–S fuzzy systems with external disturbances based on those three commonly used Lyapunov functions.

The rest of the chapter is organized as follows. First, robust H_∞ controller synthesis of T–S fuzzy systems based on common quadratic Lyapunov functions is discussed in Section 6.2. Then, the issue of H_∞ controller synthesis of T–S fuzzy systems based on piecewise quadratic Lyapunov functions is addressed in Section 6.3. Robust H_∞ controller design based on fuzzy quadratic Lyapunov functions is presented in Section 6.4, followed by a comparison of those robust stabilization results via numerical examples in Section 6.5. Some remarks conclude the chapter in Section 6.6.

6.2 ROBUST H_∞ CONTROL BASED ON COMMON QUADRATIC LYAPUNOV FUNCTIONS

In this chapter robust H_∞ control of T–S fuzzy systems is studied. For this purpose, consider a modified T–S fuzzy model with an extra item of disturbances as follows:

$$R^l: \quad \text{IF} \quad z_1 \text{ is } F_1^l, \text{ AND}, \dots z_v \text{ is } F_v^l$$
$$\text{THEN} \quad x(t+1) = A_l x(t) + B_l u(t) + D_l v(t) \tag{6.1}$$
$$q(t) = H_l x(t) + G_l u(t)$$
$$l \in L := \{1, 2, \dots, m\},$$

where most variables are the same as those defined in the previous chapters except for the disturbance term and the controlled output term; that is, R^l denotes the lth fuzzy inference rule, m the number of inference rules, F_j^l, $j = 1, 2, \ldots n$ the fuzzy sets, $x(t) \in \Re^n$ the state vector, $u(t) \in \Re^g$ the control vector, $q(t) \in \Re^r$ the controlled output vector, $v(t) \in \Re^s$ the disturbance vector that belongs to $l_2[0, \infty)$, $z(t) := [z_1, z_2, \ldots, z_v]$ the premise variables, which are some measurable variables of the system, and $(A_l, B_l, D_l, H_l, G_l)$ the lth local model of the fuzzy system (6.1).

By using a center-average defuzzifier, product inference, and singleton fuzzifier, the T–S fuzzy model (6.1) can be expressed by the following global model:

$$x(t+1) = A(\mu)x(t) + B(\mu)u(t) + D(\mu)v(t) \tag{6.2}$$

$$q(t) = H(\mu)x(t) + G(\mu)u(t)$$

where

$$A(\mu) = \sum_{l=1}^{m} \mu_l A_l, \quad B(\mu) = \sum_{l=1}^{m} \mu_l B_l, \quad D(\mu) = \sum_{l=1}^{m} \mu_l D_l,$$

$$H(\mu) = \sum_{l=1}^{m} \mu_l H_l, \quad G(\mu) = \sum_{l=1}^{m} \mu_l G_l.$$

The objective of the robust H_∞ controller synthesis is to design a suitable controller for the system (6.1) or equivalently (6.2) such that the closed-loop control system is globally exponentially stable with a guaranteed performance in the H_∞ sense. That is, given a prescribed level of disturbance attenuation $\gamma > 0$, find a controller such that the closed-loop control system is globally exponentially stable and the induced l_2-norm of the operator from $v(t)$ to the controlled output $q(t)$ is less than γ under zero initial conditions,

$$\|q(t)\|_2 < \gamma \|v(t)\|_2 \tag{6.3}$$

for all nonzero $v(t) \in l_2$. In this case, the closed-loop control system is said to be globally exponentially stable with H_∞ performance γ.

With the continuous fuzzy controller (5.3), or equivalently (5.4), the closed-loop system can be described by the equation

$$x(t+1) = A_c(\mu)x(t) + D_c(\mu)v(t) \tag{6.4}$$

$$q(t) = H_c(\mu)x(t),$$

where

$$A_c(\mu) = \sum_{l=1}^{m} \sum_{j=1}^{m} \mu_l \mu_j (A_l + B_l K_j), \quad D_c(\mu) = D(\mu) = \sum_{l=1}^{m} \mu_l D_l,$$

$$H_c(\mu) = \sum_{l=1}^{m} \sum_{j=1}^{m} \mu_l \mu_j (H_l + G_l K_j).$$

The following lemma is introduced before the main results in this section are presented.

Lemma 6.1

Given a constant $\gamma > 0$, the fuzzy system (6.4) is globally exponentially stable with H_∞ performance γ, if there exists a positive definite matrix X such that the following matrix inequalities are satisfied:

$$\gamma^2 I - D_c^T X^{-1} D_c > 0, \tag{6.5}$$

$$A_c^T X^{-1} A_c - X^{-1} + A_c^T X^{-1} D_c \left(\gamma^2 I - D_c^T X^{-1} D_c\right)^{-1} D_c^T X^{-1} A_c + H_c^T H_c < 0. \tag{6.6}$$

Proof: It is easily seen that Equations (6.5) and (6.6) imply the inequality

$$A_c^T X^{-1} A_c - X^{-1} < 0, \tag{6.7}$$

thus it follows from Lyapunov stability theory that the closed-loop system is globally exponentially stable.

Now we show the disturbance attenuation performance. Consider the Lyapunov function candidate,

$$V(x) = x^T X^{-1} x. \tag{6.8}$$

Then along the trajectories of the system, one has

$$\begin{aligned}
\Delta V(t) &:= V(t+1) - V(t) \\
&= x^T(t+1)X^{-1}x(t+1) - x^T(t)X^{-1}x(t) \\
&= x(t)^T \left(A_c^T X^{-1} A_c - X^{-1}\right)x(t) + v(t)^T D_c^T X^{-1} A_c x(t) \\
&\quad + x(t)^T A_c^T X^{-1} D_c v(t) + v(t)^T D_c^T X^{-1} D_c v(t) \\
&\leq x(t)^T \left[-A_c^T X^{-1} D_c \left(\gamma^2 I - D_c^T X^{-1} D_c\right)^{-1} D_c^T X^{-1} A_c - H_c^T H_c\right]x(t) \\
&\quad + v(t)^T D_c^T X^{-1} A_c x(t) + x(t)^T A_c^T X^{-1} D_c v(t) + v(t)^T D_c^T X^{-1} D_c v(t) \\
&= -q(t)^T q(t) + \gamma^2 v(t)^T v(t) - w(t)^T M(t)w(t), \tag{6.9}
\end{aligned}$$

where $M(t) = \gamma^2 I - D_c^T X^{-1} D_c$, $w(t) = v(t)^T - M(t)^{-1} D_c^T X^{-1} A_c x$.
Then it follows from (6.9) that

$$\Delta V(t) \leq -q(t)^T q(t) + \gamma^2 v(t)^T v(t), \tag{6.10}$$

which implies that

$$V(x(\infty)) - V(x(0)) \le -\sum_{t=0}^{\infty} q(t)^T q(t) + \sum_{t=0}^{\infty} \gamma^2 v(t)^T v(t);$$

that is, with $x(0) = 0$, $\|q\|_2 \le \gamma \|v\|_2$, and the proof is thus completed. ☐

Then based on Lemma 6.1, one has the following result.

Theorem 6.1
Given a constant $\gamma > 0$, the closed-loop control system (6.4) is globally exponentially stable with H_∞ performance γ, if there exist a positive definite matrix X and a set of matrices $Q_l, l \in L$ such that the following LMIs are satisfied:

$$\begin{bmatrix} -X & (A_i X + B_i Q_j)^T & (H_i X + G_i Q_j)^T \\ A_i X + B_i Q_j & -X + \gamma^{-2} D_i D_i^T & 0 \\ H_i X + G_i Q_j & 0 & -I \end{bmatrix} < 0, \quad \forall l, j \in L. \quad (6.11)$$

Moreover, the controller gains are given by

$$K_l = Q_l X^{-1}, \quad l \in L. \tag{6.12}$$

Proof: According to Lemma 6.1, one knows that the system (6.4) is globally exponentially stable with H_∞ performance γ, if the conditions (6.5) and (6.6) are satisfied. We show LMIs in (6.11) imply those conditions.

It follows from (6.11) that $X - \gamma^{-2} D_l D_l^T > 0$. We first show that the inequality $X - \gamma^{-2} D_l D_l^T > 0$ implies (6.5). By the Schur complements, it follows that

$$X - \gamma^{-2} D_l D_l^T > 0$$

is equivalent to the LMI

$$\begin{bmatrix} X & D_l \\ D_l^T & \gamma^2 I \end{bmatrix} > 0. \tag{6.13}$$

Then it follows from (6.13) that

$$\sum_{l=1}^{m} \mu_l \begin{bmatrix} X & D_l \\ D_l^T & \gamma^2 I \end{bmatrix} = \begin{bmatrix} X & \sum_{l=1}^{m} \mu_l D_l \\ \sum_{l=1}^{m} \mu_l D_l^T & \gamma^2 I \end{bmatrix} = \begin{bmatrix} X & D_c \\ D_c^T & \gamma^2 I \end{bmatrix} > 0. \tag{6.14}$$

And by using the Schur complement again with respect to the term X one can conclude that $\gamma^2 I - D_c^T X^{-1} D_c > 0$, which is exactly (6.5).

We then show that the inequality (6.11) implies the inequality (6.6). Note that via the matrix inversion lemma the inequality (6.6) can be expressed as

$$A_c^T \left(X - \gamma^{-2} D_c D_c^T \right)^{-1} A_c - X^{-1} + H_c^T H_c < 0. \tag{6.15}$$

Multiplying X in both sides of (6.15) leads to

$$X A_c^T \left(X - \gamma^{-2} D_c D_c^T \right)^{-1} A_c X - X + X H_c^T H_c X < 0. \tag{6.16}$$

Using the Schur complement, it follows from (6.16) that

$$\begin{bmatrix} -X + X H_c^T H_c X & X A_c^T \\ A_c X & -X + \gamma^{-2} D_c D_c^T \end{bmatrix} < 0, \tag{6.17}$$

which, via the Schur complement again, is equivalent to

$$\begin{bmatrix} -X & X A_c^T & X H_c^T \\ A_c X & -X + \gamma^{-2} D_c D_c^T & 0 \\ H_c X & 0 & -I \end{bmatrix} < 0. \tag{6.18}$$

Substituting A_c, D_c, and H_c into (6.18) leads to

$$\sum_{l=1}^{m} \sum_{j=1}^{m} \mu_l \mu_j \begin{bmatrix} -X & X(A_l + B_l K_j)^T & X(H_l + G_l K_j)^T \\ (A_l + B_l K_j) X & -X + \gamma^{-2} D_l D_j^T & 0 \\ (H_l + G_l K_j) X & 0 & -I \end{bmatrix} < 0. \tag{6.19}$$

Then by noting $D_l D_j^T \leq \frac{1}{2}(D_l D_l^T + D_j D_j^T)$ it follows that the next matrix inequalities would imply (6.19),

$$\begin{bmatrix} -X & X(A_l + B_l K_j)^T & X(H_l + G_l K_j)^T \\ (A_l + B_l K_j) X & -X + \gamma^{-2} D_l D_l^T & 0 \\ (H_l + G_l K_j) X & 0 & -I \end{bmatrix} < 0, \quad \forall l, j \in L. \tag{6.20}$$

Letting $Q_j = K_j X$, it follows from (6.20) that

$$\begin{bmatrix} -X & (A_l X + B_l Q_j)^T & (H_l X + G_l Q_j)^T \\ A_l X + B_l Q_j & -X + \gamma^{-2} D_l D_l^T & 0 \\ H_l X + G_l Q_j & 0 & -I \end{bmatrix} < 0, \quad \forall l, j \in L, \tag{6.21}$$

which is (6.11). Thus we have shown that the inequality (6.11) implies the inequality (6.6). Therefore, it can be concluded from Lemma 6.1 that the closed-loop control system is globally exponentially stable with H_∞ performance γ and the proof is thus completed. ❑

On the other hand, with the switching control law (5.5) the closed-loop fuzzy control system can be described in each region as

$$x(t+1) = A_{cl}(\mu)x(t) + D_{cl}(\mu)v(t)$$

$$q(t) = H_{cl}(\mu)x(t) \tag{6.22}$$

$$z(t) \in S_l,$$

where

$$A_{cl}(\mu) = A_l + \Delta A_l(\mu) + (B_l + \Delta B_l(\mu))K_l, \quad D_{cl} = D_l + \Delta D_l(\mu),$$

$$H_{cl}(\mu) = H_l + \Delta H_l(\mu) + (G_l + \Delta G_l(\mu))K_l. \tag{6.23}$$

Similarly, the following upper bounds for the uncertainty terms of the fuzzy system (6.23) can be introduced as in Chapters 4 and 5.

$$[\Delta A_l(\mu)]^T[\Delta A_l(\mu)] \le E_{lA}^T E_{lA}, \quad [\Delta B_l(\mu)]^T[\Delta B_l(\mu)] \le E_{lB}^T E_{lB},$$

$$[\Delta D_l(\mu)][\Delta D_l(\mu)]^T \le E_{lD}E_{lD}^T, \quad [\Delta H_l(\mu)]^T[\Delta H_l(\mu)] \le E_{lH}^T E_{lH}, \tag{6.24}$$

$$[\Delta G_l(\mu)]^T[\Delta G_l(\mu)] \le E_{lG}^T E_{lG}.$$

Lemma 6.2
Given a constant $\gamma > 0$, the fuzzy system (6.22) is globally exponentially stable with H_∞ performance γ, if there exists a positive definite matrix X such that the following matrix inequalities are satisfied:

$$\gamma^2 I - D_{cl}^T X^{-1} D_{cl} < 0, \quad l \in L, \tag{6.25}$$

$$A_{cl}^T X^{-1} A_{cl} - X^{-1} + A_{cl}^T X^{-1} D_{cl} \left(\gamma^2 I - D_{cl}^T X^{-1} D_{cl} \right)^{-1} D_{cl}^T X^{-1} A_{cl} + H_{cl}^T H_{cl} < 0, l \in L. \tag{6.26}$$

Proof: The proof is similar to that of Lemma 6.1 and thus omitted. ❑

Then based on Lemma 6.2, one has the following result.

Theorem 6.2
Given a constant $\gamma > 0$, the system (6.22) is globally exponentially stable with H_∞ performance γ, if there exist a set of constants $\varepsilon_l, l = 1, 2, \ldots, m$, a positive

definite matrix X, and a set of matrices $Q_l, l \in L$ such that the following LMIs are satisfied,

$$\begin{bmatrix} -X & (A_lX + B_lQ_l)^T & \Gamma_l^T \\ A_lX + B_lQ_l & -\Pi_l & 0 \\ \Gamma_l & 0 & -\Xi_l \end{bmatrix} < 0, \quad l \in L, \qquad (6.27)$$

where

$$\Pi_l := X - 2\gamma^{-2}D_lD_l^T - 2\gamma^{-2}E_{lD}E_{lD}^T - \varepsilon_lI,$$

$$\Gamma_l^T = \begin{bmatrix} XE_{lA}^T & XH_l^T & XE_{lH}^T & Q_l^TE_{lB}^T & Q_l^TG_l^T & Q_l^TE_{lG}^T \end{bmatrix},$$

$$\Xi_l = \mathrm{Diag}\{\varepsilon_lI/2 \quad I/4 \quad I/4 \quad \varepsilon_lI/2 \quad I/4 \quad I/4\}.$$

Moreover, the controller gains are given by

$$K_l = Q_lX^{-1}, \quad l \in L. \qquad (6.28)$$

Proof: According to Lemma 6.2, one knows that the system (6.22) is globally exponentially stable with H_∞ performance γ, if the conditions (6.25) and (6.26) are satisfied.

It follows from Equation (6.27) that $\Pi_l > 0$. We first show that the inequality $\Pi_l > 0$ implies (6.25). It follows from $\Pi_l > 0$ that

$$X - \gamma^{-2}D_{cl}D_{cl}^T > 0,$$

which, by the Schur complement, is equivalent to the following LMI:

$$\begin{bmatrix} X & D_{cl} \\ D_{cl}^T & \gamma^2I \end{bmatrix} > 0.$$

And by using the Schur complement again with respect to the other term one can conclude $\gamma^2I - D_{cl}^TX^{-1}D_{cl} > 0$.

We then show that (6.27) implies (6.26). It is noted that via the matrix inversion lemma the right-hand side of (6.26) can be expressed as

$$RH := A_{cl}^T\left(X - \gamma^{-2}D_{cl}D_{cl}^T\right)^{-1}A_{cl} - X^{-1} + H_{cl}^TH_{cl}$$

$$= [A_l + \Delta A_l + (B_l + \Delta B_l)K_l]^T\left[X - \gamma^{-2}(D_l + \Delta D_l)(D_l + \Delta D_l)^T\right]^{-1}$$

$$\times [A_l + \Delta A_l + (B_l + \Delta B_l)K_l] - X^{-1} + [H_l + \Delta H_l + (G_l + \Delta G_l)K_l]^T$$

$$\times [H_l + \Delta H_l + (G_l + \Delta G_l)K_l].$$

Letting $\Theta = [X - 2\gamma^{-2}(D_l D_l^T + E_{lD}E_{lD}^T)]^{-1}$, using Lemma 4.2, one has

$$RH \le (A_l + B_l K_l)^T \Theta (A_l + B_l K_l) + (A_l + B_l K_l)^T \Theta (\Delta A_l + \Delta B_l K_l)$$

$$+ (\Delta A_l + \Delta B_l K_l)^T \Theta (A_l + B_l K_l) + (\Delta A_l + \Delta B_l K_l)^T \Theta (\Delta A_l + \Delta B_l K_l) - X^{-1}$$

$$+ 2(H_l + \Delta H_l)^T (H_l + \Delta H_l) + 2[(G_l + \Delta G_l)K_l]^T [(\Delta G_l + \Delta G_l)K_l]$$

$$\le (A_l + B_l K_l)^T \Theta (A_l + B_l K_l) + (A_l + B_l K_l)^T \Theta \left(\frac{1}{\varepsilon_l} I - \Theta\right)^{-1} \Theta (A_l + B_l K_l)$$

$$+ \frac{1}{\varepsilon_l}(\Delta A_l + \Delta B_l K_l)^T (\Delta A_l + \Delta B_l K_l) - X^{-1} + 4\left(H_l^T H_l + \Delta H_l^T \Delta H_l\right)$$

$$+ 4K_l^T \left(G_l^T G_l + \Delta G_l^T G_l\right)K_l$$

$$\le (A_l + B_l K_l)^T \left[\Theta^{-1} - \varepsilon_l I\right]^{-1} (A_l + B_l K_l) + \frac{2}{\varepsilon_l}\left(\Delta A_l^T \Delta A_l + K_l^T \Delta B_l^T \Delta B_l K_l\right) - X^{-1}$$

$$+ 4\left(H_l^T H_l + \Delta H_l^T \Delta H_l\right) + 4K_l^T \left(G_l^T G_l + \Delta G_l^T \Delta G_l\right)K_l$$

$$\le (A_l + B_l K_l)^T \Pi_l^{-1}(A_l + B_l K_l) + \frac{2}{\varepsilon_l}\left(E_{lA}^T E_{lA} + K_l^T E_{lB}^T E_{lB} K_l\right) - X^{-1}$$

$$+ 4\left(H_l^T H_l + E_{lH}^T E_{lH}\right) + 4K_l^T \left(G_l^T G_l + E_{lG}^T E_{lG}\right)K_l.$$

Then the following inequality implies (6.26),

$$(A_l + B_l K_l)^T \Pi_l^{-1}(A_l + B_l K_l) - X^{-1} + \frac{2}{\varepsilon_l}\left(E_{lA}^T E_{lA} + K_l^T E_{lB}^T E_{lB} K_l\right)$$

$$+ 4\left(H_l^T H_l + E_{lH}^T E_{lH}\right) + 4K_l^T \left(G_l^T G_l + E_{lG}^T E_{lG}\right)K_l < 0,$$

or equivalently,

$$X(A_l + B_l K_l)^T \Pi_l^{-1}(A_l + B_l K_l)X - X + X\left(\frac{2}{\varepsilon_l} E_{lA}^T E_{lA} + 4H_l^T H_l + 4E_{lH}^T E_{lH}\right)X$$

$$(6.29)$$

$$+ XK_l^T \left(\frac{2}{\varepsilon_l} E_{lB}^T E_{lB} + 4G_l^T G_l + 4E_{lG}^T E_{lG}\right)K_l X < 0.$$

Let $Q_l = K_l X$; using the Schur complement formula a few times, one can easily show that (6.29) is equivalent to (6.27). Thus we have shown that (6.27) implies (6.26). Therefore, it can be concluded from Lemma 6.2 that the closed-loop control system is globally exponentially stable with H_∞ performance γ and thus the proof is completed. ❑

6.3 ROBUST H_∞ CONTROL BASED ON PIECEWISE QUADRATIC LYAPUNOV FUNCTIONS

In this section robust H_∞ control of T–S fuzzy systems based on piecewise quadratic Lyapunov functions is studied. With the space partitions of the first kind, the same T–S fuzzy system with external disturbance defined in (6.1), or equivalently (6.2), can be described in each region as

$$x(t+1) = (A_l + \Delta A_l(\mu))x(t) + (B_l + \Delta B_l(\mu))u(t) + (D_l + \Delta D_l(\mu))v(t)$$

$$q(t) = (H_l + \Delta H_l(\mu))x(t) + (G_l + \Delta G_l(\mu))u(t) \tag{6.30}$$

$$z(t) \in S_l, \quad l \in L := \{1, 2, \ldots, m\}.$$

With the piecewise controller,

$$u(t) = K_l x(t) \quad z(t) \in S_l, \quad l \in L, \tag{6.31}$$

the closed-loop system can be described in each local region by the equation

$$x(t+1) = A_{cl}(\mu)x(t) + D_{cl}(\mu)v(t)$$

$$q(t) = H_{cl}(\mu)x(t) \tag{6.32}$$

$$z(t) \in S_l, \quad l \in L := \{1, 2, \ldots, m\},$$

where

$$A_{cl}(\mu) = A_l + \Delta A_l(\mu) + (B_l + \Delta B_l(\mu))K_l, \quad D_{cl} = D_l + \Delta D_l(\mu),$$

$$H_{cl}(\mu) = H_l + \Delta H_l(\mu) + (G_l + \Delta G_l(\mu))K_l.$$

Define the set Ω as in (4.15) representing all possible system transitions among regions. The following lemma is first introduced.

Lemma 6.3

Given a constant $\gamma > 0$, the fuzzy system (6.32) is globally exponentially stable with H_∞ performance γ, if there exists a set of positive definite matrices $X_l, l \in L$ such that the following matrix inequalities are satisfied,

$$\gamma^2 I - D_{cl}^T X_l^{-1} D_{cl} < 0, \quad l \in L, \tag{6.33}$$

$$A_{cl}^T X_l^{-1} A_{cl} - X_l^{-1} + A_{cl}^T X_l^{-1} D_{cl} \left(\gamma^2 I - D_{cl}^T X_l^{-1} D_{cl} \right)^{-1} D_{cl}^T X_l^{-1} A_{cl} + H_{cl}^T H_{cl} < 0, \quad l \in L, \tag{6.34}$$

$$\gamma^2 I - D_{cl}^T X_j^{-1} D_{cl} < 0, \quad l, j \in \Omega, \tag{6.35}$$

$$A_{cl}^T X_j^{-1} A_{cl} - X_l^{-1} + A_{cl}^T X_j^{-1} D_{cl} \left(\gamma^2 I - D_{cl}^T X_j^{-1} D_{cl} \right)^{-1} D_{cl}^T X_j^{-1} A_{cl} + H_{cl}^T H_{cl} < 0, \quad l, j \in \Omega. \tag{6.36}$$

Proof: It is easily seen that Equations (6.33)–(6.36) imply the following inequalities, respectively,

$$A_{cl}^T X_l^{-1} A_{cl} - X_l^{-1} < 0, \quad l \in L, \tag{6.37}$$

$$A_{cl}^T X_j^{-1} A_{cl} - X_l^{-1} < 0, \quad l, j \in \Omega. \tag{6.38}$$

Thus it follows from Theorem 4.2 and its proof that the closed-loop system is globally exponentially stable.

Now we show the disturbance attenuation performance. Consider the Lyapunov function candidate,

$$V(x) = x^T X_l^{-1} x, \, z(t) \in S_l, \ l \in L. \tag{6.39}$$

Then along the trajectories of the system, one has

$$\begin{aligned} \Delta V(t) &:= V(t+1) - V(t) \\ &= x^T(t+1)X_j^{-1}x(t+1) - x^T(t)X_l^{-1}x(t) \\ &= x(t)^T \left(A_{cl}^T X_j^{-1} A_{cl} - X_l^{-1} \right)x(t) + v(t)^T D_{cl}^T X_j^{-1} A_{cl}x(t) \\ &\quad + x(t)^T A_{cl}^T X_j^{-1} D_{cl} v(t) + v(t)^T D_{cl}^T X_j^{-1} D_{cl} v(t) \\ &\leq x(t)^T \left[-A_{cl}^T X_j^{-1} D_{cl} \left(\gamma^2 I - D_{cl}^T X_j^{-1} D_{cl} \right)^{-1} D_{cl}^T X_j^{-1} A_{cl} - H_{cl}^T H_{cl} \right]x(t) \\ &\quad + v(t)^T D_{cl}^T X_j^{-1} A_{cl}x(t) + x(t)^T A_{cl}^T X_j^{-1} D_{cl} v(t) + v(t)^T D_{cl}^T X_j^{-1} D_{cl} v(t) \\ &= -q(t)^T q(t) + \gamma^2 v(t)^T v(t) - w(t)^T M(t)w(t), \end{aligned} \tag{6.40}$$

where $M(t) = \gamma^2 I - D_{cl}^T X_j^{-1} D_{cl}$, $w(t) = v(t)^T - M(t)^{-1} D_{cl}^T X_j^{-1} A_{cl}x$.

Then it follows from (6.40) that

$$\Delta V(t) \leq -q(t)^T q(t) + \gamma^2 v(t)^T v(t), \tag{6.41}$$

which implies that

$$V(x(\infty)) - V(x(0)) \leq -\sum_{t=0}^{\infty} q(t)^T q(t) + \sum_{t=0}^{\infty} \gamma^2 v(t)^T v(t);$$

that is, with $x(0) = 0$, $\|q\|_2 \leq \gamma \|v\|_2$, and the proof is thus completed. $\square$

Then based on Lemma 6.3, one has the following result.

Theorem 6.3

Given a constant $\gamma > 0$, the system (6.32) is globally exponentially stable with H_∞ performance γ, if there exist a set of constants $\varepsilon_l, l = 1, 2, \cdots m$, a set of positive definite matrices $X_l, l \in L$, and a set of matrices $Q_l, l \in L$ such that the following LMIs are satisfied:

$$\begin{bmatrix} -X_l & (A_l X_l + B_l Q_l)^T & \Gamma_l^T \\ A_l X_l + B_l Q_l & -\Pi_l & 0 \\ \Gamma_l & 0 & -\Xi_l \end{bmatrix} < 0, \quad l \in L, \tag{6.42}$$

$$\begin{bmatrix} -X_l & (A_l X_l + B_l Q_l)^T & \Gamma_l^T \\ A_l X_l + B_l Q_l & -\Pi_{lj} & 0 \\ \Gamma_l & 0 & -\Xi_l \end{bmatrix} < 0, \quad l, j \in \Omega, \tag{6.43}$$

where

$$\Pi_l := X_l - 2\gamma^{-2} D_l D_l^T - 2\gamma^{-2} E_{lD} E_{lD}^T - \varepsilon_l I,$$

$$\Pi_{lj} := X_j - 2\gamma^{-2} D_l D_l^T - 2\gamma^{-2} E_{lD} E_{lD}^T - \varepsilon_l I,$$

$$\Gamma_l^T = \begin{bmatrix} X_l E_{lA}^T & X_l H_l^T & X_l E_{lH}^T & Q_l^T E_{lB}^T & Q_l^T G_l^T & Q_l^T E_{lG}^T \end{bmatrix},$$

$$\Xi_l = \mathrm{Diag}\{\varepsilon_l I/2 \quad I/4 \quad I/4 \quad \varepsilon_l I/2 \quad I/4 \quad I/4\}.$$

Moreover, the controller gains are given by

$$K_l = Q_l X_l^{-1}, \quad l \in L. \tag{6.44}$$

Proof: According to Lemma 6.3, one knows that the system (6.32) is globally exponentially stable with H_∞ performance γ, if the conditions (6.33)–(6.36) are satisfied.

It follows from Equation (6.42) that $\Pi_l > 0$. We first show that the inequality $\Pi_l > 0$ implies (6.33). It follows from $\Pi_l > 0$ that

$$X_l - \gamma^{-2} D_{cl} D_{cl}^T > 0,$$

which, by the Schur complement, is equivalent to the LMI

$$\begin{bmatrix} X_l & D_{cl} \\ D_{cl}^T & \gamma^2 I \end{bmatrix} > 0.$$

And by using the Schur complement again with respect to the other term one can conclude $\gamma^2 I - D_{cl}^T X_l^{-1} D_{cl} > 0$. Similarly, one can easily show that $\Pi_{lj} > 0$ implies (6.36).

We then show that (6.42) implies (6.34). It is noted that via the matrix inversion lemma the left-hand side of (6.34) can be expressed as

$$LH := A_{cl}^T \left(X_l - \gamma^{-2} D_{cl} D_{cl}^T \right)^{-1} A_{cl} - X_l^{-1} + H_{cl}^T H_{cl}$$

$$= [A_l + \Delta A_l + (B_l + \Delta B_l) K_l]^T \left[X_l - \gamma^{-2} (D_l + \Delta D_l)(D_l + \Delta D_l)^T \right]^{-1}$$

$$\times [A_l + \Delta A_l + (B_l + \Delta B_l) K_l] - X_l^{-1} + [H_l + \Delta H_l + (G_l + \Delta G_l) K_l]^T$$

$$\times [H_l + \Delta H_l + (G_l + \Delta G_l) K_l].$$

Let $\Theta = [X_l - 2\gamma^{-2}(D_l D_l^T + E_{lD} E_{lD}^T)]^{-1}$; using Lemma 4.2, one has

$$LH \leq (A_l + B_l K_l)^T \Theta (A_l + B_l K_l) + (A_l + B_l K_l)^T \Theta (\Delta A_l + \Delta B_l K_l)$$

$$+ (\Delta A_l + \Delta B_l K_l)^T \Theta (A_l + B_l K_l) + (\Delta A_l + \Delta B_l K_l)^T \Theta (\Delta A_l + \Delta B_l K_l) - X_l^{-1}$$

$$+ 2(H_l + \Delta H_l)^T (H_l + \Delta H_l) + 2[(G_l + \Delta G_l) K_l]^T [(\Delta G_l + \Delta G_l) K_l]$$

$$\leq (A_l + B_l K_l)^T \Theta (A_l + B_l K_l) + (A_l + B_l K_l)^T \Theta \left(\frac{1}{\varepsilon_l} I - \Theta \right)^{-1} \Theta (A_l + B_l K_l)$$

$$+ \frac{1}{\varepsilon_l} (\Delta A_l + \Delta B_l K_l)^T (\Delta A_l + \Delta B_l K_l) - X_l^{-1} + 4 \left(H_l^T H_l + \Delta H_l^T \Delta H_l \right)$$

$$+ 4 K_l^T \left(G_l^T G_l + \Delta G_l^T G_l \right) K_l$$

$$\leq (A_l + B_l K_l)^T \left[\Theta^{-1} - \varepsilon_l I \right]^{-1} (A_l + B_l K_l) + \frac{2}{\varepsilon_l} \left(\Delta A_l^T \Delta A_l + K_l^T \Delta B_l^T \Delta B_l K_l \right) - X_l^{-1}$$

$$+ 4 \left(H_l^T H_l + \Delta H_l^T \Delta H_l \right) + 4 K_l^T \left(G_l^T G_l + \Delta G_l^T \Delta G_l \right) K_l$$

$$\leq (A_l + B_l K_l)^T \Pi_{cl}^{-1} (A_l + B_l K_l) + \frac{2}{\varepsilon_l} \left(E_{lA}^T E_{lA} + K_l^T E_{lB}^T E_{lB} K_l \right) - X_l^{-1}$$

$$+ 4 \left(H_l^T H_l + E_{lH}^T E_{lH} \right) + 4 K_l^T \left(G_l^T G_l + E_{lG}^T E_{lG} \right) K_l.$$

It then follows that the following inequality implies (6.34),

$$(A_l + B_l K_l)^T \Pi_l^{-1} (A_l + B_l K_l) - X_l^{-1} + \frac{2}{\varepsilon_l} \left(E_{lA}^T E_{lA} + K_l^T E_{lB}^T E_{lB} K_l \right)$$

$$+ 4 \left(H_l^T H_l + E_{lH}^T E_{lH} \right) + 4 K_l^T \left(G_l^T G_l + E_{lG}^T E_{lG} \right) K_l < 0,$$

or equivalently,

$$X_l(A_l + B_lK_l)^T \Pi_l^{-1}(A_l + B_lK_l)X_l - P_l + X_l\left(\frac{2}{\varepsilon_l}E_{lA}^T E_{lA} + 4H_l^T H_l + 4E_{lH}^T E_{lH}\right)X_l$$

$$+ X_l K_l^T \left(\frac{2}{\varepsilon_l}E_{lB}^T E_{lB} + 4G_l^T G_l + 4E_{lG}^T E_{lG}\right)K_l X_l < 0. \tag{6.45}$$

Let $Q_l = K_l P_l$. Using the Schur complement formula a few times, one can easily show that (6.45) is equivalent to (6.42). Thus we have shown that (6.42) implies (6.34). Following a similar procedure, one can also show that (6.44) implies (6.36). Therefore, it can be concluded from Lemma 6.3 that the closed-loop control system is globally exponentially stable with H_∞ performance γ and the proof is thus completed. ❑

With a space partition of the second kind and the switching controller defined as in (5.45) the closed-loop fuzzy control system with disturbances can be described by the following equations,

$$x(t+1) = \sum_{k\in\aleph(l)} \mu_k(z) \cdot \{A_{clk}x(t) + D_k v(t)\}$$

$$q(t) = \sum_{k\in\aleph(l)} \mu_k(z) \cdot \{H_{clk}x(t)\} \tag{6.46}$$

$$z(t) \in S_l, \quad k \in \aleph(l), \quad l \in \bar{L},$$

where

$$A_{clk} = A_k + B_k K_l, \quad H_{clk} = H_k + G_k K_l.$$

Define the set $\bar{\Omega}$ as in (4.32) representing all possible transitions among regions. Then one can easily obtain the following lemma, which is similar to Lemma 6.3.

Lemma 6.4
Given a constant $\gamma > 0$, the fuzzy system (6.46) is globally exponentially stable with H_∞ performance γ, if there exists a set of positive definite matrices $X_l, l \in L$ such that the following matrix inequalities are satisfied:

$$\gamma^2 I - D_k^T X_l^{-1} D_k < 0, \quad l \in \bar{L}, \quad k \in \aleph(l), \tag{6.47}$$

$$A_{clk}^T X_l^{-1} A_{clk} - X_l^{-1} + A_{clk}^T X_l^{-1} D_k$$

$$\times \left(\gamma^2 I - D_k^T X_l^{-1} D_k\right)^{-1} D_k^T X_l^{-1} A_{clk} + H_{clk}^T H_{clk} < 0, \quad l \in \bar{L}, \quad k \in \aleph(l), \tag{6.48}$$

$$\gamma^2 I - D_k^T X_j^{-1} D_k < 0, \quad l, j \in \bar{\Omega}, \quad k \in \aleph(l), \tag{6.49}$$

$$A_{clk}^T X_j^{-1} A_{clk} - X_l^{-1} + A_{clk}^T X_j^{-1} D_k$$
$$\times \left(\gamma^2 I - D_k^T X_j^{-1} D_k \right)^{-1} D_k^T X_j^{-1} A_{clk} + H_{clk}^T H_{clk} < 0, \quad l, j \in \bar{\Omega}, \quad k \in \aleph(l). \tag{6.50}$$

Proof: The proof is similar to that of Lemma 6.3 and thus omitted. ❑

Then one has the following robust H_∞ fuzzy controller design results.

Theorem 6.4
Given a scalar $\gamma > 0$, the system (6.46) is globally exponentially stable with H_∞ performance γ, if there exist a set of positive definite matrices X_l, $l \in \bar{L}$ and a set of matrices Q_l, $l \in \bar{L}$ such that the following LMIs are satisfied:

$$\begin{bmatrix} -X_l & (A_k X_l + B_k Q_l)^T & (H_k X_l + G_k Q_l)^T \\ (A_k X_l + B_k Q_l) & -X_l + \gamma^{-2} D_k D_k^T & 0 \\ (H_k X_l + G_k Q_l) & 0 & -I \end{bmatrix} < 0, \quad l \in \bar{L}, \quad k \in \aleph(l), \tag{6.51}$$

$$\begin{bmatrix} -X_l & (A_k X_l + B_k Q_l)^T & (H_k X_l + G_k Q_l)^T \\ (A_k X_l + B_k Q_l) & -X_j + \gamma^{-2} D_k D_k^T & 0 \\ (H_k X_l + G_k Q_l) & 0 & -I \end{bmatrix} < 0, \quad (l, j) \in \bar{\Omega}, \quad k \in \aleph(l). \tag{6.52}$$

Moreover, the controller gains are given by

$$K_l = Q_l X_l^{-1}, \quad l \in \bar{L}. \tag{6.53}$$

Proof: Based on Lemma 6.4, one knows that the system (6.46) is globally exponentially stable with H_∞ performance γ, if the conditions (6.47)–(6.50) are satisfied.

It follows from Equation (6.51) that $X_l - \gamma^{-2} D_k D_k^T > 0$. We first show that the inequality $X_l - \gamma^{-2} D_k D_k^T > 0$ implies (6.47). Via the Schur complement it is easy to see that $X_l - \gamma^{-2} D_k D_k^T > 0$ is equivalent to the following LMI:

$$\begin{bmatrix} X_l & D_k \\ D_k^T & \gamma^2 I \end{bmatrix} > 0.$$

By using the Schur complement again with respect to the other term one can conclude $\gamma^2 I - D_k^T X_l^{-1} D_k > 0$, which is (6.46). Similarly, one can easily show that (6.52) implies (6.49).

We then show that (6.52) also implies (6.50). Note that via the matrix inversion lemma (6.50) can be expressed as

$$A_{clk}^T \left(X_j - \gamma^{-2} D_k D_k^T \right)^{-1} A_{clk} - X_l^{-1} + H_{clk}^T H_{clk} < 0. \tag{6.54}$$

Multiplying by X_l in both sides of (6.54) leads to

$$X_l A_{clk}^T \left(X_j - \gamma^{-2} D_k D_k^T \right)^{-1} A_{clk} X_l - X_l + X_l H_{clk}^T H_{clk} X_l < 0. \tag{6.55}$$

Using the Schur complement, it follows from (6.55) that

$$\begin{bmatrix} -X_l + X_l H_{clk}^T H_{clk} X_l & X_l A_{clk}^T \\ A_{clk} X_l & -X_j + \gamma^{-2} D_k D_k^T \end{bmatrix} < 0,$$

which, via the Schur complement again, is equivalent to

$$\begin{bmatrix} -X_l & X_l A_{clk}^T & X_l H_{clk}^T \\ A_{clk} X_l & -X_j + \gamma^{-2} D_k D_k^T & 0 \\ H_{clk} X_l & 0 & -I \end{bmatrix} < 0. \tag{6.56}$$

Substituting A_{clk}, D_k, and H_{clk} into (6.56) leads to

$$\begin{bmatrix} -X_l & X_l (A_k + B_k K_l)^T & X_l (H_k + G_k K_l)^T \\ (A_k + B_k K_l) X_l & -X_j + \gamma^{-2} D_k D_k^T & 0 \\ (H_k + G_k K_l) X_l & 0 & -I \end{bmatrix} < 0. \tag{6.57}$$

Letting $Q_l = K_l X_l$, it follows from (6.57) that

$$\begin{bmatrix} -X_l & (A_k X_l + B_k Q_l)^T & (H_k X_l + G_k Q_l)^T \\ (A_k X_l + B_k Q_l) & -X_j + \gamma^{-2} D_k D_k^T & 0 \\ (H_k X_l + G_k Q_l) & 0 & -I \end{bmatrix} < 0,$$

which is exactly (6.52). Thus we have shown that (6.52) implies (6.50). Similarly, one can also show that (6.51) implies (6.48). Therefore, it can be concluded from Lemma 6.4 that the closed-loop control system is globally exponentially stable with H_∞ performance γ and the proof is thus completed. ❑

Remark 6.1

It is noted that similar to the stabilization case in Chapter 5, if the positive definite matrices in (6.51) and (6.52) are chosen as a common one, that is, $X_1 = X_2 = \cdots X_m = X$, then the result of Theorem 6.4 reduces to that of Theorem 6.1. It thus can be easily seen that the robust H_∞ control result based on piecewise quadratic Lyapunov functions is less conservative than that based on common quadratic Lyapunov functions.

6.4 ROBUST H_∞ CONTROL BASED ON FUZZY QUADRATIC LYAPUNOV FUNCTIONS

In this section robust H_∞ control of T–S fuzzy systems based on fuzzy Lyapunov functions is studied. Consider the same T–S fuzzy model with external disturbance as shown in (6.1), which can be written in global form as in (6.2). With the continuous controller (5.3) or equivalently (5.4), the closed-loop system can be described by Equation (6.4).

The following lemma is first introduced.

Lemma 6.5

Given a constant $\gamma > 0$, the fuzzy system (6.4) is globally exponentially stable with H_∞ performance γ, if there exists a set of positive definite matrices X_l, $l \in L$ such that the following matrix inequalities are satisfied:

$$\gamma^2 I - D_{cl}^T X_i^{-1} D_{cl} < 0, \quad i, l \in L, \tag{6.58}$$

$$A_{clj}^T X_i^{-1} A_{clj} - X_j^{-1} + A_{clj}^T X_i^{-1} D_{cl} \left(\gamma^2 I - D_{cl}^T X_i^{-1} D_{cl} \right)^{-1} D_{cl}^T X_i^{-1} A_{clj} + H_{clj}^T H_{clj} < 0 \quad i, l, j \in L \tag{6.59}$$

Proof: The proof can be established by applying the Schur complement to Theorem 4 in Zhou et al. (2005), and thus omitted. ❏

Then based on Lemma 6.5, one has the following result.

Theorem 6.5

Given a constant $\gamma > 0$, the system (6.4) is globally exponentially stable with H_∞ performance γ, if there exist a set of positive definite matrices $X_l, l \in L$ and a set of matrices $Q_l, l \in L$ such that the following LMIs are satisfied:

$$\begin{bmatrix} -X_j & (A_l X_j + B_l Q_j)^T & (H_l X_j + G_l Q_j)^T \\ (A_l X_j + B_l Q_j) & -X_i + \gamma^{-2} D_l D_l^T & 0 \\ (H_l X_j + G_l Q_j) & 0 & -I \end{bmatrix} < 0, \quad i, l, j \in L. \tag{6.60}$$

Moreover, the controller gains are given by

$$K_l = Q_l X_l^{-1}, \quad l \in L. \tag{6.61}$$

Proof: Based on Lemma 6.5, one knows that the system (6.4) is globally exponentially stable with H_∞ performance γ, if the conditions (6.58) and (6.59) are satisfied.

It follows from Equation (6.60) that $X_i - \gamma^{-2} D_{cl} D_{cl}^T > 0$. We first show that the inequality $X_i - \gamma^{-2} D_{cl} D_{cl}^T > 0$ implies (6.58). Via the Schur complement it is easy to see that $X_i - \gamma^{-2} D_{cl} D_{cl}^T > 0$ is equivalent to the following LMI:

$$\begin{bmatrix} X_i & D_{cl} \\ D_{cl}^T & \gamma^2 I \end{bmatrix} > 0. \tag{6.62}$$

By using the Schur complement again with respect to the other term one can conclude $\gamma^2 I - D_{cl}^T X_i^{-1} D_{cl} > 0$, which is (6.58).

We then show that (6.60) also implies (6.59). It is noted that via the matrix inversion lemma (6.59) can be expressed as

$$A_{clj}^T \left(X_i - \gamma^{-2} D_{cl} D_{cl}^T \right)^{-1} A_{clj} - X_j^{-1} + H_{clj}^T H_{clj} < 0. \tag{6.63}$$

Multiplying X_j in both sides of (6.63) leads to

$$X_j A_{clj}^T \left(X_i - \gamma^{-2} D_{cl} D_{cl}^T \right)^{-1} A_{clj} X_j - X_j + X_j H_{clj}^T H_{clj} X_j < 0. \tag{6.64}$$

Using the Schur complement, it follows from (6.64) that

$$\begin{bmatrix} -X_j + X_j H_{clj}^T H_{clj} X_j & X_j A_{clj}^T \\ A_{clj} X_j & -X_i + \gamma^{-2} D_{cl} D_{cl}^T \end{bmatrix} < 0, \tag{6.65}$$

which, via the Schur complement again, is equivalent to

$$\begin{bmatrix} -X_j & X_j A_{clj}^T & X_j H_{clj}^T \\ A_{clj} X_j & -X_i + \gamma^{-2} D_{cl} D_{cl}^T & 0 \\ H_{clj} X_j & 0 & -I \end{bmatrix} < 0. \tag{6.66}$$

Substituting A_{clj}, D_{cl}, and H_{clj} into (6.66) leads to

$$\begin{bmatrix} -X_j & X_j (A_l + B_l K_j)^T & X_j (H_l + G_l K_j)^T \\ (A_l + B_l K_j) X_j & -X_i + \gamma^{-2} D_l D_l^T & 0 \\ (H_l + G_l K_j) X_j & 0 & -I \end{bmatrix} < 0. \tag{6.67}$$

Letting $Q_j = K_j X_j$, it follows from (6.67) that

$$
\begin{bmatrix}
-X_j & (A_l X_j + B_l Q_j)^T & (H_l X_j + G_l Q_j)^T \\
(A_l X_j + B_l Q_j) & -X_i + \gamma^{-2} D_l D_l^T & 0 \\
(H_l X_j + G_l Q_j) & 0 & -I
\end{bmatrix} < 0,
$$

which is exactly (6.60). Thus we have shown that (6.60) implies (6.59). Therefore, it can be concluded from Lemma 6.5 that the closed-loop control system is globally exponentially stable with H_∞ performance γ and the proof is thus completed. $\square$

By using the same technique as in de Oliveira, Bernussou, and Geromel (1999), one can obtain the following improved result.

Theorem 6.6

Given a constant $\gamma > 0$, the system (6.4) is globally exponentially stable with H_∞ performance γ, if there exist a set of positive definite matrices $X_l, l \in L$ and two sets of matrices $Q_l, l \in L$, $V_l, l \in L$ such that the following LMIs are satisfied:

$$
\begin{bmatrix}
X_l - V_j^T - V_j & (A_l V_j + B_l Q_j)^T & (H_l V_j + G_l Q_j)^T \\
(A_l V_j + B_l Q_j) & -X_i + \gamma^{-2} D_l D_l^T & 0 \\
(H_l V_j + G_l Q_j) & 0 & -I
\end{bmatrix} < 0, \quad i, l, j \in L. \tag{6.68}
$$

Moreover, the controller gains are given by

$$
K_l = Q_l V_l^{-1}, \quad l \in L. \tag{6.69}
$$

Proof: By noting $(X_l - V_j)^T X_l^{-1} (X_l - V_j) = X_l - V_j^T - V_j + V_j^T X_l^{-1} V_j \geq 0$, it follows from (6.68) that

$$
\begin{bmatrix}
-V_j^T X_l^{-1} V_j & (A_l V_j + B_l Q_j)^T & (H_l V_j + G_l Q_j)^T \\
(A_l V_j + B_l Q_j) & -X_i + \gamma^{-2} D_l D_l^T & 0 \\
(H_l V_j + G_l Q_j) & 0 & -I
\end{bmatrix} < 0. \tag{6.70}
$$

Premultiplying diag $\{V_j^{-T}, I, I\}$ and postmultiplying diag $\{V_j^{-1}, I, I\}$ to (6.70) leads to

$$
\begin{bmatrix}
-X_l^{-1} & (A_l + B_l K_j)^T & (H_l + G_l K_j)^T \\
(A_l + B_l K_j) & -X_i + \gamma^{-2} D_l D_l^T & 0 \\
(H_l + G_l K_j) & 0 & -I
\end{bmatrix} < 0, \tag{6.71}
$$

which can be rewritten as

$$\begin{bmatrix} -X_l^{-1} & A_{clj}^T & H_{clj}^T \\ A_{clj} & -X_i + \gamma^{-2}D_{cl}D_{cl}^T & 0 \\ H_{clj} & 0 & -I \end{bmatrix} < 0. \qquad (6.72)$$

Applying the Schur complement twice, one can obtain

$$A_{clj}^T \left(X_i - \gamma^{-2}D_{cl}D_{cl}^T \right)^{-1} A_{clj} - X_l^{-1} + H_{clj}^T H_{clj} < 0, \qquad (6.73)$$

which is exactly (6.59) via the matrix inversion lemma. Thus it follows from Lemma 6.5 that the closed-loop fuzzy control system (6.4) is globally exponentially stable with H_∞ performance γ, and the proof is thus completed. □

Remark 6.2

Similar to the stabilization case in Chapter 5, if the positive definite matrices in (6.60) or (6.68) are chosen as a common one, that is, $X_1 = X_2 = \cdots X_m = X$, or $X_1 = X_2 = \cdots = X_m = X = V_1 = V_2 = \cdots = V_m$, respectively, then the result of Theorems 6.5 or 6.6 reduces to that of Theorem 6.1. It thus can be easily seen that the robust H_∞ control result based on fuzzy quadratic Lyapunov functions is less conservative than that based on common quadratic Lyapunov functions.

6.5 COMPARISON OF ROBUST H_∞ CONTROL RESULTS VIA NUMERICAL EXAMPLES

In this section, the robust H_∞ controller synthesis results discussed in the previous sections for T–S fuzzy systems based on common, piecewise, and fuzzy quadratic Lyapunov functions, respectively, are compared via numerical examples. For this purpose, the following optimization algorithms can be developed based on Theorems 6.1, 6.3, 6.4, 6.5, and 6.6, respectively.

Algorithm 6.1: $\min_{X,Q_l,l\in L} \gamma$, subject to LMIs (6.11)

Algorithm 6.2: $\min_{X_l,\varepsilon_l,Q_l,l\in L} \gamma$, subject to LMIs (6.42) and (6.43)

Algorithm 6.3: $\min_{X_l,Q_l,l\in \bar{L}} \gamma$, subject to LMIs (6.51) and (6.52)

Algorithm 6.4: $\min_{X_l,Q_l,l\in L} \gamma$, subject to LMIs (6.60)

Algorithm 6.5: $\min_{X_l,Q_l,V_l,l\in L} \gamma$, subject to LMIs (6.68)

Example 6.1

Consider the T–S fuzzy system defined in the form of (6.1) with matrix parameters as follows:

$$A_1 = \begin{bmatrix} 0.90 & 0.09 \\ -0.45 & 0.90 \end{bmatrix}, \quad A_2 = \begin{bmatrix} 0.45 & -0.54 \\ 0.54 & 0.45 \end{bmatrix}, \quad A_3 = \begin{bmatrix} 0.90 & 0.45 \\ -0.09 & 0.90 \end{bmatrix},$$

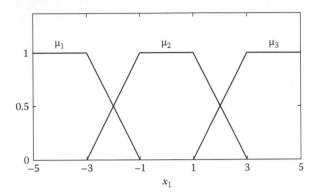

FIGURE 6.1 Membership functions in Example 6.1.

$$B_1 = \begin{bmatrix} 0 \\ 1 \end{bmatrix}, \quad B_2 = \begin{bmatrix} 0 \\ 1 \end{bmatrix}, \quad B_3 = \begin{bmatrix} 1 \\ 0 \end{bmatrix}, \quad D_1 = \begin{bmatrix} 0 \\ 0.1 \end{bmatrix}, \quad D_2 = \begin{bmatrix} 0 \\ 0.1 \end{bmatrix}, \quad D_3 = \begin{bmatrix} 0.1 \\ 0 \end{bmatrix},$$

$$H_1 = [0 \quad 1], \quad H_2 = [1 \quad 0], \quad H_3 = [0 \quad 1], \quad G_1 = 1, \quad G_2 = -1, \quad G_3 = 1,$$

and the membership functions described as in Figure 6.1.

By applying Algorithms 6.1–6.5, respectively, one obtains the optimal disturbance attenuation performances as summarized in Table 6.1.

Example 6.2

Consider the T–S fuzzy system defined in the form of (6.1) with matrix parameters as follows:

$$A_1 = \begin{bmatrix} 0.9 & 0.1 \\ -0.5 & 0.9 \end{bmatrix}, \quad A_2 = \begin{bmatrix} 0.3 & -0.5 \\ 0.5 & -0.3 \end{bmatrix}, \quad A_3 = \begin{bmatrix} 1 & 0.5 \\ -0.1 & 1 \end{bmatrix},$$

$$B_1 = \begin{bmatrix} 0 \\ 1 \end{bmatrix}, \quad B_2 = \begin{bmatrix} 0 \\ 1 \end{bmatrix}, \quad B_3 = \begin{bmatrix} 1 \\ 0 \end{bmatrix}, \quad D_1 = \begin{bmatrix} 0 \\ 0.1 \end{bmatrix}, \quad D_2 = \begin{bmatrix} 0 \\ 0.1 \end{bmatrix}, \quad D_3 = \begin{bmatrix} 0.1 \\ 0 \end{bmatrix},$$

$$H_1 = [0 \quad 1], \quad H_2 = [1 \quad 0], \quad H_3 = [0 \quad 1], \quad G_1 = 1, \quad G_2 = -1, \quad G_3 = 1,$$

TABLE 6.1

Disturbance Attenuation Performance in Example 6.1

Algorithm	6.1	6.2	6.3	6.4	6.5
Minimum γ	0.3238	0.6158	0.1855	0.3238	0.3212

TABLE 6.2
Disturbance Attenuation Performance in Example 6.2

Algorithm	6.1	6.2	6.3	6.4	6.5
Minimum γ	Infeasible	2.2013	0.2446	Infeasible	1.8036

and the membership functions described as in Figure 6.1. By applying Algorithms 6.1–6.5, respectively, one obtains the corresponding optimal disturbance attenuation performances as summarized in Table 6.2, where it is noted that the approach based on common quadratic Lyapunov functions (Theorem 6.1, Algorithm 6.1) and the approach based on fuzzy quadratic Lyapunov functions without any improvement (Theorem 6.5, Algorithm 6.4) have no feasible solutions.

It can be observed from these examples that the robust H_∞ control approaches based on piecewise quadratic Lyapunov functions or fuzzy quadratic Lyapunov functions are, in general, less conservative than those based on common quadratic Lyapunov functions. It can also be observed that the approach based on fuzzy quadratic Lyapunov functions without any modification (Theorem 6.4) is of similar conservatism as the approach based on common quadratic Lyapunov functions (Theorem 6.1) in the sense that they lead to the same solution or infeasibility for these two particular examples. However, whether the piecewise Lyapunov function based robust H_∞ control approach is less conservative than the fuzzy Lyapunov function based approach or vice versa, is in general dependent on the system under study. It should also be noted that Algorithm 6.2, which is based on piecewise quadratic Lyapunov functions, might lead to significant conservatism in some cases due to the introduction of uncertainties as demonstrated in Example 6.1.

6.6 CONCLUSIONS

This chapter has presented a number of approaches to robust H_∞ controller synthesis of T–S fuzzy systems based on common, piecewise, and fuzzy quadratic Lyapunov functions, respectively. It is shown that controller gains can be obtained by solving linear matrix inequalities. It is also shown via analysis and numerical examples that the robust H_∞ control approaches based on piecewise quadratic Lyapunov functions or fuzzy quadratic Lyapunov functions are less conservative than their counterparts based on common quadratic Lyapunov functions.

7 Observer and Output Feedback Controller Synthesis of T–S Fuzzy Systems

7.1 INTRODUCTION

Chapters 5 and 6 were devoted to the study of the stabilization controller and robust H_∞ controller synthesis, respectively. Those controllers are based on state feedback under the assumption that all the state variables are available for measurement. However, not all the state variables are measurable in most cases in practice. In such cases, controllers based on state feedback cannot be implemented, and thus output feedback control has to be considered. In this chapter, a number of results regarding observer and output feedback controller synthesis of T–S fuzzy systems are presented.

The rest of the chapter is organized as follows. First, the issue of observer and output feedback controller synthesis of T–S fuzzy systems based on common quadratic Lyapunov functions is addressed in Section 7.2. Then, observer and output feedback controller synthesis of T–S fuzzy systems based on piecewise quadratic Lyapunov functions is discussed in Section 7.3. Observer and output feedback controller design of T–S fuzzy systems based on fuzzy quadratic Lyapunov functions is presented in Section 7.4, followed by comparison of these observer design results via numerical examples in Section 7.5. Some remarks conclude the chapter in Section 7.6.

7.2 OBSERVER AND OUTPUT FEEDBACK CONTROLLER SYNTHESIS BASED ON COMMON QUADRATIC LYAPUNOV FUNCTIONS

Consider the same T–S fuzzy system as described in (5.1) with the same definitions of parameters and variables, rewritten as follows,

$$R^l: \quad \text{IF} \quad z_1 \text{ is } F_1^l, \text{AND}, \dots z_v \text{ is } F_v^l$$
$$\text{THEN} \quad x(t+1) = A_l x(t) + B_l u(t) \tag{7.1}$$
$$y_l(t) = C_l x(t)$$
$$l \in L := \{1, 2, \dots, m\},$$

or equivalently in global form,

$$x(t+1) = A(\mu)x(t) + B(\mu)u(t)$$

$$y(t) = C(\mu)x(t).$$

(7.2)

Consider the following observer rule sharing the same rule of the T–S fuzzy system:

$$O^l = R^l : \quad \text{IF} \quad z_1 \text{ is } F_1^l, \text{AND}, \ldots z_v \text{ is } F_v^l$$

$$\text{THEN} \quad \hat{x}(t+1) = A_l\hat{x}(t) + B_lu(t) + F_l(\hat{y}(t) - y(t))$$

(7.3)

$$\hat{y}_l(t) = C_l\hat{x}(t)$$

$$l \in L := \{1, 2, \ldots, m\},$$

which can be rewritten as

$$\hat{x}(t+1) = \sum_{l=1}^{m} \mu_l(A_l\hat{x}(t) + B_lu(t) + F_l(\hat{y}(t) - y(t))),$$

(7.4)

$$\hat{y}(t) = \sum_{l=1}^{m} \mu_l C_l x(t),$$

where $\hat{x}(t)$ is the estimated state vector, $\hat{y}(t)$ the estimated output vector, and $F_l, l \in L$, the observer gains to be determined.

Then the fuzzy observer error dynamic equation consisting of (7.2) and (7.4) can be given as

$$\tilde{x}(t+1) = \sum_{j=1}^{m}\sum_{l=1}^{m} \mu_j\mu_l(A_l + F_lC_j)\tilde{x}(t),$$

(7.5)

where $\tilde{x}(t) = \hat{x}(t) - x(t)$, and the following result is readily obtained (Ma, Sun, and He, 1998).

Theorem 7.1

The fuzzy observer error system (7.5) is globally exponentially stable if there exist a positive definite matrix P and a set of matrices $R_l, l \in L$ such that the following LMIs are satisfied:

$$\begin{bmatrix} -P & A_l^T P + C_j^T R_l^T \\ PA_l + R_lC_j & -P \end{bmatrix} < 0, \quad l, j \in L.$$

(7.6)

Moreover, the observer gains are given by

$$F_l = P^{-1} R_l, \quad l \in L. \tag{7.7}$$

Proof: Let $A_{clj} = A_l + F_l C_j$. It follows from Theorem 4.1 that the following matrix inequalities would imply the global exponential stability of the error system (7.5),

$$A_{clj}^T P A_{clj} - P < 0, \quad l, j \in L, \tag{7.8}$$

which, via the Schur complement, is equivalent to

$$\begin{bmatrix} -P & A_{clj}^T P \\ P A_{clj} & -P \end{bmatrix} < 0, \quad l, j \in L.$$

That is,

$$\begin{bmatrix} -P & A_l^T P + C_j^T F_l^T P \\ P A_l + P F_l C_j & -P \end{bmatrix} < 0, \quad l, j \in L. \tag{7.9}$$

Define $R_l = P F_l$. Then (7.9) can be rewritten as

$$\begin{bmatrix} -P & A_l^T P + C_j^T R_l^T \\ P A_l + R_l C_j & -P \end{bmatrix} < 0, \quad l, j \in L,$$

which is exactly (7.6). Therefore we have shown that (7.6) implies (7.8) and it then follows from Theorem 4.1 that the observer error system (7.5) is globally exponentially stable and the observer gains can be obtained by (7.7). The proof is thus completed. ❑

Similar to the switching controller, one can also design a switching observer for the T–S fuzzy system (7.2) as follows:

$$\hat{x}(t+1) = (A_l + \Delta A_l)\hat{x}(t) + (B_l + \Delta B_l)u(t) + F_l(\hat{y}(t) - y(t)), \quad z(t) \in S_l, \quad l \in L. \tag{7.10}$$

Then the observer error system can be described in this case as

$$\tilde{x}(t+1) = A_{cl}\tilde{x}(t), \quad z(t) \in S_l, \quad l \in L, \tag{7.11}$$

where $A_{cl} := A_l + \Delta A_l + F_l(C_l + \Delta C_l)$.

Similar to the previous chapters, define the following upper bounds for ΔA_l and ΔC_l, respectively,

$$[\Delta A_l(\mu)][\Delta A_l(\mu)]^T \leq E_{lA}E_{lA}^T, \quad l \in L,$$
$$[\Delta C_l(\mu)][\Delta C_l(\mu)]^T \leq E_{lC}E_{lC}^T, \quad l \in L. \tag{7.12}$$

Then one has the following result.

Theorem 7.2

The fuzzy observer error system (7.11) is globally exponentially stable if there exist a positive definite matrix P and two sets of matrices $R_l, V_l, l \in L$ such that the following LMIs are satisfied:

$$\begin{bmatrix} P - V_l - V_l^T & V_l A_l + R_l C_l & V_l & R_l \\ (V_l A_l + R_l C_l)^T & -P + \varepsilon_l \left(E_{lA}^T E_{lA} + E_{lC}^T E_{lC} \right) & 0 & 0 \\ V_l^T & 0 & -\varepsilon_l I & 0 \\ R_l^T & 0 & 0 & -\varepsilon_l I \end{bmatrix} < 0, \quad l \in L. \tag{7.13}$$

Moreover, the switching observer gains are given by

$$F_l = P^{-1} R_l, \quad l \in L. \tag{7.14}$$

Proof: Based on the result in Theorem 4.1 and its proof, one learns that the system (7.11) is globally exponentially stable if there exists a positive definite matrix P satisfying the following inequalities:

$$A_{cl}^T P_l A_{cl} - P_l < 0, \quad l \in L. \tag{7.15}$$

It can be easily seen that the following inequalities would imply (7.15).

$$\begin{bmatrix} P - V_l - V_l^T & V_l A_{cl} \\ A_{cl}^T V_l^T & -P \end{bmatrix} < 0, \quad l \in L. \tag{7.16}$$

In fact, multiplying $W = [A_{cl}^T \ \ I]$ on the left-hand side and W^T on the right-hand side of (7.16) leads exactly to (7.15).

We then show that the inequalities (7.13) imply (7.16). Substituting A_{cl} into the left-hand side of (7.16) leads to

$$\begin{bmatrix} P - V_l - V_l^T & V_l A_l + V_l \Delta A_l + V_l F_l C_l + V_l F_l \Delta C_l \\ (V_l A_l + V_l \Delta A_l + V_l F_l C_l + V_l F_l \Delta C_l)^T & -P \end{bmatrix}$$

$$= \begin{bmatrix} P - V_l - V_l^T & V_l A_l + V_l F_l C_l \\ (V_l A_l + V_l F_l C_l)^T & -P \end{bmatrix} + \begin{bmatrix} 0 & V_l \Delta A_l \\ \Delta A_l^T V_l^T & 0 \end{bmatrix} + \begin{bmatrix} 0 & V_l F_l \Delta C_l \\ \Delta C_l^T F_l^T V_l^T & 0 \end{bmatrix}.$$

$$\tag{7.17}$$

Now consider the last two terms in (7.17); one has, respectively,

$$
\begin{bmatrix} 0 & V_l \Delta A_l \\ \Delta A_l^T V_l^T & 0 \end{bmatrix} = \begin{bmatrix} 0 & V_l \\ 0 & 0 \end{bmatrix} \begin{bmatrix} 0 & 0 \\ 0 & \Delta A_l \end{bmatrix} + \begin{bmatrix} 0 & 0 \\ 0 & \Delta A_l^T \end{bmatrix} \begin{bmatrix} 0 & 0 \\ V_l^T & 0 \end{bmatrix},
$$

$$
\leq \frac{1}{\varepsilon_l} \begin{bmatrix} V_l V_l^T & 0 \\ 0 & 0 \end{bmatrix} + \varepsilon_l \begin{bmatrix} 0 & 0 \\ 0 & \Delta A_l^T \Delta A_l \end{bmatrix}
$$

$$
\leq \frac{1}{\varepsilon_l} \begin{bmatrix} V_l V_l^T & 0 \\ 0 & 0 \end{bmatrix} + \varepsilon_l \begin{bmatrix} 0 & 0 \\ 0 & E_{lA}^T E_{lA} \end{bmatrix},
$$

$$
\begin{bmatrix} 0 & V_l F_l \Delta C_l \\ \Delta C_l^T F_l^T V_l^T & 0 \end{bmatrix} = \begin{bmatrix} 0 & V_l F_l \\ 0 & 0 \end{bmatrix} \begin{bmatrix} 0 & 0 \\ 0 & \Delta C_l \end{bmatrix} + \begin{bmatrix} 0 & 0 \\ 0 & \Delta C_l^T \end{bmatrix} \begin{bmatrix} 0 & 0 \\ F_l^T V_l^T & 0 \end{bmatrix}
$$

$$
\leq \frac{1}{\varepsilon_l} \begin{bmatrix} V_l F_l F_l^T V_l^T & 0 \\ 0 & 0 \end{bmatrix} + \varepsilon_l \begin{bmatrix} 0 & 0 \\ 0 & \Delta C_l^T \Delta C_l \end{bmatrix}
$$

$$
\leq \frac{1}{\varepsilon_l} \begin{bmatrix} V_l F_l F_l^l V_l^l & 0 \\ 0 & 0 \end{bmatrix} + \varepsilon_l \begin{bmatrix} 0 & 0 \\ 0 & E_{lC}^T E_{lC} \end{bmatrix}.
$$

It then follows from (7.17) that

$$
\begin{bmatrix} P - V_l - V_l^T & V_l A_l + V_l F_l C_l \\ (V_l A_l + V_l F_l C_l)^T & -P \end{bmatrix} + \begin{bmatrix} 0 & V_l \Delta A_l \\ \Delta A_l^T V_l^T & 0 \end{bmatrix} + \begin{bmatrix} 0 & V_l F_l \Delta C_l \\ \Delta C_l^T F_l^T V_l^T & 0 \end{bmatrix}
$$

$$
\leq \begin{bmatrix} P - V_l - V_l^T & V_l A_l + V_l F_l C_l \\ (V_l A_l + V_l F_l C_l)^T & -P \end{bmatrix} + \frac{1}{\varepsilon_l} \begin{bmatrix} V_l V_l^T + V_l F_l F_l^T V_l^T & 0 \\ 0 & 0 \end{bmatrix}
$$

$$
+ \varepsilon_l \begin{bmatrix} 0 & 0 \\ 0 & E_{lA}^T E_{lA} + E_{lC}^T E_{lC} \end{bmatrix}
$$

$$
= \begin{bmatrix} P - V_l - V_l^T & V_l A_l + V_l F_l C_l \\ (V_l A_l + V_l F_l C_l)^T & -P + \varepsilon_l (E_{lA}^T E_{lA} + E_{lC}^T E_{lC}) \end{bmatrix} + \frac{1}{\varepsilon_l} \begin{bmatrix} V_l V_l^T + V_l F_l F_l^T V_l^T & 0 \\ 0 & 0 \end{bmatrix}.
$$

Then the following inequality would imply (7.16),

$$
\begin{bmatrix} P - V_l - V_l^T & V_l A_l + V_l F_l C_l \\ (V_l A_l + V_l F_l C_l)^T & -P + \varepsilon_l (E_{lA}^T E_{lA} + E_{lC}^T E_{lC}) \end{bmatrix} + \frac{1}{\varepsilon_l} \begin{bmatrix} V_l V_l^T + V_l F_l F_l^T V_l^T & 0 \\ 0 & 0 \end{bmatrix} < 0.
$$

$$
(7.18)
$$

Noting (7.14) and applying the Schur complement formula several times, it is easily shown that (7.18) is in turn equivalent to (7.13). Thus, we have shown that the inequalities (7.13) imply (7.16). Therefore, it can be concluded that the observer error system is globally exponentially stable, and the proof is thus completed. ❑

If the estimated state is used for control implementation, one gets the observer-based output feedback controller. For the resulting closed-loop output feedback control system with the continuous controller (5.3), one has the following result.

Theorem 7.3 (Separation Principle)

Suppose that there exist a positive definite matrix X and a set of matrices $Q_l, l \in L$ satisfying (5.8), and there exist a positive definite matrix P and a set of matrices $R_l, l \in L$ satisfying (7.6), with the controller gains given by (5.9) and the observer gains given by (7.7). There exists a positive constant α large enough such that the following positive definite matrix

$$Y := \begin{bmatrix} X^{-1} & 0 \\ 0 & \alpha P \end{bmatrix}$$

(7.19)

is a solution to the following matrix inequalities,

$$\bar{A}_{clj}^T Y \bar{A}_{clj} - Y < 0, \quad l, j \in L,$$

(7.20)

where

$$\bar{A}_{clj} = \begin{bmatrix} A_l + B_l K_j & B_l K_j \\ 0 & A_l + F_l C_j \end{bmatrix}.$$

(7.21)

In other words, the closed-loop system with the combined controller and observer is globally exponentially stable.

Proof: It follows from (5.2) and (7.6) with $u = \sum_{l=1}^m \mu_l K_l \hat{x}(t)$ that the closed-loop output feedback control system can be described as

$$\begin{bmatrix} x(t+1) \\ \tilde{x}(t+1) \end{bmatrix} = \sum_{l=1}^m \sum_{j=1}^m \mu_l \mu_j \bar{A}_{clj} \begin{bmatrix} x(t) \\ \tilde{x}(t) \end{bmatrix}.$$

(7.22)

It then follows from Theorem 4.1 and its proof that if there exists a positive definite matrix Y such that (7.20) is satisfied, then the closed-loop system (7.22) is globally exponentially stable. Now we show that (7.19) with large enough α is a solution to (7.20). Substituting (7.19) and (7.21) into (7.20) leads to

$$\begin{bmatrix} A_l + B_l K_j & B_l K_j \\ 0 & A_l + F_l C_j \end{bmatrix}^T \begin{bmatrix} X^{-1} & 0 \\ 0 & \alpha P \end{bmatrix} \begin{bmatrix} A_l + B_l K_j & B_l K_j \\ 0 & A_l + F_l C_j \end{bmatrix} - \begin{bmatrix} X^{-1} & 0 \\ 0 & \alpha P \end{bmatrix} < 0,$$

(7.23)

which is equivalent to

$$
\begin{bmatrix}
(A_l + B_l K_j)^T X^{-1}(A_l + B_l K_j) - X^{-1} & (A_l + B_l K_j)^T X^{-1} B_l K_j \\
K_j^T B_l^T X^{-1}(A_l + B_l K_j) & \Phi_{lj}
\end{bmatrix} < 0, \qquad (7.24)
$$

where $\Phi_{lj} := \alpha(A_l + F_l C_j)^T P(A_l + F_l C_j) - \alpha P + K_j^T B_l^T X^{-1} B_l K_j$.
Using the Schur complements, (7.24) is equivalent to

$$
(A_l + B_l K_j)^T X^{-1}(A_l + B_l K_j) - X^{-1} < 0, \qquad (7.25)
$$

and

$$
\alpha(A_l + F_l C_j)^T P(A_l + F_l C_j) - \alpha P + K_j^T B_l^T X^{-1} B_l K_j - K_j^T B_l^T X^{-1}(A_l + B_l K_j)
$$
$$
[(A_l + B_l K_j)^T X^{-1}(A_l + B_l K_j) - X^{-1}]^{-1}(A_l + B_l K_j)^T X^{-1} B_l K_j < 0. \qquad (7.26)
$$

Using the matrix inversion lemma, (7.26) is equivalent to

$$
\alpha(A_l + F_l C_j)^T P(A_l + F_l C_j) - \alpha P - K_j^T B_l^T[(A_l + B_l K_j)X(A_l + B_l K_j)^T - X]^{-1} B_l K_j < 0; \qquad (7.27)
$$

that is,

$$
\alpha[(A_l + F_l C_j)^T P(A_l + F_l C_j) - P] < K_j^T B_l^T[(A_l + B_l K_j)X(A_l + B_l K_j)^T - X]^{-1} B_l K_j. \qquad (7.28)
$$

Define

$$
\Omega_{lj} := K_j^T B_l^T[(A_l + B_l K_j)X(A_l + B_l K_j)^T - X]^{-1} B_l K_j, \qquad (7.29)
$$

$$
\Xi_{lj} := [(A_l + F_l C_j)^T P(A_l + F_l C_j) - P]. \qquad (7.30)
$$

Then it follows that there exists a large enough positive constant α such that

$$
\alpha z^T \Xi_{lj} z < z^T \Omega_{lj} z, \quad z \neq 0, \qquad (7.31)
$$

because Ξ_{lj} is negative definite. It is easily seen that (7.31) implies (7.28). Thus, we have shown that with a large enough α, (7.19) is a solution to (7.20). Therefore, the claim of the theorem is established and the proof is thus completed. ❑

Remark 7.1

A different version of the separation principle has also been presented in Ma, Sun, and He (1998) and proved in a different way. However, these separation principles are essentially the same in the sense that controller design and observer design can be independently carried out based on common quadratic Lyapunov functions, and the resulting closed-loop fuzzy control system, with estimated state variables to be used for state feedback control, is globally exponentially stable.

If the switching controller (5.5) and switching observer (7.10) are used, one can also obtain the similar result of separation principle as follows.

Theorem 7.4 (Separation Principle)

Suppose that there exist a positive definite matrix X and a set of matrices $Q_l, l \in L$ satisfying (5.26), and there exist a positive definite matrix P and a set of matrices $R_l, l \in L$ satisfying (7.13), with the controller gains given by (5.27) and the observer gains given by (7.14). There exists a positive constant α large enough such that the positive definite matrix

$$Y := \begin{bmatrix} X^{-1} & 0 \\ 0 & \alpha P \end{bmatrix} \tag{7.32}$$

is a solution to the matrix inequalities

$$\bar{A}_{cl}^T Y \bar{A}_{cl} - Y < 0, \quad l \in L, \tag{7.33}$$

where

$$\bar{A}_{cl} = \begin{bmatrix} A_l + \Delta A_l + (B_l + \Delta B_l)K_l & (B_l + \Delta B_l)K_l \\ 0 & A_l + \Delta A_l + F_l(C_l + \Delta C_l) \end{bmatrix}. \tag{7.34}$$

In other words, the closed-loop system with the combined controller and observer is globally exponentially stable.

Proof: With the switching controller as follows,

$$u(t) = K_l \hat{x}(t), \quad z(t) \in S_l, \quad l \in L,$$

Equation (5.23) becomes

$$x(t+1) = (A_l + \Delta A_l(\mu))x(t) + (B_l + \Delta B_l(\mu))K_l \hat{x}(t). \tag{7.35}$$

And with (7.11) the closed-loop output feedback control system can be described as

$$
\begin{bmatrix} x(t+1) \\ \tilde{x}(t+1) \end{bmatrix} = \bar{A}_{cl} \begin{bmatrix} x(t) \\ \tilde{x}(t) \end{bmatrix}, \quad z(t) \in S_l, \quad l \in L. \tag{7.36}
$$

Then the proof can be established following a similar procedure as in the proof of Theorem 7.3, and thus it is omitted. ❑

Example 7.1

Consider the T–S fuzzy system of the form (7.1) with the system parameters (A_l, B_l) defined in Example 5.1 and the output matrices given as follows:

$$
C_1 = [0.2 \quad 0.3], \quad C_2 = [0.1 \quad 0.1].
$$

By applying Theorem 7.1, one obtains the solution to those LMIs as

$$
P = \begin{bmatrix} 106.9787 & 10.6549 \\ 10.6549 & 38.9046 \end{bmatrix}, \quad F_1 = \begin{bmatrix} -2.1895 \\ -3.6198 \end{bmatrix}, \quad F_2 = \begin{bmatrix} -2.6108 \\ -3.8025 \end{bmatrix}.
$$

Then the observer error response can be illustrated in Figure 7.1 where initial conditions are given as $x = [-0.2 \quad 0.1]^T$ and $\tilde{x} = [-0.5 \quad 0.5]^T$. It can be easily observed that the observer errors converge to zero as time goes to infinity.

Moreover, with the controller obtained in Example 5.1, the observer-based output feedback control system can be obtained and the closed-loop system responses to a couple of initial conditions are illustrated in Figure 7.2. It can be easily observed that

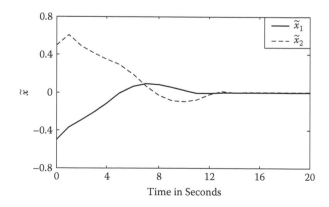

FIGURE 7.1   Observer error responses for Example 7.1.

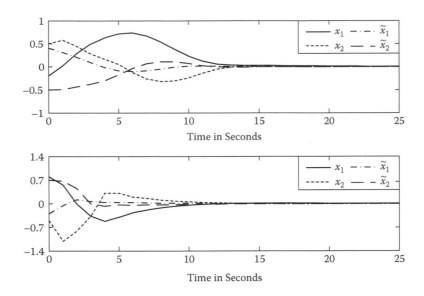

FIGURE 7.2   Closed-loop responses with output feedback for Example 7.1.

all the trajectories converge to the origin, and thus the separation principle is verified in this case.

7.3   OBSERVER AND OUTPUT FEEDBACK CONTROLLER SYNTHESIS BASED ON PIECEWISE QUADRATIC LYAPUNOV FUNCTIONS

Consider the switching observer in (7.10) and the observer error system described in (7.11). Define the set Ω as in (4.15), representing all possible system trajectory transitions among regions.

Then one has the following result.

Theorem 7.5

The fuzzy observer error system (7.11) is globally exponentially stable if there exist a set of positive definite matrices $P_l, l \in L$, two sets of matrices $V_l, R_l, l \in L$, and a set of constant parameters $\varepsilon_l, l \in L$ such that the following LMIs are satisfied:

$$\begin{bmatrix} P_l - V_l - V_l^T & V_l A_l + R_l C_l & V_l & R_l \\ (V_l A_l + R_l C_l)^T & -P_l + \varepsilon_l \left(E_{lA}^T E_{lA} + E_{lC}^T E_{lC} \right) & 0 & 0 \\ V_l^T & 0 & -\varepsilon_l I & 0 \\ R_l^T & 0 & 0 & -\varepsilon_l I \end{bmatrix} < 0, \quad l \in L,$$

(7.37)

$$
\begin{bmatrix}
P_j - V_l - V_l^T & V_l A_l + R_l C_l & V_l & R_l \\
(V_l A_l + R_l C_l)^T & -P_l + \varepsilon_l \left(E_{lA}^T E_{lA} + E_{lC}^T E_{lC} \right) & 0 & 0 \\
V_l^T & 0 & -\varepsilon_l I & 0 \\
R_l^T & 0 & 0 & -\varepsilon_l I
\end{bmatrix} < 0, \quad l, j \in \Omega.
$$

(7.38)

Moreover, the switching observer gains are given by

$$
F_l = V_l^{-1} R_l, \quad l \in L.
$$

(7.39)

Proof: Based on the result in Theorem 4.2 and its proof, one learns that the system (7.11) is globally exponentially stable if there exists a set of positive definite matrices $P_l, l \in L$ satisfying the following inequalities:

$$
A_{cl}^T P_l A_{cl} - P_l < 0, \quad l \in L,
$$

(7.40)

$$
A_{cl}^T P_j A_{cl} - P_l < 0, \quad l, j \in \Omega.
$$

(7.41)

It can be easily seen that the following inequalities would imply (7.40) and (7.41), respectively.

$$
\begin{bmatrix}
P_l - V_l - V_l^T & V_l A_{cl} \\
A_{cl}^T V_l^T & -P_l
\end{bmatrix} < 0,
$$

(7.42)

$$
\begin{bmatrix}
P_j - V_l - V_l^T & V_l A_{cl} \\
A_{cl}^T V_l^T & -P_l
\end{bmatrix} < 0.
$$

(7.43)

In fact, multiplying $W = [A_{cl}^T \quad I]$ on the left-hand side and W^T on the right-hand side of (7.42) leads exactly to (7.40). Doing the same to (7.43) also leads to (7.41).

We then show that the inequalities (7.38) imply (7.43). Substituting A_{cl} into the left-hand side of (7.43) leads to

$$
\begin{bmatrix}
P_j - V_l - V_l^T & V_l A_l + V_l \Delta A_l + V_l F_l C_l + V_l F_l \Delta C_l \\
(V_l A_l + V_l \Delta A_l + V_l F_l C_l + V_l F_l \Delta C_l)^T & -P_l
\end{bmatrix}
$$

$$
= \begin{bmatrix}
P_j - V_l - V_l^T & V_l A_l + V_l F_l C_l \\
(V_l A_l + V_l F_l C_l)^T & -P_l
\end{bmatrix} + \begin{bmatrix}
0 & V_l \Delta A_l \\
\Delta A_l^T V_l^T & 0
\end{bmatrix} + \begin{bmatrix}
0 & V_l F_l \Delta C_l \\
\Delta C_l^T F_l^T V_l^T & 0
\end{bmatrix}.
$$

(7.44)

Now consider the last two terms in (7.44); one has, respectively,

$$
\begin{bmatrix} 0 & V_l \Delta A_l \\ \Delta A_l^T V_l^T & 0 \end{bmatrix} = \begin{bmatrix} 0 & V_l \\ 0 & 0 \end{bmatrix} \begin{bmatrix} 0 & 0 \\ 0 & \Delta A_l \end{bmatrix} + \begin{bmatrix} 0 & 0 \\ 0 & \Delta A_l^T \end{bmatrix} \begin{bmatrix} 0 & 0 \\ V_l^T & 0 \end{bmatrix},
$$

$$
\leq \frac{1}{\varepsilon_l} \begin{bmatrix} V_l V_l^T & 0 \\ 0 & 0 \end{bmatrix} + \varepsilon_l \begin{bmatrix} 0 & 0 \\ 0 & \Delta A_l^T \Delta A_l \end{bmatrix}
$$

$$
\leq \frac{1}{\varepsilon_l} \begin{bmatrix} V_l V_l^T & 0 \\ 0 & 0 \end{bmatrix} + \varepsilon_l \begin{bmatrix} 0 & 0 \\ 0 & E_{lA}^T E_{lA} \end{bmatrix},
$$

$$
\begin{bmatrix} 0 & V_l F_l \Delta C_l \\ \Delta C_l^T F_l^T V_l^T & 0 \end{bmatrix} = \begin{bmatrix} 0 & V_l F_l \\ 0 & 0 \end{bmatrix} \begin{bmatrix} 0 & 0 \\ 0 & \Delta C_l \end{bmatrix} + \begin{bmatrix} 0 & 0 \\ 0 & \Delta C_l^T \end{bmatrix} \begin{bmatrix} 0 & 0 \\ F_l^T V_l^T & 0 \end{bmatrix}
$$

$$
\leq \frac{1}{\varepsilon_l} \begin{bmatrix} V_l F_l F_l^T V_l^T & 0 \\ 0 & 0 \end{bmatrix} + \varepsilon_l \begin{bmatrix} 0 & 0 \\ 0 & \Delta C_l^T \Delta C_l \end{bmatrix}
$$

$$
\leq \frac{1}{\varepsilon_l} \begin{bmatrix} V_l F_l F_l^T V_l^T & 0 \\ 0 & 0 \end{bmatrix} + \varepsilon_l \begin{bmatrix} 0 & 0 \\ 0 & E_{lC}^T E_{lC} \end{bmatrix}.
$$

It then follows from (7.44) that

$$
\begin{bmatrix} P_j - V_l - V_l^T & V_l A_l + V_l F_l C_l \\ (V_l A_l + V_l F_l C_l)^T & -P_l \end{bmatrix} + \begin{bmatrix} 0 & V_l \Delta A_l \\ \Delta A_l^T V_l^T & 0 \end{bmatrix} + \begin{bmatrix} 0 & V_l F_l \Delta C_l \\ \Delta C_l^T F_l^T V_l^T & 0 \end{bmatrix}
$$

$$
\leq \begin{bmatrix} P_j - V_l - V_l^T & V_l A_l + V_l F_l C_l \\ (V_l A_l + V_l F_l C_l)^T & -P_l \end{bmatrix} + \frac{1}{\varepsilon_l} \begin{bmatrix} V_l V_l^T + V_l F_l F_l^T V_l^T & 0 \\ 0 & 0 \end{bmatrix}
$$

$$
+ \varepsilon_l \begin{bmatrix} 0 & 0 \\ 0 & E_{lA}^T E_{lA} + E_{lC}^T E_{lC} \end{bmatrix}
$$

$$
= \begin{bmatrix} P_j - V_l - V_l^T & V_l A_l + V_l F_l C_l \\ (V_l A_l + V_l F_l C_l)^T & -P_l + \varepsilon_l (E_{lA}^T E_{lA} + E_{lC}^T E_{lC}) \end{bmatrix} + \frac{1}{\varepsilon_l} \begin{bmatrix} V_l V_l^T + V_l F_l F_l^T V_l^T & 0 \\ 0 & 0 \end{bmatrix}.
$$

Then the following inequality would imply (7.43):

$$
\begin{bmatrix} P_j - V_l - V_l^T & V_l A_l + V_l F_l C_l \\ (V_l A_l + V_l F_l C_l)^T & -P_l + \varepsilon_l (E_{lA}^T E_{lA} + E_{lC}^T E_{lC}) \end{bmatrix} + \frac{1}{\varepsilon_l} \begin{bmatrix} V_l V_l^T + V_l F_l F_l^T V_l^T & 0 \\ 0 & 0 \end{bmatrix} < 0.
$$

$$
\tag{7.45}
$$

Noting (7.39) and applying the Schur complement formula several times, it is easily shown that (7.45) is in turn equivalent to (7.38). Thus, we have shown that the inequalities (7.38) imply (7.41). Similarly, one can also show that the inequalities (7.37) imply (7.40). Therefore, it can be concluded that the observer error system is globally exponentially stable, and the proof is thus completed. ❏

If the estimated state is used for the control implementation in (5.33), one gets the observer-based output feedback controller. In this case, the closed-loop system can be described as

$$x(t+1) = (A_l + \Delta A_l)x(t) + (B_l + \Delta B_l)K_l \hat{x}(t), \quad z(t) \in S_l, \quad l \in L. \quad (7.46)$$

Combining (7.46) and (7.11) leads to

$$\begin{bmatrix} x(t+1) \\ \tilde{x}(t+1) \end{bmatrix} = \bar{A}_{cl} \begin{bmatrix} x(t) \\ \tilde{x}(t) \end{bmatrix} := \begin{bmatrix} A_{cl} & (B_l + \Delta B_l)K_l \\ 0 & A_{ol} \end{bmatrix} \begin{bmatrix} x(t) \\ \tilde{x}(t) \end{bmatrix}, \quad z(t) \in S_l, \quad l \in L,$$

$$(7.47)$$

where

$$A_{cl} = A_l + \Delta A_l + (B_l + \Delta B_l)K_l,$$

$$A_{ol} = A_l + \Delta A_l + F_l(C_l + \Delta C_l).$$

For the resulting closed-loop output feedback control system (7.47), one has the following result.

Theorem 7.6 (Separation Principle)

Suppose that there exist a set of positive definite matrices $X_l, l \in L$ and a set of matrices $Q_l, l \in L$ satisfying (5.36) and (5.37), and there exist a set of positive definite matrices $P_l, l \in L$ and a set of matrices $R_l, l \in L$, satisfying (7.37) and (7.38), respectively, with the controller gains given by (5.38) and the observer gains given by (7.39). There exists a positive constant α large enough such that the following set of positive definite matrices

$$Y_l := \begin{bmatrix} X_l^{-1} & 0 \\ 0 & \alpha P_l \end{bmatrix}, \quad l \in L \quad (7.48)$$

is a solution to the following matrix inequalities:

$$\bar{A}_{cl}^T Y_l \bar{A}_{cl} - Y_l < 0, \quad l \in L, \quad (7.49)$$

$$\bar{A}_{cl}^T Y_j \bar{A}_{cl} - Y_l < 0, \quad l, j \in \Omega. \quad (7.50)$$

In other words, the closed-loop system with the combined controller and observer is globally exponentially stable.

Proof: It follows from Theorem 4.2 and its proof that if there exists a set of positive definite matrices $Y_l, l \in L$ such that (7.49) and (7.50) are satisfied, then the closed-loop system (7.47) is globally exponentially stable. Now we show that (7.48) with large enough α is a solution to (7.49) and (7.50). Substituting (7.48) into (7.50) and noting (7.47) leads to

$$\begin{bmatrix} A_{cl} & (B_l + \Delta B_l)K_l \\ 0 & A_{ol} \end{bmatrix}^T \begin{bmatrix} X_j^{-1} & 0 \\ 0 & \alpha P_j \end{bmatrix} \begin{bmatrix} A_{cl} & (B_l + \Delta B_l)K_l \\ 0 & A_{ol} \end{bmatrix} - \begin{bmatrix} X_l^{-1} & 0 \\ 0 & \alpha P_l \end{bmatrix} < 0,$$

(7.51)

which is equivalent to

$$\begin{bmatrix} A_{cl}^T X_j^{-1} A_{cl} - X_l^{-1} & A_{cl}^T X_j^{-1}(B_l + \Delta B_l)K_l \\ K_l^T(B_l + \Delta B_l)^T X_j^{-1} A_{cl} & \Phi_l \end{bmatrix} < 0,$$

(7.52)

where $\Phi_l := \alpha A_{ol}^T P_j A_{ol} - \alpha P_l + K_l^T(B_l + \Delta B_l)^T X_j^{-1}(B_l + \Delta B_l)K_l$.

Using Schur complements, (7.52) is equivalent to

$$A_{cl}^T X_j^{-1} A_{cl} - X_l^{-1} < 0,$$

(7.53)

and

$$\alpha A_{ol}^T P_j A_{ol} - \alpha P_l + K_l^T(B_l + \Delta B_l)^T X_j^{-1}(B_l + \Delta B_l)K_l$$

$$- K_l^T(B_l + \Delta B_l)^T X_j^{-1} A_{cl} \left(A_{cl}^T X_j^{-1} A_{cl} - X_l^{-1} \right)^{-1} A_{cl}^T X_j^{-1}(B_l + \Delta B_l)K_l < 0.$$

(7.54)

Using the matrix inversion lemma, (7.54) is equivalent to

$$\alpha A_{ol}^T P_j A_{ol} - \alpha P_l + - K_l^T(B_l + \Delta B_l)^T [A_{cl} X_l A_{cl}^T - X_j]^{-1}(B_l + \Delta B_l)K_l < 0; \quad (7.55)$$

that is,

$$\alpha(A_{ol}^T P_j A_{ol} - P_l) < K_l^T(B_l + \Delta B_l)^T [A_{cl} X_l A_{cl}^T - X_j]^{-1}(B_l + \Delta B_l)K_l. \quad (7.56)$$

Define

$$\Omega_{lj} := K_l^T(B_l + \Delta B_l)^T [A_{cl} X_l A_{cl}^T - X_j]^{-1}(B_l + \Delta B_l)K_l,$$

$$\Xi_{lj} := A_{ol}^T P_j A_{ol} - P_l.$$

Then it follows that there exists a large enough positive constant α such that

$$\alpha z^T \Xi_{lj} z < z^T \Omega_{lj} z, \quad z \neq 0 \tag{7.57}$$

because Ξ_{lj} is negative definite. It is easily seen that (7.57) implies (7.56). Thus, we have shown that with a large enough α (7.48) is a solution to (7.50). Similarly, it can also be shown that with a large enough α (7.48) is also a solution to (7.49). Therefore, the claim of the theorem is established, and the proof is thus completed. ❑

With the space partition of the second kind, observer design of T–S fuzzy systems can also be developed based on piecewise quadratic Lyapunov functions. Consider the following piecewise fuzzy observer of the form

$$\hat{x}(t+1) = \sum_{k \in \aleph(l)} \mu_k(z(t))\{A_k\hat{x}(t) + B_k u(t) + F_l(\hat{y}(t) - y(t))\}$$

$$\hat{y}(t) = \sum_{k \in \aleph(l)} \mu_k(z(t))C_k\hat{x}(t) \tag{7.58}$$

$$z(t) \in S_l, \quad l \in \bar{L}.$$

The fuzzy observer error dynamic equation can be described as

$$\tilde{x}(t+1) = \sum_{k \in \aleph(l)} \mu_k(z(t))A_{clk}\tilde{x}(t), \quad z(t) \in S_l, \quad l \in \bar{L}, \tag{7.59}$$

where $\tilde{x}(t) = \hat{x}(t) - x(t)$, $A_{clk} = A_k + F_l C_k$.

Define the set $\bar{\Omega}$ as in (4.32) representing all possible trajectory transitions among regions. Then the following result is readily obtained.

Theorem 7.7

The fuzzy observer error system (7.59) is globally exponentially stable if there exist a set of positive definite matrices P_l, $l \in \bar{L}$ and two sets of matrices V_l, R_l, $l \in \bar{L}$ such that the following LMIs are satisfied:

$$\begin{bmatrix} P_l - V_l - V_l^T & V_l A_k + R_l C_k \\ (V_l A_k + R_l C_k)^T & -P_l \end{bmatrix} < 0, \quad l \in \bar{L}, \quad k \in \aleph(l), \tag{7.60}$$

$$\begin{bmatrix} P_j - V_l - V_l^T & V_l A_k + R_l C_k \\ (V_l A_k + R_l C_k)^T & -P_l \end{bmatrix} < 0, \quad (l,j) \in \bar{\Omega}, \quad k \in \aleph(l). \tag{7.61}$$

Moreover, the observer gains are given by

$$F_l = V_l^{-1} R_l, \quad l \in \bar{L}. \tag{7.62}$$

Proof: Based on the result in Theorem 4.3 and its proof, one learns that the fuzzy control system (7.59) is globally exponentially stable if there exists a set of positive definite matrices $P_l, l \in L$, satisfying the following inequalities,

$$A_{clk}^T P_l A_{clk} - P_l < 0, \quad l \in \bar{L}, \quad k \in \aleph(l), \tag{7.63}$$

$$A_{clk}^T P_j A_{clk} - P_l < 0, \quad (l, j) \in \bar{\Omega}, \quad k \in \aleph(l). \tag{7.64}$$

It can be easily seen that the following inequalities would imply (7.63) and (7.64), respectively.

$$\begin{bmatrix} P_l - V_l - V_l^T & V_l A_{clk} \\ A_{clk}^T V_l^T & -P_l \end{bmatrix} < 0, \tag{7.65}$$

$$\begin{bmatrix} P_j - V_l - V_l^T & V_l A_{clk} \\ A_{clk}^T V_l^T & -P_l \end{bmatrix} < 0. \tag{7.66}$$

In fact, multiplying $W = [A_{clk}^T \quad I]$ on the left-hand side and W^T on the right-hand side of (7.65) leads exactly to (7.63). Doing the same to (7.66) also leads to (7.64).

Substituting A_{clk} into (7.65) and (7.66) and noting (7.62) lead to, respectively,

$$\begin{bmatrix} P_l - V_l - V_l^T & V_l A_k + R_l C_k \\ (V_l A_k + R_l C_k)^T & -P_l \end{bmatrix} < 0,$$

$$\begin{bmatrix} P_j - V_l - V_l^T & V_l A_k + R_l C_k \\ (V_l A_k + R_l C_k)^T & -P_l \end{bmatrix} < 0,$$

which are exactly (7.60) and (7.61), respectively. Therefore, it can be concluded that the observer error system is globally exponentially stable, and the proof is thus completed. ❏

Similarly, if the estimated state is used for controller implementation, one obtains the observer based output feedback controller. In this case, the closed-loop system can be described as

$$x(t+1) = \sum_{k \in \aleph(l)} \mu_k(z(t))[A_k x(t) + B_k K_l \hat{x}(t)], \quad z(t) \in S_l, \quad l \in \bar{L}. \tag{7.67}$$

Combining (7.67) and (7.59) leads to

$$
\begin{bmatrix} x(t+1) \\ \tilde{x}(t+1) \end{bmatrix} = \sum_{k \in \aleph(l)} \mu_k(z(t)) \bar{A}_{clk} \begin{bmatrix} x(t) \\ \tilde{x}(t) \end{bmatrix}, \quad z(t) \in S_l, \quad l \in \bar{L},
$$

(7.68)

where

$$
\bar{A}_{clk} = \begin{bmatrix} A_k + B_k K_l & B_k K_l \\ 0 & A_k + F_l C_k \end{bmatrix}.
$$

For the resulting closed-loop output feedback control system (7.68), one has the following result.

Theorem 7.8 (Separation Principle)
Suppose that there exist a set of positive definite matrices $X_l, l \in \bar{L}$ and a set of matrices $Q_l, l \in \bar{L}$ satisfying (5.47) and (5.48), and there exist a set of positive definite matrices $P_l, l \in \bar{L}$ and a set of matrices $R_l, l \in \bar{L}$ satisfying (7.60) and (7.61), respectively, with the controller gains given by (5.49) and the observer gains given by (7.62), respectively. There exists a positive constant α large enough such that the following set of positive definite matrices,

$$
Y_l := \begin{bmatrix} X_l^{-1} & 0 \\ 0 & \alpha P_l \end{bmatrix}, \quad l \in \bar{L},
$$

(7.69)

is a solution to the following matrix inequalities,

$$
\bar{A}_{clk}^T Y_l \bar{A}_{clk} - Y_l < 0, \quad l \in \bar{L}, \quad k \in \aleph(l),
$$

(7.70)

$$
\bar{A}_{clk}^T Y_j \bar{A}_{clk} - Y_l < 0, \quad (l,j) \in \bar{\Omega}, \quad k \in \aleph(l).
$$

(7.71)

In other words, the closed-loop system with the combined controller and observer is globally exponentially stable.

Proof: The proof is similar to that of Theorem 7.6 and is thus omitted. ❑

Example 7.2
Consider the T–S fuzzy system of the form (7.1) with the system parameters (A_l, B_l) defined in Example 5.3 and the output matrices given as follows:

$$
C_1 = [0.2 \quad 0.3], \quad C_2 = [0.1 \quad 0.1].
$$

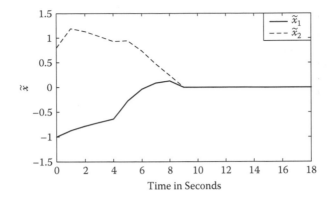

FIGURE 7.3   Observer error responses for Example 7.2.

By applying Theorem 7.5, one obtains the following solution to those LMIs:

$$P_1 = \begin{bmatrix} 91.7960 & 2.1478 \\ 2.1478 & 15.4211 \end{bmatrix}, \quad P_2 = \begin{bmatrix} 74.7836 & 8.8341 \\ 8.8341 & 30.3289 \end{bmatrix},$$

$$V_1 = \begin{bmatrix} 168.7767 & 69.9204 \\ 65.3174 & 74.3440 \end{bmatrix}, \quad V_2 = \begin{bmatrix} 99.0400 & -2.9673 \\ -1.8673 & 39.3458 \end{bmatrix},$$

$$F_1 = \begin{bmatrix} -0.8187 \\ -2.7101 \end{bmatrix}, \quad F_2 = \begin{bmatrix} -3.5081 \\ -3.9347 \end{bmatrix}.$$

Then the observer error responses can be illustrated in Figure 7.3 where initial conditions are given as $x = [-0.2 \quad 0.1]^T$ and $\tilde{x} = [-1.0 \quad 0.8]^T$. It can be easily observed that the observer errors converge to zero as time goes to infinity.

Moreover, with the controller obtained in Example 5.3, the observer-based output feedback control system can be obtained and the closed-loop system responses to a couple of initial conditions are illustrated in Figure 7.4. It can be easily observed that all the trajectories converge to the origin, and thus the separation principle is verified in this case.

7.4   OBSERVER AND OUTPUT FEEDBACK CONTROLLER SYNTHESIS BASED ON FUZZY QUADRATIC LYAPUNOV FUNCTIONS

Consider the smooth observer defined in (7.3) and the resulting closed-loop observer error system defined in (7.5). Then the following result on observer design of T–S fuzzy systems based on fuzzy quadratic Lyapunov functions can be easily established.

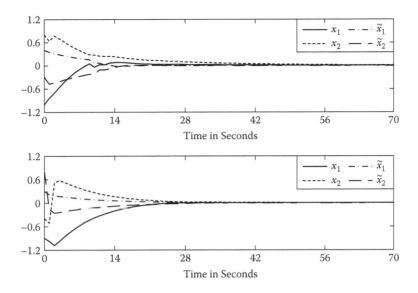

FIGURE 7.4   Closed-loop responses with output feedback for Example 7.2.

Theorem 7.9

The fuzzy observer error system (7.5) is globally exponentially stable if there exist a set of positive definite matrices $P_l, l \in L$, and two sets of matrices $R_l, V_l, l \in L$, such that the following LMIs are satisfied:

$$\begin{bmatrix} -P_j & A_l^T V_j^T + C_l^T R_j^T \\ V_j A_l + R_j C_l & P_i - V_j - V_j^T \end{bmatrix} < 0, \quad i, l, j \in L. \tag{7.72}$$

Moreover, the observer gains are given by

$$F_l = V_l^{-1} R_l, \quad l \in L. \tag{7.73}$$

Proof: By noting $(P_i - V_j)P_i^{-1}(P_i - V_j)^T = P_i - V_j - V_j^T + V_j P_i^{-1} V_j^T \geq 0$, it follows from (7.72) that

$$\begin{bmatrix} -P_j & A_l^T V_j^T + C_l^T R_j^T \\ V_j A_l + R_j C_l & -V_j^T P_i^{-1} V_j \end{bmatrix} < 0. \tag{7.74}$$

Premultiplying diag $\{I, V_j^{-1}\}$ and postmultiplying diag $\{I, V_j^{-T}\}$ in (7.74) lead to

$$\begin{bmatrix} -P_j & A_l^T + C_l^T R_j^T V_j^{-T} \\ A_l + V_j^{-1} R_j C_l & -P_i^{-1} \end{bmatrix} < 0. \tag{7.75}$$

By noting (7.73), (7.75) can be rewritten as

$$
\begin{bmatrix} -P_j & A_l^T + C_l^T F_j^T \\ A_l + F_j C_l & -P_i^{-1} \end{bmatrix} < 0;
$$

that is,

$$
\begin{bmatrix} -P_j & A_{clj}^T \\ A_{clj} & -P_i^{-1} \end{bmatrix} < 0,
\tag{7.76}
$$

which, via the Schur complement, is equivalent to

$$
A_{clj}^T P_i A_{clj} - P_j < 0.
\tag{7.77}
$$

Then it follows from Theorem 4.4 that the fuzzy observer error system (7.3) is globally exponentially stable, and the proof is thus completed. ❑

Similar to the technique in Kim and Lee (2000) to reduce the number of LMIs, the following result is straightforward.

Theorem 7.10
The fuzzy observer error system (7.5) is globally exponentially stable if there exist a set of positive definite matrices $P_l, l \in L$, and two sets of matrices $R_l, V_l, l \in L$, such that the following LMIs are satisfied:

$$
\begin{bmatrix} -P_l & A_i^T V_l^T + C_l^T R_l^T \\ V_l A_l + R_l C_l & P_i - V_l - V_l^T \end{bmatrix} < 0, \quad l, i \in L,
\tag{7.78}
$$

$$
\begin{bmatrix} -P_l & A_i^T V_j^T + C_i^T R_j^T \\ V_j A_l + R_j C_l & P_i - V_j^T - V_j \end{bmatrix} + \begin{bmatrix} -P_j & A_j^T V_l^T + C_j^T R_l^T \\ V_l A_j + R_l C_j & P_i - V_l^T - V_l \end{bmatrix} < 0, \; l < j, \; l, j, i \in L.
\tag{7.79}
$$

Moreover, the observer gains are given by

$$
F_l = V_l^{-1} R_l, \quad l \in L.
\tag{7.80}
$$

If the estimated state is used for control implementation, one obtains the observer-based output feedback controller. In this case, the closed-loop system can be described as in (7.22). Then one has the following result.

Theorem 7.11 (Separation Principle)

Suppose that there exist a set of positive definite matrices $X_l, l \in L$ and a set of matrices $Q_l, l \in L$ satisfying (5.61), and a set of positive definite matrices $P_l, l \in L$ and a set of matrices $R_l, l \in L$ satisfying (7.72), respectively, with the controller gains given by (5.62) and the observer gains given by (7.73), respectively. There exists a positive constant α large enough such that the following set of positive definite matrices

$$Y_l := \begin{bmatrix} X_l^{-1} & 0 \\ 0 & \alpha P_l \end{bmatrix}, \quad l \in L \tag{7.81}$$

is a solution to the following matrix inequalities,

$$\bar{A}_{clj}^T Y_i \bar{A}_{clj} - Y_j < 0, \quad i, l, j \in L. \tag{7.82}$$

In other words, the closed-loop system with the combined controller and observer is globally exponentially stable.

Proof: It follows from Theorem 4.4 and its proof that if there exists a set of positive definite matrices $Y_l, l \in L$ such that (7.82) is satisfied, then the closed-loop system (7.22) is globally exponentially stable. Now we are going to show that (7.81) is a solution to (7.82). Substituting (7.22) and (7.81) into (7.82) leads to

$$\begin{bmatrix} A_{clj} & B_l K_j \\ 0 & A_{olj} \end{bmatrix}^T \begin{bmatrix} X_i^{-1} & 0 \\ 0 & \alpha P_i \end{bmatrix} \begin{bmatrix} A_{clj} & B_l K_j \\ 0 & A_{olj} \end{bmatrix} - \begin{bmatrix} X_j^{-1} & 0 \\ 0 & \alpha P_j \end{bmatrix} < 0, \tag{7.83}$$

which is equivalent to

$$\begin{bmatrix} A_{clj}^T X_i^{-1} A_{clj} - X_j^{-1} & A_{clj}^T X_i^{-1} B_l K_j \\ K_j^T B_l^T X_i^{-1} A_{clj} & \Phi_{ilj} \end{bmatrix} < 0, \tag{7.84}$$

where $\Phi_{ilj} := \alpha A_{olj}^T P_i A_{olj} - \alpha P_j + K_j^T B_l^T X_i^{-1} B_l K_l$.

Using the Schur complements, (7.84) is equivalent to

$$A_{clj}^T X_i^{-1} A_{clj} - X_j^{-1} < 0 \tag{7.85}$$

and

$$\alpha A_{olj}^T P_i A_{olj} - \alpha P_j + K_j^T B_l^T X_i^{-1} B_l K_l - K_j^T B_l^T X_i^{-1} A_{clj} \left(A_{clj}^T X_i^{-1} A_{clj} - X_j^{-1} \right)^{-1} A_{clj}^T X_i^{-1} B_l K_l < 0. \tag{7.86}$$

Using the matrix inversion lemma, (7.86) is equivalent to

$$\alpha A_{olj}^T P_i A_{olj} - \alpha P_j - K_l^T B_l^T \left[A_{clj} X_j A_{clj}^T - X_i \right]^{-1} B_l K_l < 0; \tag{7.87}$$

that is,

$$\alpha \left(A_{olj}^T P_i A_{olj} - P_j \right) < K_l^T B_l^T \left[A_{clj} X_j A_{clj}^T - X_i \right]^{-1} B_l K_l. \tag{7.88}$$

Define

$$\Omega_{ilj} := K_l^T B_l^T \left[A_{clj} X_j A_{clj}^T - X_i \right]^{-1} B_l K_l,$$

$$\Xi_{ilj} := A_{olj}^T P_i A_{olj} - P_j;$$

then it follows that there exists a large enough positive constant α such that

$$\alpha z^T \Xi_{ilj} z < z^T \Omega_{ilj} z, z \neq 0 \tag{7.89}$$

because Ξ_{ilj} is negative definite. It is easily seen that (7.89) implies (7.88). Thus, we have shown that with a large enough α, (7.81) is a solution to (7.82). Therefore, the claim of the theorem is established, and the proof is thus completed. ❑

Remark 7.2
Note that the separation principle can also be established for the other controller and observer design schemes.

Example 7.3
Reconsider the T–S fuzzy system defined in Example 7.2. By applying Theorem 7.9, one obtains the solution to those LMIs as follows:

$$P_1 = \begin{bmatrix} 1.3272 & 0.3051 \\ 0.3051 & 0.5599 \end{bmatrix}, \quad P_2 = \begin{bmatrix} 1.3272 & 0.3051 \\ 0.3051 & 0.5599 \end{bmatrix},$$

$$V_1 = \begin{bmatrix} 1.1693 & 0.2598 \\ 0.2476 & 0.6154 \end{bmatrix}, \quad V_2 = \begin{bmatrix} 1.1693 & 0.2598 \\ 0.2476 & 0.6154 \end{bmatrix},$$

$$F_1 = \begin{bmatrix} -1.1389 \\ -5.2195 \end{bmatrix}, \quad F_2 = \begin{bmatrix} -1.1389 \\ -5.2195 \end{bmatrix}.$$

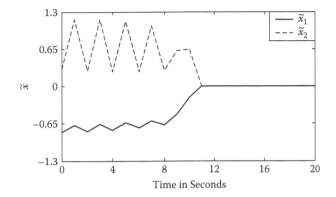

FIGURE 7.5   Observer error responses for Example 7.3.

Then the observer error responses can be illustrated in Figure 7.5 where initial conditions are given as $x = [-0.2 \quad 0.1]^T$ and $\tilde{x} = [-0.8 \quad 0.3]^T$. It can be easily observed that the observer errors converge to zero as time goes to infinity.

Moreover, with the controller obtained in Example 5.4, the observer-based output feedback control system can be obtained and the closed-loop system responses to a couple of initial conditions are illustrated in Figure 7.6. It can be easily observed that all the trajectories converge to the origin, and thus the separation principle is verified in this case.

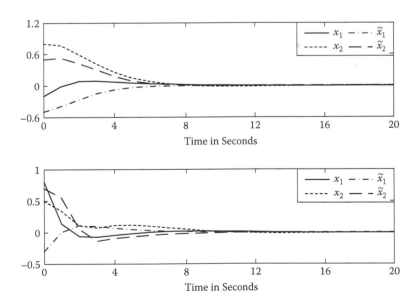

FIGURE 7.6   Closed-loop responses with output feedback for Example 7.3.

7.5 COMPARISON OF OBSERVER DESIGN RESULTS VIA NUMERICAL EXAMPLES

In this section, the observer synthesis results discussed in the previous sections for T–S fuzzy models are compared via numerical examples. For this purpose, the following modified observer design results in terms of exponential decay rate parameter λ are adopted, where the parameter λ is used for the performance indicator for comparison, and the proofs of these results are straightforward and thus omitted.

Theorem 7.12

If there exist two positive constants η, ξ, a positive definite matrix P, and two sets of matrices R_l, V_l $l \in L$ such that the following LMIs are satisfied,

$$\eta I < P < I, \tag{7.90}$$

$$\begin{bmatrix} P - V_l - V_l^T & V_l A_l + R_l C_j \\ (V_l A_l + R_l C_j)^T & -P + \xi I \end{bmatrix} < 0, \quad l, j \in L, \tag{7.91}$$

then the fuzzy observer error system (7.5) is globally exponentially stable with the decay rate λ defined as follows:

$$\|x(t)\| \le \alpha e^{-\lambda t} \|x(0)\|, \tag{7.92}$$

where $\alpha = \sqrt{1/\eta}$ and $\lambda = -\ln(1 - \xi)/2$.

Moreover, the observer gains are given by

$$F_l = V_l^{-1} R_l, \quad l \in L. \tag{7.93}$$

Theorem 7.13

The fuzzy observer error system (7.11) is globally exponentially stable with the decay rate λ as defined in (7.92), if there exist two positive constants η, ξ, a set of positive constants ε_l, $l \in L$, a set of positive definite matrices P_l, $l \in L$, and two sets of matrices R_l, V_l, $l \in L$ such that the following LMIs are satisfied:

$$\eta I < P_l < I, \quad l \in L, \tag{7.94}$$

$$\begin{bmatrix} P_l - V_l - V_l^T & V_l A_l + R_l C_l & V_l & R_l \\ (V_l A_l + R_l C_l)^T & -P_l + \xi I + \varepsilon_l \left(E_{lA}^T E_{lA} + E_{lC}^T E_{lC} \right) & 0 & 0 \\ V_l^T & 0 & -\varepsilon_l I & 0 \\ R_l^T & 0 & 0 & -\varepsilon_l I \end{bmatrix} < 0, \quad l \in L, \tag{7.95}$$

$$
\begin{bmatrix}
P_j - V_l - V_l^T & V_l A_l + R_l C_l & V_l & R_l \\
(V_l A_l + R_l C_l)^T & -P_l + \xi I + \varepsilon_l \left(E_{lA}^T E_{lA} + E_{lC}^T E_{lC} \right) & 0 & 0 \\
V_l^T & 0 & -\varepsilon_l I & 0 \\
R_l^T & 0 & 0 & -\varepsilon_l I
\end{bmatrix} < 0, \quad l, j \in \Omega.
$$

$$(7.96)$$

Moreover, the switching observer gains are given by

$$
F_l = V_l^{-1} R_l, \quad l \in L.
$$

$$(7.97)$$

Theorem 7.14

The fuzzy observer error system (7.59) is globally exponentially stable with the decay rate λ as defined in (7.92) if there exist positive constants η, ξ, a set of positive definite matrices P_l, $l \in \bar{L}$, and two sets of matrices V_l, R_l, $l \in \bar{L}$ such that the following LMIs are satisfied:

$$
\eta I < P_l < I, \quad l \in \bar{L},
$$

$$(7.98)$$

$$
\begin{bmatrix}
P_l - V_l - V_l^T & V_l A_k + R_l C_k \\
(V_l A_k + R_l C_k)^T & -P_l + \xi I
\end{bmatrix} < 0, \quad l \in \bar{L}, \quad k \in \aleph(l),
$$

$$(7.99)$$

$$
\begin{bmatrix}
P_j - V_l - V_l^T & V_l A_k + R_l C_k \\
(V_l A_k + R_l C_k)^T & -P_l + \xi I
\end{bmatrix} < 0, \quad (l, j) \in \bar{\Omega}, \quad k \in \aleph(l).
$$

$$(7.100)$$

Moreover, the observer gains are given by

$$
F_l = V_l^{-1} R_l, \quad l \in \bar{L}.
$$

$$(7.101)$$

Theorem 7.15

The fuzzy observer error system (7.5) is globally exponentially stable with the decay rate λ as defined in (7.92) if there exist positive constants η, ξ, a set of positive definite matrices P_l, $l \in L$, and two sets of matrices R_l, V_l, $l \in L$ such that the following LMIs are satisfied:

$$
\eta I < P_l < I, \quad l \in L,
$$

$$(7.102)$$

$$
\begin{bmatrix}
-P_l + \xi I & A_l^T V_j^T + C_l^T R_j^T \\
V_j A_l + R_j C_l & P_i - V_j - V_j^T
\end{bmatrix} < 0, \quad i, l, j \in L.
$$

$$(7.103)$$

Moreover, the observer gains are given by

$$F_l = V_l^{-1} R_l, \quad l \in L. \tag{7.104}$$

The objective of comparison is to find the maximum decay rate for each observer design approach and then compare their performance. For this purpose, the following optimization algorithms can be developed based on Theorems 7.12–7.15, respectively.

Algorithm 7.1: $\max\limits_{P, R_l, V_l, l \in L} \xi$, subject to LMIs (7.90) and (7.91)

Algorithm 7.2: $\max\limits_{P_l, \varepsilon_l, R_l, V_l, l \in L} \xi$, subject to LMIs (7.94)–(7.96)

Algorithm 7.3: $\max\limits_{P_l, R_l, V_l, l \in \bar{L}} \xi$, subject to LMIs (7.98)–(7.100)

Algorithm 7.4: $\max\limits_{P_l, R_l, V_l, l \in L} \xi$, subject to LMIs (7.102) and (7.103)

Example 7.4

Consider the T–S fuzzy system (7.1) with parameters as follows:

$$A_1 = \begin{bmatrix} 0.85 & 0.1 \\ -0.4 & 0.85 \end{bmatrix}, \quad A_2 = \begin{bmatrix} 0.45 & -0.54 \\ 0.54 & 0.45 \end{bmatrix}, \quad A_3 = \begin{bmatrix} 0.9 & 0.45 \\ -0.1 & 0.9 \end{bmatrix}$$

$$B_1 = \begin{bmatrix} 0 \\ 1 \end{bmatrix}, \quad B_2 = \begin{bmatrix} 0 \\ 1 \end{bmatrix}, \quad B_3 = \begin{bmatrix} 1 \\ 0 \end{bmatrix} \square$$

$$C_1 = [0 \quad 1], \quad C_2 = [1 \quad 0], \quad C_3 = [0 \quad 1],$$

and the membership functions described in Figure 7.7.

By applying Algorithms 7.1–7.4, respectively, one can obtain the corresponding maximum decay rates, which are summarized in Table 7.1.

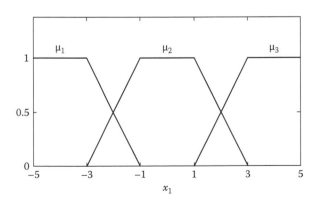

FIGURE 7.7   Membership functions in Example 7.4.

TABLE 7.1
Maximum Decay Rates for Example 7.4

Algorithm	7.1	7.2	7.3	7.4
Maximum λ	0.0245	0.0074	0.1103	0.1141

This example clearly demonstrates the advantage of the approaches based on piecewise or fuzzy quadratic Lyapunov functions to the approaches based on common quadratic Lyapunov functions.

7.6 CONCLUSIONS

This chapter has presented a number of approaches to observer and output feedback controller synthesis for T–S fuzzy systems based on common, piecewise, and fuzzy quadratic Lyapunov functions. It has been shown that the observer gains can be obtained by solving linear matrix inequalities. It has also been shown that the separation principle holds for the observer-based output feedback control, in the sense that the controller design and observer design can be carried out separately and the resulting observer-based output feedback closed-loop control system is still globally exponentially stable.

8 Robust Controller Synthesis of Uncertain T–S Fuzzy Systems

8.1 INTRODUCTION

T–S fuzzy models described in the previous chapters are assumed to be accurate in describing the underlying systems. However, in practice, many complex non-linear dynamic systems are subject to varieties of uncertainties such as parameter uncertainties or even structure variations. Quite often these uncertainties degrade the performance of the resulting closed-loop control systems, or even worse, lead to instability if they are not dealt with adequately in the controller design process. In this chapter, a class of uncertain T–S fuzzy systems is introduced, and the issue of their stabilization and H_∞ controller synthesis is addressed based on piecewise and fuzzy quadratic Lyapunov functions, respectively. It is also noted that the common quadratic Lyapunov functions are special cases of piecewise and fuzzy quadratic Lyapunov functions and thus the corresponding results based on common quadratic Lyapunov functions are presented as corollaries.

The rest of the chapter is organized as follows. Section 8.2 is devoted to the introduction of uncertain T–S fuzzy models. Sections 8.3 and 8.4 present controller synthesis of uncertain T–S fuzzy systems based on piecewise and fuzzy quadratic Lyapunov functions, respectively. A simulation example is given in Section 8.5, followed by some conclusions in Section 8.6.

8.2 MODEL OF UNCERTAIN T–S FUZZY SYSTEMS

Consider the following uncertain T–S fuzzy model,

$$R^l : \quad \text{IF} \quad z_1 \text{ is } F_1^l \text{ and } \dots z_v \text{ is } F_v^l$$

$$\text{THEN} \quad x(t+1) = A_l(t)x(t) + B_{1l}(t)u(t) + D_{1l}(t)v(t) \tag{8.1}$$

$$y(t) = C_l(t)x(t) + B_{2l}(t)u(t) + D_{2l}(t)v(t)$$

$$q(t) = H_l(t)x(t) + B_{3l}(t)u(t) + D_{3l}(t)v(t)$$

$$l \in L := \{1, 2, \dots, m\},$$

where most variables are the same as those defined in the previous chapters except for uncertain matrices; that is, R^l denotes the lth inference rule, m the number of inference rules, $F_j^l (j = 1, 2, \ldots, v)$ the fuzzy sets, $x(t) \in \Re^n$ the system state vector, $u(t) \in \Re^g$ the control variable vector, $v(t) \in \Re^s$ the disturbance input vector which belongs to $l_2[0, \infty)$, $y(t) \in \Re^p$ the measurement output vector, $q(t) \in \Re^r$ the controlled output vector, and $z(t) := [z_1(t), z_2(t), \ldots, z_v(t)]$ some measurable variables of the system, for example, the state variables. $A_l(t)$, $C_l(t)$, $H_l(t)$, $B_{\eta l}(t)$, $D_{\eta l}(t)$, $\eta \in \{1, 2, 3\}$ are appropriately dimensioned system matrices with time-varying parametric uncertainties, which are assumed to be of the form

$$
\begin{bmatrix}
A_l(t) & B_{1l}(t) & D_{1l}(t) \\
C_l(t) & B_{2l}(t) & D_{2l}(t) \\
H_l(t) & B_{3l}(t) & D_{3l}(t)
\end{bmatrix}
=
\begin{bmatrix}
A_l & B_{1l} & D_{1l} \\
C_l & B_{2l} & D_{2l} \\
H_l & B_{3l} & D_{3l}
\end{bmatrix}
+
\begin{bmatrix}
W_{1l} \\
W_{2l} \\
W_{3l}
\end{bmatrix}
\Delta(t)[E_{1l} \quad E_{2l} \quad E_{3l}]
$$

$$(8.2a)$$

$$\Delta(t) = F(t)[I - JF(t)]^{-1} \tag{8.2b}$$

$$I - JJ^T > 0, \tag{8.2c}$$

with A_l, C_l, H_l, $B_{\eta l}$, $D_{\eta l}$, $W_{\eta l}$, $E_{\eta l}$, $\eta \in \{1, 2, 3\}$, and J being known real constant matrices of appropriate dimensions. $F(t) \in \Re^{s_1 \times s_2}$ is an unknown real-valued matrix function with Lesbesgue measurable elements satisfying

$$F^T(t)F(t) \leq I. \tag{8.2d}$$

The parameter uncertainties are said to be admissible if (8.2a)–(8.2d) hold.

Remark 8.1
The parametric uncertainties are assumed to have a structured linear fractional form. This kind of uncertainty has been fairly well investigated in the area of robust control. Notice that when $J = 0$, the linear fractional form uncertainties reduce to the norm-bounded ones. Notice also that the condition (8.2c) guarantees that $I - JF(t)$ is invertible for all $F(t)$ satisfying (8.2d).

Similar to the standard fuzzy inference method as described in Chapter 3, that is, using a singleton fuzzifier, product fuzzy inference, and center-average defuzzifier, the following globally uncertain T–S fuzzy model can be obtained,

$$x(t+1) = A(\mu, t)x(t) + B_1(\mu, t)u(t) + D_1(\mu, t)v(t)$$

$$y(t) = C(\mu, t)x(t) + B_2(\mu, t)u(t) + D_2(\mu, t)v(t) \tag{8.3}$$

$$q(t) = H(\mu, t)x(t) + B_3(\mu, t)u(t) + D_3(\mu, t)v(t),$$

where

$$A(\mu, t) = \sum_{l=1}^{m} \mu_l A_l(t), \quad B_1(\mu, t) = \sum_{l=1}^{m} \mu_l B_{1l}(t), \quad D_1(\mu, t) = \sum_{l=1}^{m} \mu_l D_{1l}(t),$$

$$C(\mu,t) = \sum_{l=1}^{m} \mu_l C_l(t), \quad B_2(\mu,t) = \sum_{l=1}^{m} \mu_l B_{2l}(t), \quad D_2(\mu,t) = \sum_{l=1}^{m} \mu_l D_{2l}(t), \quad (8.4)$$

$$H(\mu,t) = \sum_{l=1}^{m} \mu_l H_l(t), \quad B_3(\mu,t) = \sum_{l=1}^{m} \mu_l B_{3l}(t), \quad D_3(\mu,t) = \sum_{l=1}^{m} \mu_l D_{3l}(t),$$

$$\sum_{l=1}^{m} \mu_l(z(t)) = 1.$$

The objective of this chapter, similar to that of Chapter 6, is to design a suitable controller for system (8.1) or equivalently (8.3) such that the resulting closed-loop control system is globally exponentially stable with a guaranteed performance in the H_∞ sense. In other words, given a prescribed level of noise attenuation $\gamma > 0$, find a suitable controller such that the induced l_2-norm of the operator from v to the controlled output q is less than γ under zero initial conditions; that is, $\|q(t)\|_2 < \gamma \|v(t)\|_2$, for all nonzero $v \in l_2[0,\infty)$ and all admissible uncertainties. In this case, the closed-loop control system is said to be globally exponentially stable with H_∞ performance γ.

8.3 CONTROLLER SYNTHESIS BASED ON PIECEWISE QUADRATIC LYAPUNOV FUNCTIONS

In this section, we consider a robust H_∞ control problem of the uncertain T–S fuzzy model (8.1) based on a piecewise quadratic Lyapunov function.

With the space partition of the second kind described in Chapter 4, the global uncertain fuzzy system (8.3) can be expressed in each region as

$$x(t+1) = A_i(t)x(t) + B_{1i}(t)u(t) + D_{1i}(t)v(t)$$

$$y(t) = C_i(t)x(t) + B_{2i}(t)u(t) + D_{2i}(t)v(t)$$

$$q(t) = H_i(t)x(t) + B_{3i}(t)u(t) + D_{3i}(t)v(t) \quad (8.5)$$

$$z(t) \in S_i, i \in \bar{L},$$

where

$$A_i(t) = \sum_{k \in \aleph(i)} \mu_k A_k(t), \quad B_{1i}(t) = \sum_{k \in \aleph(i)} \mu_k B_{1k}(t), \quad D_{1i}(t) = \sum_{k \in \aleph(i)} \mu_k D_{1k}(t),$$

$$C_i(t) = \sum_{k \in \aleph(i)} \mu_k C_k(t), \quad B_{2i}(t) = \sum_{k \in \aleph(i)} \mu_k B_{2k}(t), \quad D_{2i}(t) = \sum_{k \in \aleph(i)} \mu_k D_{2k}(t), \quad (8.6)$$

$$H_i(t) = \sum_{k \in \aleph(i)} \mu_k H_k(t), \quad B_{3i}(t) = \sum_{k \in \aleph(i)} \mu_k B_{3k}(t), \quad D_{3i}(t) = \sum_{k \in \aleph(i)} \mu_k D_{3k}(t),$$

with $0 \le \mu_k(z(t)) \le 1$, $\sum_{k \in \aleph(i)} \mu_k(z(t)) = 1$. Similar to Chapter 4, for each region S_i, the set $\aleph(i)$ contains the indices for those subsystem matrices used in the interpolation within that region. For a crisp region, $\aleph(i)$ contains a single element.

The piecewise controller design is considered for the fuzzy system (8.3) or equivalently (8.5). Thus it is assumed that when the state of the system transits from the region S_i to S_j at the time t, the dynamics of the system is governed by the dynamics of the blended model in the region S_i at that time. For future use, we also define a set $\bar{\Omega}$ as in Chapter 4 to represent all possible region transitions,

$$\bar{\Omega} := \{(i, j) \mid z(t) \in S_i, z(t+1) \in S_j, \forall i, j \in \bar{L}\}, \tag{8.7}$$

where $j = i$ when $z(t)$ stays in the same region S_i, and $j \ne i$ when $z(t)$ transits from the region S_i to S_j.

Several approaches to robust control analysis and synthesis for the uncertain fuzzy system (8.3) or equivalently (8.5) are developed based on piecewise quadratic Lyapunov functions in this section. A robust H_∞ performance analysis result for the nominal unforced fuzzy system (8.3) with $u(t) \equiv 0$ is first presented in Section 8.3.1, and the issues of piecewise state feedback and output feedback controller designs are addressed in Sections 8.3.2 and 8.3.3, respectively.

8.3.1 ROBUST H_∞ PERFORMANCE ANALYSIS

Theorem 8.1

The unforced fuzzy system (8.3), or equivalently (8.5) with $u(t) \equiv 0$ is globally exponentially stable with H_∞ performance γ if there exist a set of positive definite matrices X_i, $i \in \bar{L}$, a set of matrices $\tilde{G}_i$, $i \in \bar{L}$, and a set of positive scalars $\varepsilon_{kij} > 0$, $k \in \aleph(i)$, $(i, j) \in \bar{\Omega}$, such that the following linear matrix inequalities are satisfied:

$$\begin{bmatrix} \tilde{X}_{ij} + He\{\tilde{A}_k \tilde{G}_i\} & * & * \\ \tilde{E}_k \tilde{G}_i & -\varepsilon_{kij} I & * \\ \varepsilon_{kij} \tilde{W}_k^T & \varepsilon_{kij} J^T & -\varepsilon_{kij} I \end{bmatrix} < 0, \quad k \in \aleph(i), (i, j) \in \bar{\Omega}, \tag{8.8}$$

where

$$\tilde{X}_{ij} := \begin{bmatrix} X_i & 0 & 0 & 0 \\ 0 & -X_j & 0 & 0 \\ 0 & 0 & I & 0 \\ 0 & 0 & 0 & -\gamma^2 I \end{bmatrix}, \tilde{A}_k := \begin{bmatrix} I & 0 \\ A_k & D_{1k} \\ 0 & I \\ H_k & D_{3k} \end{bmatrix}, \tilde{W}_k := \begin{bmatrix} 0 \\ W_{1k} \\ 0 \\ W_{3k} \end{bmatrix}, \tilde{E}_k := [E_{1k} \quad E_{3k}],$$

$$\tag{8.9}$$

and $He\{A\}$ is the shorthand notation for $A + A^T$.

Proof: Consider the following piecewise quadratic Lyapunov function candidate

$$V(x(t)) = x^T(t)X_i^{-1}x(t), \; z(t) \in S_i, \, i \in \bar{L}. \tag{8.10}$$

Then, it is well known that it suffices to show the following inequality,

$$V(t+1) - V(t) + \gamma^{-2}q^T(t)q(t) - v^T(t)v(t) < 0, \tag{8.11}$$

to prove that the nominal unforced fuzzy system (8.3) with $u(t) \equiv 0$ is globally exponentially stable with H_∞ performance γ under zero initial conditions for all admissible uncertainties satisfying (8.2).

Note that with the definition of the Lyapunov function candidate in (8.10) the
inequality (8.11) is equivalent to the following set of inequalities,

$$\begin{bmatrix} x(t) \\ x(t+1) \\ v(t) \\ q(t) \end{bmatrix}^T \begin{bmatrix} -X_i^{-1} & 0 & 0 & 0 \\ 0 & X_j^{-1} & 0 & 0 \\ 0 & 0 & -I & 0 \\ 0 & 0 & 0 & \gamma^{-2}I \end{bmatrix} \begin{bmatrix} x(t) \\ x(t+1) \\ v(t) \\ q(t) \end{bmatrix} < 0, \; (i,j) \in \bar{\Omega} \tag{8.12}$$

and thus it suffices to show (8.12) instead of (8.11).
   In addition, it follows from (8.5) with $u(t) \equiv 0$ that

$$\begin{bmatrix} A_i(t) & -I & D_{1i}(t) & 0 \\ H_i(t) & 0 & D_{3i}(t) & -I \end{bmatrix} \begin{bmatrix} x(t) \\ x(t+1) \\ v(t) \\ q(t) \end{bmatrix} = 0. \tag{8.13}$$

Note that

$$Ker \begin{bmatrix} A_i(t) & -I & D_{1i}(t) & 0 \\ H_i(t) & 0 & D_{3i}(t) & -I \end{bmatrix}^{\perp} = Im \begin{bmatrix} A_i^T(t) & H_i^T(t) \\ -I & 0 \\ D_{1i}^T(t) & D_{3i}^T(t) \\ 0 & -I \end{bmatrix} = Ker \begin{bmatrix} I & A_i^T(t) & 0 & H_i^T(t) \\ 0 & D_{1i}^T(t) & I & D_{3i}^T(t) \end{bmatrix}.$$

Then it follows from the dualization lemma in the appendix that (8.12) is equivalent to

$$\tilde{x}^T(t)\, diag\{X_i, -X_j, I, -\gamma^2 I\}\, \tilde{x}(t) < 0, \tag{8.14}$$

where

$$\tilde{x}(t) := \text{column vector}\{x(t), x(t+1), v(t), q(t)\} \in Ker \begin{bmatrix} I & A_i^T(t) & 0 & H_i^T(t) \\ 0 & D_{1i}^T(t) & I & D_{3i}^T(t) \end{bmatrix}. \tag{8.15}$$

Assigning $\tilde{x}(t) \to x_\eta$,

$$\text{diag}\{X_i, -X_j, I, -\gamma^2 I\} \to P, \begin{bmatrix} I & A_i^T(t) & 0 & H_i^T(t) \\ 0 & D_{1i}^T(t) & I & D_{3i}^T(t) \end{bmatrix} \to H, \tilde{G}_i^T \to N,$$

where $\tilde{G}_i^T$ is a free slack matrix, and applying Finsler's lemma in the appendix, one has that the following inequalities imply (8.14) with (8.15):

$$\begin{bmatrix} X_i & 0 & 0 & 0 \\ 0 & -X_j & 0 & 0 \\ 0 & 0 & I & 0 \\ 0 & 0 & 0 & -\gamma^2 I \end{bmatrix} + He \left\{ \begin{bmatrix} I & 0 \\ A_i(t) & D_{1i}(t) \\ 0 & I \\ H_i(t) & D_{3i}(t) \end{bmatrix} \tilde{G}_i \right\} < 0, (i, j) \in \bar{\Omega}. \text{ (8.16)}$$

On the other hand, using relations (8.2) and (8.6), one has

$$\begin{bmatrix} I & 0 \\ A_i(t) & D_{1i}(t) \\ 0 & I \\ H_i(t) & D_{3i}(t) \end{bmatrix} = \sum_{k \in \aleph(i)} \mu_k(z(t)) \begin{bmatrix} I & 0 \\ A_k(t) & D_{1k}(t) \\ 0 & I \\ H_k(t) & D_{3k}(t) \end{bmatrix},$$

$$\begin{bmatrix} I & 0 \\ A_k(t) & D_{1k}(t) \\ 0 & I \\ H_k(t) & D_{3k}(t) \end{bmatrix} = \begin{bmatrix} I & 0 \\ A_k & D_{1k} \\ 0 & I \\ H_k & D_{3k} \end{bmatrix} + \begin{bmatrix} 0 \\ W_{1k} \\ 0 \\ W_{3k} \end{bmatrix} \Delta(t) [E_{1k} \quad E_{3k}]. \quad \text{(8.17)}$$

Then, based on relationship (8.17) and by the Schur complement lemma and S-procedure in the appendix, it is clear that (8.8) guarantees (8.16). The proof is thus completed. ❑

8.3.2 PIECEWISE STATE FEEDBACK CONTROLLER DESIGN

In this section, we address the piecewise H_∞ state feedback controller synthesis problem for the uncertain fuzzy system (8.3) or equivalently (8.5). Consider the following piecewise controller defined as

$$u(t) = K_i x(t), \quad z(t) \in S_i, i \in \bar{L} \tag{8.18}$$

and the closed-loop fuzzy control system can be described by

$$x(t+1) = \sum_{k \in \aleph(i)} \mu_k(z(t))((A_k(t) + B_{1k}(t)K_i)x(t) + D_{1i}(t)v(t))$$

$$q(t) = \sum_{k \in \aleph(i)} \mu_k(z(t))((H_k(t) + B_{3k}(t)K_i)x(t) + D_{3i}(t)v(t))$$

$$z(t) \in S_i, i \in \bar{L}.$$

(8.19)

By using the same piecewise quadratic Lyapunov function as in (8.10), one has the following H_∞ controller design results for the uncertain fuzzy system described in (8.3) or equivalently (8.5).

Theorem 8.2

The closed-loop fuzzy control system (8.19) is globally exponentially stable with H_∞ performance γ if there exist a set of positive definite matrices X_i, sets of matrices $G_{\beta i}, \beta \in \{1,2,3,4,5\}, \bar{K}_i, i \in \bar{L},$ and a set of positive scalars $\varepsilon_{kij} > 0$, $k \in \aleph(i)$, $(i,j) \in \bar{\Omega}$, such that the following linear matrix inequalities are satisfied:

$$\begin{bmatrix} G_{1i} + G_{1i}^T + X_i & * & * & * & * & * \\ A_k G_{1i} + B_{1k}\bar{K}_i + D_{1k}G_{2i} & D_{1k}G_{3i} + G_{3i}^T D_{1k}^T - X_j & * & * & * & * \\ G_{2i} & G_{3i} + G_{4i}^T D_{1k}^T & G_{4i} + G_{4i}^T + I & * & * & * \\ H_k G_{1i} + B_{3k}\bar{K}_i + D_{3k}G_{2i} & D_{3k}G_{3i} + G_{5i}^T D_{1k}^T & D_{3k}G_{4i} + G_{5i}^T & D_{3k}G_{5i} + G_{5i}^T D_{3k}^T - \gamma^2 I & * & * \\ E_{1k}G_{1i} + E_{2k}\bar{K}_i + E_{3k}G_{2i} & E_{3k}G_{3i} & E_{3k}G_{4i} & E_{3k}G_{5i} & -\varepsilon_{kij}I & * \\ 0 & \varepsilon_{kij}W_{1k}^T & 0 & \varepsilon_{kij}W_{3k}^T & \varepsilon_{kij}J^T & -\varepsilon_{kij}I \end{bmatrix} < 0$$

$$k \in \aleph(i), (i,j) \in \bar{\Omega}.$$

(8.20)

Moreover, the controller gains are given by

$$K_i = \bar{K}_i G_{1i}^{-1}, i \in \bar{L}.$$

(8.21)

Proof: Based on the result in Theorem 8.1 and its proof, one learns that the closed-loop system (8.19) is globally exponentially stable with H_∞ performance γ if the following matrix inequalities are satisfied:

$$\begin{bmatrix} \tilde{X}_{ij} + He\{\tilde{A}_{ki}\tilde{G}_i\} & * & * \\ \tilde{E}_{ki}\tilde{G}_i & -\varepsilon_{kij}I & * \\ \varepsilon_{kij}\tilde{W}_k^T & \varepsilon_{kij}J^T & -\varepsilon_{kij}I \end{bmatrix} < 0, k \in \aleph(i), (i,j) \in \bar{\Omega},$$

(8.22)

where $\tilde{X}_{ij}$ and $\tilde{W}_k$ are defined in (8.9) and

$$
\tilde{A}_{ki} := \begin{bmatrix} I & 0 \\ A_k + B_{1k}K_i & D_{1k} \\ 0 & I \\ H_k + B_{3k}K_i & D_{3K} \end{bmatrix}, \tilde{E}_{ki} := [E_{1k} + E_{2k}K_i \quad E_{3k}]. \tag{8.23}
$$

For simplicity in the controller design procedure of this case, we specify the slack variables $\tilde{G}_i$ as

$$
\tilde{G}_i = \begin{bmatrix} G_{1i} & 0 & 0 & 0 \\ G_{2i} & G_{3i} & G_{4i} & G_{5i} \end{bmatrix}, i \in \bar{L}. \tag{8.24}
$$

Then, substituting the matrices $\tilde{G}_i$ defined in (8.24) into (8.22) together with consideration of (8.21) leads to exactly (8.20). The proof is thus completed. ❏

In the case of no external disturbances, that is, $v(t) \equiv 0$, in the fuzzy system (8.3) or equivalently (8.5), one can easily obtain the corresponding results on piecewise state feedback stabilization of the following fuzzy system,

$$
x(t+1) = A(\mu, t)x(t) + B_1(\mu, t)u(t), \tag{8.25}
$$

where $A(\mu, t)$ and $B_1(\mu, t)$ are defined in (8.4).

With the same space partition and the same piecewise state feedback controller defined in (8.18), the closed-loop fuzzy control system can be described by

$$
x(t+1) = \sum_{k \in \aleph(i)} \mu_k(z(t))((A_k(t) + B_{1k}(t)K_i)x(t)), z(t) \in S_i, i \in \bar{L}. \tag{8.26}
$$

By using the same piecewise quadratic Lyapunov function candidate defined in (8.10) and following a similar procedure as in the proof of Theorem 8.2, one has the following piecewise state feedback stabilization result for the uncertain fuzzy system described in (8.25).

Theorem 8.3

The closed-loop fuzzy control system (8.26) is globally exponentially stable if there exist a set of positive definite matrices X_i, $i \in \bar{L}$, two sets of matrices G_{1i}, $\bar{K}_i$, $i \in \bar{L}$, and a set of positive scalars $\varepsilon_{kij} > 0$, $k \in \aleph(i)$, $(i, j) \in \bar{\Omega}$, such that the following linear matrix inequalities are satisfied:

$$
\begin{bmatrix} X_i + G_{1i} + G_{1i}^T & * & * & * \\ A_k G_{1i} + B_{1k}\bar{K}_i & -X_j & * & * \\ E_{1k}G_{1i} + E_{2k}\bar{K}_i & 0 & -\varepsilon_{kij}I & * \\ 0 & \varepsilon_{kij}W_{1k}^T & \varepsilon_{kij}J^T & -\varepsilon_{kij}I \end{bmatrix} < 0, k \in \aleph(i), (i, j) \in \bar{\Omega}. \tag{8.27}
$$

Moreover, the controller gains are given by (8.21).

In Theorems 8.2–8.3, if the Lyapunov matrices are chosen as follows,

$$X_i \equiv X, \forall i \in \bar{L}, \tag{8.28}$$

one readily obtains the corresponding piecewise state feedback H_∞ control and stabilization results, respectively, based on the following common quadratic Lyapunov function

$$V(x(t)) = x^T(t)X^{-1}x(t), \tag{8.29}$$

where the matrix X is positive definite.

Corollary 8.1
The closed-loop fuzzy control system (8.19) is globally exponentially stable with H_∞ performance γ if the conditions (8.20) with $X_i \equiv X$, $i \in \bar{L}$ are satisfied.

Corollary 8.2
The closed-loop fuzzy control system (8.26) is globally exponentially stable if the conditions (8.27) with $X_i \equiv X, i \in \bar{L}$ are satisfied.

8.3.3 PIECEWISE OUTPUT FEEDBACK CONTROLLER DESIGN

In this section, we extend the results of Sections 8.3.1 and 8.3.2 to design a piecewise output feedback controller for the uncertain fuzzy system (8.3) or equivalently (8.5).

Consider the following piecewise dynamic output feedback controller:

$$\hat{x}(t+1) = \sum_{k \in \aleph(i)} \mu_k(z(t))\hat{A}_{ki}\hat{x}(t) + \hat{B}_i y(t)$$

$$u(t) = \hat{C}_i \hat{x}(t) \tag{8.30}$$

$$z(t) \in S_i, i \in \bar{L},$$

where $\hat{x}(t) \in \Re^n$ is the controller state vector and $\hat{A}_{ki}$, $\hat{B}_i$, $\hat{C}_i$, $i \in \bar{L}$ are appropriately dimensioned matrices to be determined.

Combining (8.30) with the system (8.5) leads to the following closed-loop fuzzy control system:

$$\bar{x}(t+1) = \sum_{k \in \aleph(i)} \mu_k(z(t))(\bar{A}_{ki}(t)\bar{x}(t) + \bar{B}_{ki}(t)v(t))$$

$$q(t) = \sum_{k \in \aleph(i)} \mu_k(z(t))(\bar{H}_{ki}(t)\bar{x}(t) + D_{3k}(t)v(t)) \tag{8.31}$$

$$z(t) \in S_i, i \in \bar{L},$$

where

$$\bar{x}(t) := [x^T(t) \quad \hat{x}^T(t)]^T, \ \bar{A}_{ki}(t) := \begin{bmatrix} A_k(t) & B_{1k}(t)\hat{C}_i \\ \hat{B}_i C_k(t) & \hat{A}_{ki} + \hat{B}_i B_{2k}(t)\hat{C}_i \end{bmatrix},$$
(8.32)

$$\bar{B}_{ki}(t) := \begin{bmatrix} D_{1k}(t) \\ \hat{B}_i D_{2k}(t) \end{bmatrix}, \ \bar{H}_{ki}(t) := [H_k(t) \quad B_{3k}(t)\hat{C}_i].$$

Then, the main result for the piecewise output feedback H_∞ control of the uncertain fuzzy system (8.3) or equivalently (8.5) is summarized in the following theorem.

Theorem 8.4

The closed-loop fuzzy control system (8.31) is globally exponentially stable with H_∞ performance γ if there exist a set of positive definite matrices,

$$\bar{X}_i := \begin{bmatrix} \bar{X}_{i1} & * \\ \bar{X}_{i2} & \bar{X}_{i3} \end{bmatrix}, \ i \in \bar{L},$$

matrices G_1, U_1, M, sets of matrices $\vec{A}_{ki}$, $\vec{B}_i$, $\vec{C}_i$, $k \in \aleph(i)$, $i \in \bar{L}$, and a set of positive scalars $\varepsilon_{kij} > 0$, $k \in \aleph(i)$, $(i,j) \in \bar{\Omega}$, such that the following linear matrix inequalities are satisfied:

$$\begin{bmatrix} G_1 + G_1^T + \bar{X}_{i1} & * & * & * & * & * & * & * & * & * \\ M + I + \bar{X}_{i2} & U_1 + U_1^T + \bar{X}_{i3} & * & * & * & * & * & * & * & * \\ A_k G_1 + B_{1k}\vec{C}_i & A_k & -\bar{X}_{j1} & * & * & * & * & * & * & * \\ \vec{A}_{ki} & U_1 A_k + \vec{B}_i C_k & -\bar{X}_{j2} & -\bar{X}_{j3} & * & * & * & * & * & * \\ 0 & 0 & -D_{1k}^T & -D_{1k}^T U_1^T - D_{2k}^T \vec{B}_i^T & -I & * & * & * & * & * \\ H_k G_1 + B_{3k}\vec{C}_i & H_k & 0 & 0 & -D_{3k} & -\gamma^2 I & * & * & * & * \\ E_{1k} G_1 + E_{2k}\vec{C}_i & E_{1k} & 0 & 0 & -E_{3k} & 0 & -\varepsilon_{kij}I & * & * & * \\ 0 & 0 & \varepsilon_{kij}W_{1k}^T & 0 & 0 & \varepsilon_{kij}W_{3k}^T & \varepsilon_{kij}J^T & -\varepsilon_{kij}I & * & * \\ 0 & 0 & 0 & \varepsilon_{kij}I & 0 & 0 & 0 & 0 & -I & * \\ 0 & 0 & 0 & 0 & 0 & 0 & 0 & U_1 W_{1k} + \vec{B}_i W_{2k} & 0 & -I \end{bmatrix} < 0,$$

$$k \in \aleph(i), \ (i,j) \in \bar{\Omega}.$$
(8.33)

Moreover, the controller gain matrices can be constructed as

$$U_2 G_2 = S.V.D(M - U_1 G_1),$$

$$\hat{A}_{ki} = U_2^{-1}\left(\vec{A}_{ki} - U_1 A_k G_1 - \vec{B}_i C_k G_1 - U_1 B_{1k}\vec{C}_i - \vec{B}_i B_{2k}\vec{C}_i\right)G_2^{-1},$$
(8.34)

$$\hat{B}_i = U_2^{-1}\vec{B}_i, \ \hat{C}_i = \vec{C}_i G_2^{-1}$$

$$k \in \aleph(i), i \in \bar{L}.$$

Proof: Using the relations (8.2) and (8.6), one has

$$
\begin{bmatrix}
I & 0 \\
\bar{A}_{ki}(t) & \bar{B}_{ki}(t) \\
0 & I \\
\bar{H}_{ki}(t) & D_{3k}(t)
\end{bmatrix}
= \tilde{A}_{ki} + \tilde{W}_{ki}\Delta(t)\tilde{E}_{ki},
\tag{8.35}
$$

where

$$
\tilde{A}_{ki} :=
\begin{bmatrix}
I & 0 & 0 \\
0 & I & 0 \\
A_k & B_{1k}\hat{C}_i & D_{1k} \\
\hat{B}_i C_k & \hat{A}_{ki} + \hat{B}_i B_{2k}\hat{C}_i & \hat{B}_i D_{2k} \\
0 & 0 & I \\
H_k & B_{3k}\hat{C}_i & D_{3k}
\end{bmatrix},
\tilde{W}_{ki} :=
\begin{bmatrix}
0 \\
0 \\
W_{1k} \\
\hat{B}_i W_{2k} \\
0 \\
W_{3k}
\end{bmatrix},
\tilde{E}_{ki} := [E_{1k} \quad E_{2k}\hat{C}_i \quad E_{3k}].
$$

$$\tag{8.36}$$

Based on the results in Theorem 8.1 and its proof, it is easy to see that the closed-loop system (8.31) is globally exponentially stable with H_∞ performance γ if the following matrix inequalities are satisfied:

$$
\begin{bmatrix}
\tilde{X}_{ij} + He\{\tilde{A}_{ki}\tilde{G}_i\} & * & * \\
\tilde{E}_{ki}\tilde{G}_i & -\varepsilon_{kij}I & * \\
\varepsilon_{kij}\tilde{W}_{ki}^T & \varepsilon_{kij}J^T & -\varepsilon_{kij}I
\end{bmatrix} < 0, \, k \in \aleph(i), (i,j) \in \bar{\Omega},
\tag{8.37}
$$

where $\tilde{X}_{ij} := \mathrm{diag}\{X_i, -X_j, I, -\gamma^2 I\}$ and $\tilde{A}_{ki}$, $\tilde{E}_{ki}$, $\tilde{W}_{ki}$ are defined in (8.36).

For simplicity of the controller design procedure in this case, we specify the slack variables $\tilde{G}_i$ as

$$
\tilde{G}_i =
\begin{bmatrix}
G & 0 & 0 & 0 \\
0 & 0 & -I & 0
\end{bmatrix}, \forall i \in \bar{L}.
\tag{8.38}
$$

Let the matrix G and its inverse be partitioned as

$$
G =
\begin{bmatrix}
G_1 & \bullet \\
G_2 & \bullet
\end{bmatrix},
G^{-1} =
\begin{bmatrix}
U_1^T & \bullet \\
U_2^T & \bullet
\end{bmatrix},
\tag{8.39}
$$

where "$\bullet$" denotes the elements that are uniquely determined from the equality $G^{-1}G = GG^{-1} = I$.

For matrix inequality linearization purposes, we further define the following matrices,

$$J := \begin{bmatrix} I & U_1^T \\ 0 & U_2^T \end{bmatrix},$$

$$\Lambda := \text{diag}\{J, J, I, I\},$$

$$\bar{X}_i := J^T X_i J := \begin{bmatrix} \bar{X}_{i1} & \bar{X}_{i2}^T \\ \bar{X}_{i2} & \bar{X}_{i3} \end{bmatrix},$$

$$M := U_1 G_1 + U_2 G_2, \tag{8.40}$$

$$\vec{A}_{ki} := U_1 A_k G_1 + \vec{B}_i C_k G_1 + U_1 B_{1k} \vec{C}_i + U_2 \hat{A}_{ki} G_2 + \vec{B}_i B_{2k} \vec{C}_i,$$

$$\vec{B}_i := U_2 \hat{B}_i,$$

$$\vec{C}_i := \hat{C}_i G_2.$$

Then, substituting the matrices $\tilde{A}_{ki}$, $\tilde{E}_{ki}$, and $\tilde{W}_{ki}$ defined in (8.36) into (8.37) and performing the congruence transformation to (8.37) by Λ together with consideration of the matrices defined in (8.40) and the following bounding inequality lead to (8.33) exactly.

$$R_1 R_2^T + R_2 R_1^T \leq [R_1 \quad R_2] \begin{bmatrix} R_1^T \\ R_2^T \end{bmatrix}, \tag{8.41}$$

where

$$R_1 := [0 \ 0 \ 0 \ \varepsilon_{kij} I \ 0 \ 0 \ 0 \ 0]^T,$$

$$R_2 := \begin{bmatrix} 0 & 0 & 0 & 0 & 0 & 0 & 0 & U_1 W_{1k} + \vec{B}_i W_{2k} \end{bmatrix}^T. \tag{8.42}$$

The proof is thus completed. ❑

In the case of no external disturbances, that is, $v(t) \equiv 0$, in the fuzzy system (8.3) or equivalently (8.5), one can easily obtain the corresponding results on piecewise output feedback stabilization of the following fuzzy system,

$$x(t+1) = A(\mu, t)x(t) + B_1(\mu, t)u(t) \tag{8.43}$$

$$y(t) = C(\mu, t)x(t) + B_2(\mu, t)u(t),$$

where $A(\mu, t)$, $B_1(\mu, t)$, $C(\mu, t)$ and $B_2(\mu, t)$ are defined in (8.4).

With the same partition method and the same output feedback piecewise controller defined in (8.30), the closed-loop fuzzy control system can be described by

$$\bar{x}(t+1) = \sum_{k\in\aleph(i)} \mu_k(z(t))\bar{A}_{ki}(t)\bar{x}(t), \; z(t) \in S_i, \; i \in \bar{L},$$ (8.44)

where $\bar{x}(t)$ and $\bar{A}_{ki}(t)$ are defined in (8.32).

By using the same piecewise quadratic Lyapunov function candidate defined in (8.10) and following a similar line as in the proof of Theorem 8.4, one has the following piecewise output feedback stabilization result for the fuzzy system described in (8.43).

Theorem 8.5

The closed-loop fuzzy control system (8.44) is globally exponentially stable if there exist a set of positive definite matrices

$$\bar{X}_i := \begin{bmatrix} \bar{X}_{i1} & * \\ \bar{X}_{i2} & \bar{X}_{i3} \end{bmatrix}, \; i \in \bar{L},$$

matrices G_1, U_1, M, sets of matrices $\vec{A}_{ki}$, $\vec{B}_i$, $\vec{C}_i$, $k \in \aleph(i)$, $i \in \bar{L}$, and a set of positive scalars $\varepsilon_{kij} > 0$, $k \in \aleph(i)$, $(i,j) \in \bar{\Omega}$, such that the following linear matrix inequalities are satisfied:

$$\begin{bmatrix}
G_1 + G_1^T + \bar{X}_{i1} & * & * & * & * & * & * & * \\
M + I + \bar{X}_{i2} & U_1 + U_1^T + \bar{X}_{i3} & * & * & * & * & * & * \\
A_k G_1 + B_{1k}\vec{C}_i & A_k & -\bar{X}_{j1} & * & * & * & * & * \\
\vec{A}_{ki} & U_1 A_k + \vec{B}_i C_k & -\bar{X}_{j2} & -\bar{X}_{j3} & * & * & * & * \\
E_{1k} G_1 + E_{2k}\vec{C}_i & E_{1k} & 0 & 0 & -\varepsilon_{kij}I & * & * & * \\
0 & 0 & \varepsilon_{kij}W_{1k}^T & 0 & \varepsilon_{kij}J^T & -\varepsilon_{kij}I & * & * \\
0 & 0 & 0 & \varepsilon_{kij}I & 0 & 0 & -I & * \\
0 & 0 & 0 & 0 & 0 & U_1 W_{1k} + \vec{B}_i W_{2k} & 0 & -I
\end{bmatrix} < 0$$

$$k \in \aleph(i), \; (i,j) \in \bar{\Omega}.$$

(8.45)

Moreover, the controller gain matrices can be constructed by (8.35).

Similarly, in Theorems 8.4 and 8.5, if the Lyapunov matrices are chosen to be the same as

$$\bar{X}_i \equiv \bar{X}, \; \forall i \in \bar{L},$$ (8.46)

one obtains the corresponding piecewise output feedback controller design results based on a common quadratic Lyapunov function, which are summarized in the following corollaries.

Corollary 8.3

The closed-loop fuzzy control system (8.31) is globally exponentially stable with H_∞ performance γ if the condition (8.33) with $\bar{X}_i \equiv \bar{X}$, $\forall i \in \bar{L}$ is satisfied.

Corollary 8.4

The closed-loop fuzzy control system (8.44) is globally exponentially stable if the condition (8.45) with $\bar{X}_i \equiv \bar{X}$, $\forall i \in \bar{L}$ is satisfied.

8.4 CONTROLLER SYNTHESIS BASED ON FUZZY QUADRATIC LYAPUNOV FUNCTIONS

In this section, we present controller design methods for uncertain fuzzy systems based on fuzzy quadratic Lyapunov functions. Following a similar structure to that in Section 8.3, a robust H_∞ performance analysis result based on fuzzy quadratic Lyapunov functions is first presented in Section 8.4.1, and then the fuzzy state feedback and output feedback controller designs are discussed in Sections 8.4.2 and 8.4.3, respectively.

8.4.1 ROBUST H_∞ PERFORMANCE ANALYSIS

Theorem 8.6

The unforced fuzzy system (8.3) with $u(t) \equiv 0$ is globally exponentially stable with H_∞ performance γ if there exist a set of positive definite matrices X_i, $i \in L$ and a set of positive scalars $\varepsilon_{il} > 0$, $i, l \in L$ such that the following linear matrix inequalities are satisfied:

$$
\begin{bmatrix}
-X_i & * & * & * & * & * \\
0 & -\gamma^2 I & * & * & * & * \\
A_i X_i & D_{1i} & -X_l & * & * & * \\
H_i X_i & D_{3i} & 0 & -I & * & * \\
E_{1i} X_i & E_{3i} & 0 & 0 & -\varepsilon_{il} I & * \\
0 & 0 & \varepsilon_{il} W_{1i}^T & \varepsilon_{il} W_{3i}^T & \varepsilon_{il} J^T & -\varepsilon_{il} I
\end{bmatrix} < 0, \; i, l \in L. \tag{8.47}
$$

Proof: Consider the following fuzzy quadratic Lyapunov function candidate,

$$
V(x(t)) = x^T(t) \left(\sum_{i=1}^m \mu_i X_i^{-1} \right) x(t), \tag{8.48}
$$

where X_i, $i \in L$ is a set of positive definite matrices to be determined.

It suffices to show the following inequality

$$
V(t+1) - V(t) + q^T(t) q(t) - \gamma^2 v^T(t) v(t) < 0 \tag{8.49}
$$

to prove that the unforced fuzzy system (8.3) is globally exponentially stable with H_∞ performance γ.

Based on the fuzzy quadratic Lyapunov function defined in (8.48), (8.49) can be rewritten as

$$
x^T(t+1)\left(\sum_{l=1}^{m}\mu_l^+ X_l^{-1}\right)x(t+1)-x^T(t)\left(\sum_{i=1}^{m}\mu_i X_i^{-1}\right)x(t)+q^T(t)q(t)-\gamma^2 v^T(t)v(t)<0,
$$

(8.50)

where $\mu^+ := \mu(z(t+1)) := [\mu_1(z(t+1)),\mu_2(z(t+1)),\ldots,\mu_m(z(t+1))]$.

Substituting the system state space equation (8.3) with $u(t)\equiv 0$ into (8.50) leads to

$$
\begin{bmatrix} x(t) \\ v(t) \end{bmatrix}^T \begin{bmatrix} A^T(\mu,t) \\ D_1^T(\mu,t) \end{bmatrix}\left(\sum_{l=1}^{m}\mu_l^+ X_l^{-1}\right)[A(\mu,t)\quad D_1(\mu,t)]\begin{bmatrix} x(t) \\ v(t) \end{bmatrix} - x^T(t)\left(\sum_{i=1}^{m}\mu_i X_i^{-1}\right)x(t)
$$

$$
+\begin{bmatrix} x(t) \\ v(t) \end{bmatrix}^T \begin{bmatrix} H^T(\mu,t) \\ D_3^T(\mu,t) \end{bmatrix}[H(\mu,t)\quad D_3(\mu,t)]\begin{bmatrix} x(t) \\ v(t) \end{bmatrix} - \gamma^2 v^T(t)v(t)<0.
$$

(8.51)

Thus, the following inequality implies (8.49):

$$
\begin{bmatrix} A^T(\mu,t) \\ D_1^T(\mu,t) \end{bmatrix}\left(\sum_{l=1}^{m}\mu_l^+ X_l^{-1}\right)[A(\mu,t)\quad D_1(\mu,t)]
$$

$$
+\begin{bmatrix} H^T(\mu,t) \\ D_3^T(\mu,t) \end{bmatrix}[H(\mu,t)\quad D_3(\mu,t)]+\begin{bmatrix} -\left(\sum_{i=1}^{m}\mu_i X_i^{-1}\right) & 0 \\ 0 & -\gamma^2 I \end{bmatrix}<0.
$$

(8.52)

By the Schur complement lemma, (8.52) is equivalent to

$$
\begin{bmatrix} -\left(\sum_{i=1}^{m}\mu_i X_i^{-1}\right) & 0 & * & * \\ & -\gamma^2 I & * & * \\ \left(\sum_{l=1}^{m}\mu_l^+ X_l^{-1}\right)A(\mu,t) & \left(\sum_{l=1}^{m}\mu_l^+ X_l^{-1}\right)D_1(\mu,t) & -\left(\sum_{l=1}^{m}\mu_l^+ X_l^{-1}\right) & * \\ H(\mu,t) & D_3(\mu,t) & 0 & -I \end{bmatrix}<0.
$$

(8.53)

On the other hand, (8.53) can be rewritten as

$$
\sum_{i=1}^{m}\sum_{l=1}^{m}\mu_i\mu_l^+
\begin{bmatrix}
-X_i^{-1} & 0 & * & * \\
0 & -\gamma^2 I & * & * \\
X_l^{-1}A_i(t) & X_l^{-1}D_{1i}(t) & -X_l^{-1} & * \\
H_i(t) & D_{3i}(t) & 0 & -I
\end{bmatrix} < 0.
\tag{8.54}
$$

Thus, it is easy to see that the following inequalities imply (8.54):

$$
\begin{bmatrix}
-X_i & 0 & * & * \\
0 & -\gamma^2 I & * & * \\
A_i(t)X_i & D_{1i}(t) & -X_l & * \\
H_i(t)X_i & D_{3i}(t) & 0 & -I
\end{bmatrix} < 0, \quad i,l \in L.
\tag{8.55}
$$

Based on the uncertainty descriptions given in (8.2), (8.55) can be further rewritten as

$$
\begin{bmatrix}
-X_i & 0 & * & * \\
0 & -\gamma^2 I & * & * \\
A_i X_i & D_{1i} & -X_l & * \\
H_i X_i & D_{3i} & 0 & -I
\end{bmatrix}
+ He\left\{
\begin{bmatrix}
0 \\ 0 \\ W_{1i} \\ W_{3i}
\end{bmatrix}
\Delta(t)[E_{1i}X_i \quad E_{3i} \quad 0 \quad 0]\right\} < 0, \quad i,l \in L.
\tag{8.56}
$$

Then, by the S-procedure given in the Appendix, it is clear that (8.47) guarantees (8.56). The proof is thus completed. ❏

8.4.2   State Feedback Controller Design

In this section, we address the H_∞ state feedback controller synthesis problem for the uncertain fuzzy system (8.3) based on a fuzzy quadratic Lyapunov function. Consider the following smooth fuzzy control scheme defined as

$$
R^l: \quad \text{IF} \qquad z_1 \text{ is } F_1^l \text{ and } \dots z_v \text{ is } F_v^l
$$

$$
\text{THEN } u(t) = K_l x(t), \quad l \in L,
\tag{8.57}
$$

which can be rewritten as

$$
u(t) = \sum_{l=1}^{m}\mu_l K_l x(t) = K(\mu)x(t).
\tag{8.58}
$$

The closed-loop fuzzy control system consisting of the T–S fuzzy system (8.3) and the smooth controller (8.58) can be described as

$$x(t+1) = (A(\mu,t) + B_1(\mu,t)K(\mu))x(t) + D_1(\mu,t)v(t) \tag{8.59}$$

$$q(t) = (H(\mu,t) + B_3(\mu,t)K(\mu))x(t) + D_3(\mu,t)v(t).$$

By using the same fuzzy quadratic Lyapunov function candidate as in (8.48), one can easily obtain the following fuzzy state feedback H_∞ controller design result for the uncertain fuzzy system described in (8.3).

Theorem 8.7

The closed-loop fuzzy control system (8.59) is globally exponentially stable with H_∞ performance γ if there exist a set of positive definite matrices X_i, $i \in L$, two sets of matrices G_i, $\bar{K}_i$, $i \in L$, and a set of positive scalars $\varepsilon_{ijl} > 0$, $i,j,l \in L$, such that the following linear matrix inequalities are satisfied:

$$\begin{bmatrix} X_i - G_j - G_j^T & 0 & * & * & * & * \\ 0 & -\gamma^2 I & * & * & * & * \\ A_i G_j + B_{1i}\bar{K}_j & D_{1i} & -X_l & * & * & * \\ H_i G_j + B_{3i}\bar{K}_j & D_{3i} & 0 & -I & * & * \\ E_{1i}G_j + E_{2i}\bar{K}_j & E_{3i} & 0 & 0 & -\varepsilon_{ijl}I & * \\ 0 & 0 & \varepsilon_{ijl}W_{1i}^T & \varepsilon_{ijl}W_{3i}^T & \varepsilon_{ijl}J^T & -\varepsilon_{ijl}I \end{bmatrix} < 0, \quad i,j,l \in L. \tag{8.60}$$

Moreover, the controller gains are given by

$$K_i = \bar{K}_i G_i^{-1}, i \in L. \tag{8.61}$$

Proof: Consider the same fuzzy quadratic Lyapunov function candidate as in (8.48). Based on the result in Theorem 8.6 and its proof, one learns that the closed-loop system (8.59) is globally exponentially stable with H_∞ performance γ if the following matrix inequality holds:

$$\begin{bmatrix} -\left(\sum_{i=1}^m \mu_i X_i^{-1}\right) & * & * & * \\ 0 & -\gamma^2 I & * & * \\ \left(\sum_{l=1}^m \mu_l^+ X_l^{-1}\right)(A(\mu,t)+B_1(\mu,t)K(\mu)) & \left(\sum_{l=1}^m \mu_l^+ X_l^{-1}\right)D_1(\mu,t) & -\left(\sum_{l=1}^m \mu_l^+ X_l^{-1}\right) & * \\ (H(\mu,t)+B_3(\mu,t)K(\mu)) & D_3(\mu,t) & 0 & -I \end{bmatrix} < 0, \tag{8.62}$$

which can be rewritten as

$$\sum_{i=1}^{m}\sum_{j=1}^{m}\sum_{l=1}^{m}\mu_i\mu_j\mu_l^+ \begin{bmatrix} -X_i^{-1} & * & * & * \\ 0 & -\gamma^2 I & * & * \\ X_i^{-1}(A_i(t)+B_{1i}(t)K_j) & X_i^{-1}D_{1i}(t) & -X_i^{-1} & * \\ H_i(t)+B_{3i}(t)K_j & D_{3i}(t) & 0 & -I \end{bmatrix} < 0. \qquad (8.63)$$

Thus, it is easy to see that the following inequalities imply (8.63):

$$\begin{bmatrix} -X_i^{-1} & * & * & * \\ 0 & -\gamma^2 I & * & * \\ A_i(t)+B_{1i}(t)K_j & D_{1i}(t) & -X_l & * \\ H_i(t)+B_{3i}(t)K_j & D_{3i}(t) & 0 & -I \end{bmatrix} < 0, \quad i,j,\, l \in L. \qquad (8.64)$$

Now, performing the congruence transformation to (8.64) by $diag\{G_j,I,I,I\}$ together with consideration of (8.61) and $-G_j^T X_i^{-1} G_j \le X_i - G_j - G_j^T$, one has that the following inequalities imply (8.64):

$$\begin{bmatrix} X_i - G_j - G_j^T & * & * & * \\ 0 & -\gamma^2 I & * & * \\ A_i(t)G_j+B_{1i}(t)\overline{K}_j & D_{1i}(t) & -X_l & * \\ H_i(t)G_j+B_{3i}(t)\overline{K}_j & D_{3i}(t) & 0 & -I \end{bmatrix} < 0, \quad i,j,\, l \in L. \qquad (8.65)$$

Based on the uncertainty descriptions given in (8.2), (8.65) can be further rewritten as

$$\begin{bmatrix} X_i - G_j - G_j^T & * & * & * \\ 0 & -\gamma^2 I & * & * \\ A_i(t)G_j+B_{1i}(t)\overline{K}_j & D_{1i}(t) & -X_l & * \\ H_i(t)G_j+B_{3i}(t)\overline{K}_j & D_{3i}(t) & 0 & -I \end{bmatrix} + He\left\{ \begin{bmatrix} 0 \\ 0 \\ W_{1i} \\ W_{3i} \end{bmatrix} \Delta(t)[E_{1i}G_j + E_{2i}\overline{K}_j \quad E_{3i} \quad 0 \quad 0] \right\} < 0$$

$$i,j,l \in L \qquad (8.66)$$

Then by the S-procedure given in the Appendix, it is clear that (8.60) guarantees (8.66). The proof is thus completed. $\square$

In the case of no external disturbances, that is, $v(t) \equiv 0$, in the fuzzy system (8.1), or equivalently (8.3), one can easily obtain the corresponding results on state feedback stabilization of the following fuzzy system,

$$x(t+1) = A(\mu, t)x(t) + B_1(\mu, t)u(t), \tag{8.67}$$

where $A(\mu, t)$ and $B_1(\mu, t)$ are defined in (8.4).

With the same smooth fuzzy controller defined in (8.57) or equivalently (8.58), the closed-loop fuzzy control system can be described by

$$x(t+1) = (A(\mu, t) + B_1(\mu, t)K(\mu))x(t). \tag{8.68}$$

By using the same fuzzy quadratic Lyapunov function candidate defined in (8.48) and following a similar procedure as in the proof of Theorem 8.7, one has the following fuzzy state feedback stabilization result for the fuzzy system described in (8.67).

Theorem 8.8

The closed-loop fuzzy control system (8.68) is globally exponentially stable if there exist a set of positive definite matrices X_i, $i \in L$, two sets of matrices G_i, $\bar{K}_i$, $i \in L$, and a set of positive scalars $\varepsilon_{ijl} > 0$, $i, j, l \in L$ such that the following linear matrix inequalities are satisfied:

$$\begin{bmatrix} X_i - G_j - G_j^T & * & * & * \\ A_i G_j + B_{1i}\bar{K}_j & -X_l & * & * \\ E_{1i}G_j + E_{2i}\bar{K}_j & 0 & -\varepsilon_{ijl}I & * \\ 0 & \varepsilon_{ijl}W_{1i}^T & \varepsilon_{ijl}J^T & -\varepsilon_{ijl}I \end{bmatrix}, \quad i, j, l \in L. \tag{8.69}$$

Moreover, the controller gains are given by (8.61).

Note that in Theorems 8.7 and 8.8, if the Lyapunov matrices are chosen to be the same as

$$X_i \equiv X, \forall i \in L, \tag{8.70}$$

one readily obtains the corresponding fuzzy state feedback H_∞ control and stabilization results, respectively, based on the following common quadratic Lyapunov function

$$V(x(t)) = x^T(t)X^{-1}x(t). \tag{8.71}$$

Corollary 8.5

The closed-loop fuzzy control system (8.59) is globally exponentially stable with H_∞ performance γ if the condition (8.60) with $X_i \equiv X$, $\forall i \in L$ is satisfied.

Corollary 8.6

The closed-loop fuzzy control system (8.68) is globally exponentially stable if the condition (8.69) with $X_i \equiv X$, $\forall i \in L$ is satisfied.

8.4.3 OUTPUT FEEDBACK CONTROLLER DESIGN

In this section, we extend the results in the Sections 8.4.1 and 8.4.2 to design an output feedback controller for the uncertain fuzzy system (8.3).

Consider the following fuzzy dynamic output feedback controller,

$$\hat{x}(t+1) = \hat{A}(\mu)\hat{x}(t) + \hat{B}(\mu)y(t) \tag{8.72}$$

$$u(t) = \hat{C}(\mu)\hat{x}(t),$$

where $\hat{x}(t) \in \Re^n$ is the controller state, and $\hat{A}(\mu)$, $\hat{B}(\mu)$, $\hat{C}(\mu)$ are the fuzzy membership function dependent matrices to be determined.

Combining (8.72) with the system (8.3) leads to the following closed-loop fuzzy control system,

$$\bar{x}(t+1) = \bar{A}(\mu,t)\bar{x}(t) + \bar{B}(\mu,t)v(t) \tag{8.73}$$

$$q(t) = \bar{H}(\mu,t)\bar{x}(t) + D_3(\mu,t)v(t),$$

where

$$\bar{x}(t) := \begin{bmatrix} x^T(t) & \hat{x}^T(t) \end{bmatrix}^T, \bar{A}(\mu,t) := \begin{bmatrix} A(\mu,t) & B_1(\mu,t)\hat{C}(\mu) \\ \hat{B}(\mu)C(\mu,t) & \hat{A}(\mu) + \hat{B}(\mu)B_2(\mu,t)\hat{C}(\mu) \end{bmatrix},$$

$$\bar{B}(\mu,t) := \begin{bmatrix} D_1(\mu,t) \\ \hat{B}(\mu)D_2(\mu,t) \end{bmatrix}, \bar{H}(\mu,t) := \begin{bmatrix} H(\mu,t) & B_3(\mu,t)\hat{C}(\mu) \end{bmatrix}. \tag{8.74}$$

Then, the main result for the output feedback H_∞ control of the uncertain fuzzy system (8.3) is summarized in the following theorem.

Theorem 8.9

The closed-loop fuzzy control system (8.73) is globally exponentially stable with H_∞ performance γ if there exist a set of positive definite matrices

$$\bar{X}_i := \begin{bmatrix} \bar{X}_{i1} & * \\ \bar{X}_{i2} & \bar{X}_{i3} \end{bmatrix}, i \in L,$$

matrices G_1, U_1, M, sets of matrices $\bar{A}_i$, $\bar{B}_i$, $\bar{C}_i$, $i \in L$, and a set of positive scalars $\varepsilon_{ijrl} > 0$, $i,j,r,l \in L$, such that the following linear matrix inequalities are satisfied:

$$
\begin{bmatrix}
\bar{X}_{i1} - G_1 - G_1^T & * & * & * & * & * & * & * & * & * \\
\bar{X}_{i2} - M - I & \bar{X}_{i3} - U_1 - U_1^T & * & * & * & * & * & * & * & * \\
0 & 0 & -\gamma^2 I & * & * & * & * & * & * & * \\
A_i G_1 + B_{1i}\vec{C}_j & A_i & D_{1i} & -\bar{X}_{i1} & * & * & * & * & * & * \\
\vec{A}_i & U_1 A_i + \bar{B}_r C_i & U_1 D_{1i} + \bar{B}_r D_{2i} & -\bar{X}_{i2} & -\bar{X}_{i3} & * & * & * & * & * \\
H_i G_1 + B_{3i}\vec{C}_j & H_i & D_{3i} & 0 & 0 & -I & * & * & * & * \\
E_{1i} G_1 + E_{2i}\vec{C}_j & E_{1i} & E_{3i} & 0 & 0 & 0 & -\varepsilon_{ijrl} I & * & * & * \\
0 & 0 & 0 & \varepsilon_{ijrl} W_{1i}^T & 0 & \varepsilon_{ijrl} W_{3i}^T & \varepsilon_{ijrl} J^T & -\varepsilon_{ijrl} I & * & * \\
0 & 0 & 0 & 0 & \varepsilon_{ijrl} I & 0 & 0 & 0 & -I & * \\
0 & 0 & 0 & 0 & 0 & 0 & 0 & U_1 W_{1i} + \bar{B}_r W_{2i} & 0 & -I
\end{bmatrix} < 0
$$

$$
i, j, r, l \in L. \tag{8.75}
$$

Moreover, the controller gain matrices can be constructed as

$$
U_2 G_2 = S.V.D(M - U_1 G_1),
$$

$$
\hat{A}(\mu) = U_2^{-1}\left(\vec{A}(\mu) - U_1 A(\mu) G_1 - \vec{B}(\mu) C(\mu) G_1 - U_1 B_1(\mu)\vec{C}(\mu) - \vec{B}(\mu) B_2(\mu)\vec{C}(\mu) \right) G_2^{-1},
$$

$$
\hat{B}(\mu) = U_2^{-1}\vec{B}(\mu), \tag{8.76}
$$

$$
\hat{C}(\mu) = \vec{C}(\mu) G_2^{-1},
$$

where

$$
\begin{bmatrix} \vec{A}(\mu) & \vec{B}(\mu) & \vec{C}(\mu) & A(\mu) & C(\mu) & B_1(\mu) & B_2(\mu) \end{bmatrix}
$$

$$
= \sum_{i=1}^{m} \mu_i \begin{bmatrix} \vec{A}_i & \vec{B}_i & \vec{C}_i & A_i & C_i & B_{1i} & B_{2i} \end{bmatrix}.
$$

Proof: Consider the following fuzzy quadratic Lyapunov function candidate,

$$
V(x(t)) = \bar{x}^T(t)\left(\sum_{i=1}^{m} \mu_i X_i^{-1} \right) \bar{x}(t), \tag{8.77}
$$

where X_i, $i \in L$ is a set of positive definite matrices to be determined.

Based on the results in Theorem 8.7 and its proof, one learns that the closed-loop system (8.73) is globally exponentially stable with H_∞ performance γ if there exist a set of positive definite matrices $X_i, i \in L$ and a matrix G such that the following matrix inequalities are satisfied:

$$
\begin{bmatrix}
X_i - G - G^T & 0 & * & * \\
0 & -\gamma^2 I & * & * \\
\overline{A}(\mu,t)G & \overline{B}(\mu,t) & -X_l & * \\
\overline{H}(\mu,t)G & D_3(\mu,t) & 0 & -I
\end{bmatrix} < 0, \ i,l \in L. \tag{8.78}
$$

Similar to the proof in Theorem 8.4, let the matrix G and its inverse be partitioned as

$$
G = \begin{bmatrix} G_1 & \bullet \\ G_2 & \bullet \end{bmatrix}, G^{-1} = \begin{bmatrix} U_1^T & \bullet \\ U_2^T & \bullet \end{bmatrix}, \tag{8.79}
$$

where "$\bullet$" denotes the elements that are uniquely determined from the equality $G^{-1}G = GG^{-1} = I$.

For matrix inequality linearization purposes, we further define the following matrices,

$$
J := \begin{bmatrix} I & U_1^T \\ 0 & U_2^T \end{bmatrix},
$$

$$
\Lambda := \mathrm{diag}\{J, I, J, I\},
$$

$$
\overline{X}_i := J^T X_i J := \begin{bmatrix} \overline{X}_{i1} & \overline{X}_{i2}^T \\ \overline{X}_{i2} & \overline{X}_{i3} \end{bmatrix},
$$

$$
M := U_1 G_1 + U_2 G_2,
$$

$$
A(\mu) = \sum_{i=1}^m \mu_i A_i, \ C(\mu) = \sum_{i=1}^m \mu_i C_i, \ B_1(\mu) = \sum_{i=1}^m \mu_i B_{1i}, \ B_2(\mu) = \sum_{i=1}^m \mu_i B_{2i},
$$

$$
\vec{A}(\mu) := U_1 A(\mu)G_1 + \overline{B}(\mu)C(\mu)G_1 + U_1 B_1(\mu)\vec{C}(\mu) + U_2 \hat{A}(\mu)G_2 + \overline{B}(\mu)B_2(\mu)\vec{C}(\mu),
$$

$$
\vec{B}(\mu) = U_2 \hat{B}(\mu) = \sum_{i=1}^m \mu_i \vec{B}_i, \ \vec{C}(\mu) = \hat{C}(\mu)G_2 = \sum_{i=1}^m \mu_i \vec{C}_i,
$$

$$
\vec{A}(\mu) = \sum_{i=1}^m \mu_i \vec{A}_i, \ \vec{B}(\mu) = \sum_{i=1}^m \mu_i \vec{B}_i, \ \vec{C}(\mu) = \sum_{i=1}^m \mu_i \vec{C}_i. \tag{8.80}
$$

Then, substituting the matrices defined in (8.80) in (8.78) and performing the congruence transformation to (8.78) by Λ together with consideration of the matrices defined in (8.80), one deduces that the following inequalities imply (8.78):

$$
\begin{bmatrix}
\bar{X}_{i1} - G_1 - G_1^T & * & * & * & * & * \\
\bar{X}_{i2} - M - I & \bar{X}_{i3} - U_1 - U_1^T & * & * & * & * \\
0 & 0 & -\gamma^2 I & * & * & * \\
A_i G_1 + B_{1i} \vec{C}_j & A_i & D_{1i} & -\bar{X}_{i1} & * & * \\
\vec{A}_i & U_1 A_i + \bar{B}_r C_i & U_1 D_{1i} + \bar{B}_r D_{2i} & -\bar{X}_{i2} & -\bar{X}_{i3} & * \\
H_i G_1 + B_{3i} \vec{C}_j & H_i & D_{3i} & 0 & 0 & -I
\end{bmatrix}
$$

$$
+ He\left\{ \begin{bmatrix} 0 \\ 0 \\ 0 \\ W_{1i} \\ U_1 W_{1i} + \bar{B}_r W_{2i} \\ W_{3i} \end{bmatrix} \Delta(t)[E_{1i} G_1 + E_{2i} \vec{C}_j \quad E_{1i} \quad E_{3i} \quad 0 \quad 0 \quad 0] \right\} < 0,
$$

$$
i,j,r,l \subset L. \tag{8.81}
$$

Then the S-procedure given in the Appendix, together with consideration of the following bounding inequality, leads to (8.75) exactly,

$$
R_1 R_2^T + R_2 R_1^T \le [R_1 \quad R_2] \begin{bmatrix} R_1^T \\ R_2^T \end{bmatrix}, \tag{8.82}
$$

where

$$
R_1 := [0 \quad 0 \quad 0 \quad 0 \quad \varepsilon_{ijrl} I \quad 0 \quad 0 \quad 0]^T,
$$

$$
R_2 := [0 \quad 0 \quad 0 \quad 0 \quad 0 \quad 0 \quad 0 \quad U_1 W_{1i} + \bar{B}_r W_{2i}]^T. \tag{8.83}
$$

The proof is thus completed. ❑

In the case of no external disturbances, that is, $v(t) \equiv 0$, in the fuzzy system (8.1), or equivalently (8.3), one can easily obtain the corresponding results on output feedback stabilization of the following fuzzy system,

$$
x(t+1) = A(\mu,t)x(t) + B_1(\mu,t)u(t) \tag{8.84}
$$

$$
y(t) = C(\mu,t)x(t) + B_2(\mu,t)u(t),
$$

where $A(\mu,t)$, $B_1(\mu,t)$, $C(\mu,t)$ and $B_2(\mu,t)$ are defined in (8.4).

With the same smooth output feedback control scheme defined in (8.72), the closed-loop fuzzy control system can be described by

$$\bar{x}(t+1) = \bar{A}(\mu,t)x(t), \tag{8.85}$$

where $\bar{x}(t)$ and $\bar{A}(\mu,t)$ are defined in (8.74). By using the same fuzzy quadratic Lyapunov function candidate defined in (8.77) and following a similar procedure as in the proof of Theorem 8.9, one has the following output feedback stabilization result for the fuzzy system described in (8.84).

Theorem 8.10

The closed-loop fuzzy control system (8.85) is globally exponentially stable if there exist a set of positive definite matrices

$$\bar{X}_i := \begin{bmatrix} \bar{X}_{i1} & * \\ \bar{X}_{i2} & \bar{X}_{i3} \end{bmatrix}, i \in L,$$

matrices G_1, U_1, M, sets of matrices $\bar{A}_i$, $\bar{B}_i$, $\bar{C}_i$, $i \in L$, and a set of positive scalars $\varepsilon_{ijrl} > 0$, $i,j,r,l \in L$, such that the following linear matrix inequalities are satisfied:

$$\begin{bmatrix} \bar{X}_{i1} - G_1 - G_1^T & * & * & * & * & * & * & * \\ \bar{X}_{i2} - M - I & \bar{X}_{i3} - U_1 - U_1^T & * & * & * & * & * & * \\ A_iG_1 + B_{1i}\bar{C}_j & A_i & -\bar{X}_{i1} & * & * & * & * & * \\ \bar{A}_i & U_1A_i + \bar{B}_rC_i & -\bar{X}_{i2} & -\bar{X}_{i3} & * & * & * & * \\ E_{1i}G_1 + E_{2i}\bar{C}_j & E_{1i} & 0 & 0 & -\varepsilon_{ijrl}I & * & * & * \\ 0 & 0 & \varepsilon_{ijrl}W_{1i}^T & 0 & \varepsilon_{ijrl}J^T & -\varepsilon_{ijrl}I & * & * \\ 0 & 0 & 0 & \varepsilon_{ijrl}I & 0 & 0 & -I & * \\ 0 & 0 & 0 & 0 & 0 & U_1W_{1i} + \bar{B}_rW_{2i} & * & -I \end{bmatrix} < 0,$$

$$i,j,r,l \in L. \tag{8.86}$$

Moreover, the controller gain matrices are given by (8.76).

Similarly, in Theorems 8.9 and 8.10, if the Lyapunov matrices are chosen to be the same as follows,

$$\bar{X}_i \equiv \bar{X}, \forall i \in L, \tag{8.87}$$

one obtains the corresponding output feedback controller design results based on a common quadratic Lyapunov function, summarized in the following corollaries.

Corollary 8.7

The closed-loop fuzzy control system (8.73) is globally exponentially stable with H_∞ performance γ if the condition (8.75) with $\bar{X}_i \equiv \bar{X}$, $\forall i \in L$ is satisfied.

Corollary 8.8

The closed-loop fuzzy control system (8.85) is globally exponentially stable if the condition (8.86) with $\bar{X}_i \equiv \bar{X}$, $\forall i \in L$ is satisfied.

8.5 AN EXAMPLE

An example is presented in this section to illustrate the design procedure and performance of the proposed control design approaches.

Example 8.1

Consider the following discrete-time chaotic Lorenz system with sampling period $T_s = 0.002s$ in the T–S fuzzy model form as described in Zhou et al. (2005),

R^l : IF $x_1(t)$ is about M_l,

THEN

$$x(t+1) = (A_l + W_{1l}\Delta(t)E_{1l})x(t) + (B_{1l} + W_{1l}\Delta(t)E_{2l})u(t) + (D_{1l} + W_{1l}\Delta(t)E_{3l})v(t)$$

$$y(t) = (C_l + W_{2l}\Delta(t)E_{1l})x(t) + (B_{2l} + W_{2l}\Delta(t)E_{2l})u(t) + (D_{2l} + W_{2l}\Delta(t)E_{3l})v(t) \quad (8.88)$$

$$q(t) = (H_l + W_{3l}\Delta(t)E_{1l})x(t) + (B_{3l} + W_{3l}\Delta(t)E_{2l})u(t) + (D_{3l} + W_{3l}\Delta(t)E_{3l})v(t)$$

$l = \{1, 2\}$,

with matrices and parameters given as follows:

$$A_1 = \begin{bmatrix} 1-\sigma T_s & \sigma T_s & 0 \\ \eta T_s & 1-T_s & -M_1 T_s \\ 0 & M_1 T_s & 1-bT_s \end{bmatrix}, \quad A_2 = \begin{bmatrix} 1-\sigma T_s & \sigma T_s & 0 \\ \eta T_s & 1-T_s & -M_2 T_s \\ 0 & M_2 T_s & 1-bT_s \end{bmatrix},$$

$$B_{11} = B_{12} = \begin{bmatrix} 1 \\ 0 \\ 0 \end{bmatrix}, D_{11} = \begin{bmatrix} 0.01 \\ 0.02 \\ 0.008 \end{bmatrix}, D_{12} = \begin{bmatrix} 0.01 \\ 0.0015 \\ 0.007 \end{bmatrix},$$

$$C_1 = \begin{bmatrix} 1 & -1 & 3 \\ -1 & 5 & 1 \end{bmatrix}, C_2 = \begin{bmatrix} 1 & -1 & 3.5 \\ -1 & 4.5 & 1 \end{bmatrix}, B_{21} = B_{22} = \begin{bmatrix} 0 \\ 0 \end{bmatrix},$$

$$D_{21} = \begin{bmatrix} 0.05 \\ 0.01 \end{bmatrix}, D_{22} = \begin{bmatrix} 0.04 \\ 0.01 \end{bmatrix},$$

$$H_1 = [-0.06 \quad -0.06 \quad 0.18], H_2 = [-0.045 \quad -0.025 \quad -0.015],$$

$$B_{31} = -1, B_{32} = 0.5, D_{31} = 0.005, D_{32} = -0.002,$$

TABLE 8.1

Disturbance Attenuation Performance via State Feedback

Algorithm	8.1	8.2	8.3
Minimum γ	0.3125	0.3107	0.3099

$$W_{11} = W_{12} = \begin{bmatrix} 0.03 & 0 & 0 \\ 0 & 0.03 & 0 \\ 0 & 0 & 0.03 \end{bmatrix}, W_{21} = W_{22} = \begin{bmatrix} 0 & 0 & 0 \\ 0 & 0 & 0 \end{bmatrix}, W_{31} = W_{32} = [0 \quad 0 \quad 0],$$

$$E_{11} = E_{12} = 10^{-3} \begin{bmatrix} -\sigma & \sigma & 0 \\ \eta & 0 & 0 \\ 0 & 0 & -b \end{bmatrix}, E_{21} = E_{22} = \begin{bmatrix} 0 \\ 0 \\ 0 \end{bmatrix}, E_{31} = E_{32} = \begin{bmatrix} 0 \\ 0 \\ 0 \end{bmatrix}, J = \begin{bmatrix} 0 & 0 & 0 \\ 0 & 0 & 0 \\ 0 & 0 & 0 \end{bmatrix},$$

$$M_1 = -20, M_2 = 30, (\sigma, \eta, b) = (10, 28, 8/3).$$

We consider the controller design methods based on common, piecewise, and fuzzy quadratic Lyapunov functions, respectively. In order to compare the performance of those design approaches, the algorithms minimizing the performance index γ based on those methods are developed and summarized as follows.

Algorithm 8.1: $\min\limits_{X_l, \bar{K}_l, l \in \bar{L}, G_{\beta l}, \varepsilon_{klj}} \gamma$, subject to LMIs (8.20) with $X_l = X, l \in \bar{L}$

Algorithm 8.2: $\min\limits_{X_l, \bar{K}_l, l \in \bar{L}, G_{\beta l}, \varepsilon_{klj}} \gamma$, subject to LMIs (8.20)

Algorithm 8.3: $\min\limits_{X_l, \bar{K}_l, l \in \bar{L}, G_{\beta l}, \varepsilon_{klj}} \gamma$, subject to LMIs (8.60)

Algorithm 8.4: $\min\limits_{\bar{X}_l, \bar{A}_l, \bar{B}_l, \bar{C}_l, l \in \bar{L}, G_1, U_1, M, \varepsilon_{klj}} \gamma$, subject to LMIs (8.33) with $\bar{X}_l = \bar{X}, l \in \bar{L}$

Algorithm 8.5: $\min\limits_{\bar{X}_l, \bar{A}_l, \bar{B}_l, \bar{C}_l, l \in \bar{L}, G_1, U_1, M, \varepsilon_{klj}} \gamma$, subject to LMIs (8.33)

Algorithm 8.6: $\min\limits_{\bar{X}_l, \bar{A}_l, \bar{B}_l, \bar{C}_l, l \in \bar{L}, G_1, U_1, M, \varepsilon_{iklj}} \gamma$, subject to LMIs (8.75)

Algorithms 8.1–8.3 correspond to state feedback control schemes based on common, piecewise, and fuzzy quadratic Lyapunov functions, respectively, whereas Algorithms 8.4–8.6 correspond to output feedback control schemes based on common, piecewise, and fuzzy quadratic Lyapunov functions, respectively.

By applying all these algorithms to the system, the optimal disturbance attenuation performances are summarized in Table 8.1 for state feedback control and Table 8.2 for output feedback control, respectively.

TABLE 8.2

Disturbance Attenuation Performance via Output Feedback

Algorithm	8.4	8.5	8.6
Minimum γ	0.5522	0.5486	0.3140

It can be observed that both state feedback and output feedback can achieve robust stabilization of chaotic systems with optimized disturbance attenuation performance, and that the approaches based on piecewise or fuzzy quadratic Lyapunov functions achieve better disturbance attenuation performance than the approaches based on common quadratic Lyapunov functions.

8.6 CONCLUSIONS

This chapter has presented a number of approaches to controller synthesis for both stabilization and H_∞ control of uncertain T–S fuzzy systems based on common, piecewise, and fuzzy quadratic Lyapunov functions, respectively. Both state feedback and output feedback control are considered. It has been shown that the controller gains can be obtained by solving linear matrix inequalities. An example is given to illustrate the effectiveness of the proposed approaches and it is observed that the approaches based on piecewise or fuzzy quadratic Lyapunov functions achieve better disturbance attenuation performance than the approaches based on common quadratic Lyapunov functions.

9 Controller Synthesis of T–S Fuzzy Systems with Time-Delay

9.1 INTRODUCTION

Time-delays are frequently encountered in various complex nonlinear systems, such as chemical processes, mechanical systems, and communication networks. It has been well recognized that the presence of time-delays may result in poor performance, chaotic mode, or even instability of control systems. Control of dynamic systems with time-delay is a subject of great practical and theoretical significance. In this chapter, control of discrete-time T–S fuzzy systems with time-delay is studied by using the so-called Lyapunov–Krasovskii functional approaches. Both delay-independent and delay-dependent controller design approaches are developed.

The rest of the chapter is organized as follows. Section 9.2 is devoted to introduction of T–S fuzzy models with time-delay. Sections 9.3 and 9.4 present controller synthesis of delayed T–S fuzzy systems based on piecewise and fuzzy quadratic Lyapunov functionals, respectively. It is also noted that the common quadratic Lyapunov functionals are special cases of the more general piecewise and fuzzy quadratic Lyapunov functionals, and thus the corresponding results can be readily obtained. Simulation examples are given to illustrate the effectiveness of the proposed design approaches in Section 9.5, followed by some conclusions in Section 9.6.

9.2 MODEL OF T–S FUZZY SYSTEMS WITH TIME-DELAY

A discrete-time T–S fuzzy dynamic model with time-delay can be described as

$$
\begin{aligned}
R^l: \quad &\text{IF} \quad z_1 \text{ is } F_1^l \text{ and } \dots \text{ and } z_v \text{ is } F_1^l \\
&\text{THEN} \quad x(t+1) = A_l x(t) + A_{dl} x(t - \tau(t)) + B_{1l} u(t) + D_{1l} v(t) \\
&\qquad\qquad q(t) = C_l x(t) + C_{dl} x(t - \tau(t)) + B_{2l} u(t) + D_{2l} v(t) \\
&\qquad\qquad x(t) = \varphi(t), -\tau_2 \le t \le 0 \\
&l \in L := \{1, 2, \cdots, m\},
\end{aligned}
$$

(9.1)

where most variables are the same as those defined in the previous chapters except for time-delay variables and matrices; that is, R^l denotes the lth fuzzy inference rule, m the number of inference rules, F_j^l ($j = 1, 2, \ldots, v$) the fuzzy sets, $x(t) \in \mathfrak{R}^n$ the state vector, $u(t) \in \mathfrak{R}^g$ the control input vector, $q(t) \in \mathfrak{R}^r$ the controlled output vector, and $v(t) \in \mathfrak{R}^s$ the disturbance input vector which is assumed to belong to $l_2[0,\infty)$, $z(t) = [z_1(t), z_2(t), \ldots, z_v(t)]$ some measurable variables of the system. $\tau(t)$ is a positive integer function representing the time-varying state delay of the system (9.1) and satisfying the following assumption,

$$\tau_1 \le \tau(t) \le \tau_2, \tag{9.2}$$

with τ_1 and τ_2 being two constant positive integers representing the minimum and maximum time-delay, respectively. In this case, $\tau(t)$ is called an interval-like or range-like time-varying delay. $\varphi(t)$ is a real-valued initial condition sequence on $[-\tau_2, 0]$. A_l, A_{dl}, B_{1l}, D_{1l}, C_l, C_{dl}, B_{2l} and D_{2l}, $l \in L$ are known constant matrices with appropriate dimensions.

Similar to the standard fuzzy inference method described in Chapter 3, that is, using a center-average defuzzifier, product fuzzy inference, and singleton fuzzifier, the following global T–S fuzzy dynamic model with time-delay can be obtained:

$$x(t+1) = A(\mu)x(t) + A_d(\mu)x(t - \tau(t)) + B_1(\mu)u(t) + D_1(\mu)v(t)$$

$$q(t) = C(\mu)x(t) + C_d(\mu)x(t - \tau(t)) + B_2(\mu)u(t) + D_2(\mu)v(t) \tag{9.3}$$

$$x(t) = \varphi(t), -\tau_2 \le t \le 0,$$

where

$$A(\mu) = \sum_{l=1}^m \mu_l A_l, \quad A_d(\mu) = \sum_{l=1}^m \mu_l A_{dl}, \quad B_1(\mu) = \sum_{l=1}^m \mu_l B_{1l}, \quad D_1(\mu) = \sum_{l=1}^m \mu_l D_{1l},$$

$$C(\mu) = \sum_{l=1}^m \mu_l C_l, \quad C_d(\mu) = \sum_{l=1}^m \mu_l C_{dl}, \quad B_2(\mu) = \sum_{l=1}^m \mu_l B_{2l}, \quad D_2(\mu) = \sum_{l=1}^m \mu_l D_{2l}, \tag{9.4}$$

$$\sum_{l=1}^m \mu_l[z(t)] = 1.$$

The objective of this chapter, similar to that of Chapter 6, is to design a suitable controller for system (9.1) or equivalently (9.3) such that the resulting closed-loop control system is globally exponentially stable with a guaranteed performance in the H_∞ sense; that is, given a prescribed level of noise attenuation $\gamma > 0$, find a suitable controller such that the induced l_2-norm of the operator from v to the controlled output q is less than γ under zero initial conditions. In other words, $\|q(t)\|_2 < \gamma \|v(t)\|_2$, for all nonzero $v \in l_2[0,\infty]$.

9.3   CONTROLLER SYNTHESIS BASED ON PIECEWISE QUADRATIC LYAPUNOV FUNCTIONALS

In this section, we consider the robust H_∞ control problem of the T–S fuzzy model (9.1) based on piecewise quadratic Lyapunov functionals. Based on two different Lyapunov–Krasovskii functionals and some matrix inequality convexifying procedures, both delay-independent and delay-dependent approaches are developed.

With a space partition of the second kind as described in Chapter 4, the global time-delay fuzzy system (9.3) can be expressed in each region as

$$x(t+1) = A_i x(t) + A_{di} x(t - \tau(t)) + B_{1i} u(t) + D_{1i} v(t)$$

$$q(t) = C_i x(t) + C_{di} x(t - \tau(t)) + B_{2i} u(t) + D_{2i} v(t) \tag{9.5}$$

$$z(t) \in S_i, i \in \bar{L},$$

where

$$A_i = \sum_{k \in \aleph(i)} \mu_k A_k, \quad A_{di} = \sum_{k \in \aleph(i)} \mu_k A_{dk}, \quad B_{1i} = \sum_{k \in \aleph(i)} \mu_k B_{1k}, \quad D_{1i} = \sum_{k \in \aleph(i)} \mu_k D_{1k},$$

$$C_i = \sum_{k \in \aleph(i)} \mu_k C_k, \quad C_{di} = \sum_{k \in \aleph(i)} \mu_k C_{dk}, \quad B_{2i} = \sum_{k \in \aleph(i)} \mu_k B_{2k}, \quad D_{2i} = \sum_{k \in \aleph(i)} \mu_k D_{2k}, \tag{9.6}$$

with $0 \le \mu_k(z(t)) \le 1$, $\Sigma_{k \in \aleph(i)} \mu_k(z(t)) = 1$. As discussed in Chapter 4, for each region S_i, the set $\aleph(i)$ contains the indices for those subsystem matrices used in the interpolation within that region. For a crisp region, $\aleph(i)$ contains a single element.

The piecewise controller design of the fuzzy system (9.1) or equivalently (9.5) is considered. Thus it is assumed that when the state of the system transits from region S_i to S_j at time t, the dynamics of the system is governed by the dynamics of the blended model in region S_i at that time. For future use, we also define a set $\bar{\Omega}$ as in Chapter 4 to represent all possible region transitions,

$$\bar{\Omega} := \{(i,j) \mid z(t) \in S_i, z(t+1) \in S_j, \forall i, j \in \bar{L}\}, \tag{9.7}$$

where $j = i$ when $z(t)$ stays in the same region S_i, and $j \neq i$ when $z(t)$ transits from the region S_i to S_j.

Now, consider the following piecewise controller,

$$u(t) = K_i x(t), z(t) \in S_i, i \in \bar{L}; \tag{9.8}$$

then the closed-loop system can be described as

$$x(t+1) = \bar{A}_i x(t) + A_{di} x(t - \tau(t)) + D_{1i} v(t)$$

$$q(t) = \bar{C}_i x(t) + C_{di} x(t - \tau(t)) + D_{2i} v(t) \tag{9.9}$$

$$z(t) \in S_i, i \in \bar{L},$$

where

$$\bar{A}_i := \sum_{k \in \aleph(i)} \mu_k (A_k + B_{1k} K_i), \quad \bar{C}_i := \sum_{k \in \aleph(i)} \mu_k (C_k + B_{2k} K_i). \tag{9.10}$$

9.3.1 Delay-Independent H_∞ Controller Design

In this section, based on a piecewise Lyapunov–Krasovskii functional, the delay-independent H_∞ control problem is studied for the fuzzy system (9.5) with the piecewise control law (9.8). The result is summarized in the following theorem.

Theorem 9.1

The closed-loop system (9.9) is globally exponentially stable with H_∞ performance γ if there exist two sets of matrices $U_i = U_i^T > 0$, $\bar{K}_i$, $i \in \bar{L}$, matrices $Q_1 = Q_1^T \geq 0$, and G such that the following linear matrix inequalities are satisfied:

$$\begin{bmatrix} -U_j & 0 & A_k G + B_{1k} \bar{K}_i & A_{dk} G & D_{1k} \\ * & -I & C_k G + B_{2k} \bar{K}_i & C_{dk} G & D_{2k} \\ * & * & U_i - G - G^T + (\tau_2 - \tau_1 + 1) Q_1 & 0 & 0 \\ * & * & * & -Q_1 & 0 \\ * & * & * & * & -\gamma^2 I \end{bmatrix} \quad k \in \aleph(i), (i,j) \in \bar{\Omega}. \tag{9.11}$$

Moreover, the controller gains are given by

$$K_i = \bar{K}_i G^{-1}, \ i \in \bar{L}. \tag{9.12}$$

Proof: It is well known that it suffices to find a Lyapunov function $V(t, x(t)) > 0$, $\forall x(t) \neq 0$ satisfying the following inequality,

$$V(t+1, x(t+1)) - V(t, x(t)) + \| q(t) \|_2^2 - \gamma^2 \| v(t) \|_2^2 < 0, \tag{9.13}$$

to prove that the closed-loop system (9.9) is globally exponentially stable with H_∞ performance γ under zero initial conditions for any nonzero $v \in l_2[0, \infty)$.

Consider the following piecewise Lyapunov–Krasovskii functional candidate

$$V(t) := V_1(t) + V_2(t)$$

$$V_1(t) := x^T(t) P_i x(t), \ z(t) \in S_i, i \in \bar{L} \tag{9.14}$$

$$V_2(t) := \sum_{s=-\tau_2}^{-\tau_1} \sum_{v=t+s}^{t-1} x^T(v) Q_1 x(v),$$

where $P_i = P_i^T > 0$, $i \in \bar{L}$, $Q_1 = Q_1^T \geq 0$ are Lyapunov matrices to be determined.

Define $\Delta V(t) := V(t+1) - V(t)$, where along the trajectory of system (9.9), one has

$$\Delta V_1(t) = x^T(t+1) P_j x(t+1) - x^T(t) P_i x(t), \ (i,j) \in \bar{\Omega} \tag{9.15}$$

$$\Delta V_2(t) \le (\tau_2 - \tau_1 + 1)x^T(t)Q_1 x(t) - x^T(t - \tau(t))Q_1 x(t - \tau(t)). \tag{9.16}$$

Thus, based on (9.14)–(9.16), it is easy to see that the following inequality implies (9.13)

$$\begin{bmatrix} x(t+1) \\ q(t) \end{bmatrix}^T \begin{bmatrix} P_j & 0 \\ 0 & I \end{bmatrix} \begin{bmatrix} x(t+1) \\ q(t) \end{bmatrix}$$

$$- \begin{bmatrix} x(t) \\ x(t - \tau(t)) \\ v(t) \end{bmatrix}^T \begin{bmatrix} -P_i + (\tau_2 - \tau_1 + 1)Q_1 & 0 & 0 \\ 0 & -Q_1 & 0 \\ 0 & 0 & -\gamma^2 I \end{bmatrix} \begin{bmatrix} x(t) \\ x(t - \tau(t)) \\ v(t) \end{bmatrix} < 0 \tag{9.17}$$

and thus it suffices to show (9.17) instead of (9.13).

On the other hand, it follows from (9.9) that

$$\begin{bmatrix} x(t+1) \\ q(t) \end{bmatrix} = \begin{bmatrix} \bar{A}_i & A_{di} & D_{1i} \\ \bar{C}_i & C_{di} & D_{2i} \end{bmatrix} \begin{bmatrix} x(t) \\ x(t - \tau(t)) \\ v(t) \end{bmatrix}. \tag{9.18}$$

Then, substituting (9.18) into (9.17), together with consideration of (9.10) and the Schur complement lemma, yields

$$\begin{bmatrix} -P_j & 0 & P_j(A_k + B_{1k}K_i) & P_j A_{dk} & P_j D_{1k} \\ * & -I & C_k + B_{2k}K_i & C_{dk} & D_{2k} \\ * & * & -P_i + (\tau_2 - \tau_1 + 1)Q_1 & 0 & 0 \\ * & * & * & -Q_1 & 0 \\ * & * & * & * & -\gamma^2 I \end{bmatrix} < 0, \, k \in \aleph(i), \, (i,j) \in \bar{\Omega}. \tag{9.19}$$

The Lyapunov matrix P_j in (9.19) is associated with the system matrices, and thus (9.19) is not convex. To make it more convenient for controller synthesis purposes, we decouple the Lyapunov matrix by introducing a slack variable $G \in \Re^{n \times n}$. To this end, define the following matrices:

$$U_j := P_j^{-1}, \, \bar{K}_i := K_i G, \, \bar{Q}_1 := G^T Q_1 G. \tag{9.20}$$

Pre- and postmultiplying (9.19) by $\text{diag}\{P_j^{-1}, I, G, G, I\}$ and its transpose, respectively, lead to

$$\begin{bmatrix} -U_j & 0 & A_k G + B_{1k}\bar{K}_i & A_{dk}G & D_{1k} \\ * & -I & C_k G + B_{2k}\bar{K}_i & C_{dk}G & D_{2k} \\ * & * & -G^T U_j^{-1} G + (\tau_2 - \tau_1 + 1)\bar{Q}_1 & 0 & 0 \\ * & * & * & -\bar{Q}_1 & 0 \\ * & * & * & * & -\gamma^2 I \end{bmatrix} < 0, \, k \in \aleph(i), \, (i,j) \in \bar{\Omega}.$$

$$\tag{9.21}$$

Note that

$$U_i - G - G^T + G^T U_i^{-1} G = (U_i - G)^T U_i^{-1}(U_i - G) \geq 0 \tag{9.22}$$

implies

$$-G^T U_i^{-1} G \leq U_i - G - G^T. \tag{9.23}$$

It then follows from (9.23) that (9.11) implies (9.21). The proof is thus completed. ❑

9.3.2 DELAY-DEPENDENT H_∞ CONTROLLER DESIGN

It is well known that the delay-independent results are usually more conservative than the delay-dependent ones especially when the size of the time-delay is small. In this section, delay-dependent H_∞ control of the fuzzy system (9.5) with the control law (9.8) is considered by constructing a new piecewise Lyapunov–Krasovskii functional.

Theorem 9.2

The closed-loop system (9.9) is globally exponentially stable with H_∞ performance γ if there exist matrices

$$P_i = P_i^T > 0, \, U_i = U_i^T > 0, \, M_i = \begin{bmatrix} M_{1i} \\ M_{2i} \end{bmatrix}, N_i = \begin{bmatrix} N_{1i} \\ N_{2i} \end{bmatrix}, \, R_i = \begin{bmatrix} R_{1i} \\ R_{2i} \end{bmatrix},$$

$$X_i = X_i^T = \begin{bmatrix} X_{11i} & X_{12i} \\ X_{12i}^T & X_{22i} \end{bmatrix}, \, Y_i = Y_i^T = \begin{bmatrix} Y_{11i} & Y_{12i} \\ Y_{12i}^T & Y_{22i} \end{bmatrix}, \, K_i, \, i \in \bar{L}, \, Q_\alpha = Q_\alpha^T > 0, \, \alpha \in \{1, 2, 3\}$$

$$Z_\beta = Z_\beta^T > 0,$$

and $S_\beta = S_\beta^T > 0, \, \beta \in \{1, 2\}$ such that

$$\begin{bmatrix} -U_j & 0 & 0 & 0 & A_k + B_{1k}K_i & A_{dk} & D_{1k} & 0 & 0 \\ * & -S_1 & 0 & 0 & \sqrt{\tau_2}(A_k + B_{1k}K_i - I) & \sqrt{\tau_2}A_{dk} & \sqrt{\tau_2}D_{1k} & 0 & 0 \\ * & * & -S_2 & 0 & \sqrt{\tau_2 - \tau_1}(A_k + B_{1k}K_i - I) & \sqrt{\tau_2 - \tau_1}A_{dk} & \sqrt{\tau_2 - \tau_1}D_{1k} & 0 & 0 \\ * & * & * & -I & C_k + B_{2k}K_i & C_{dk} & D_{2k} & 0 & 0 \\ * & * & * & * & \Phi_{11i} & \Phi_{12i} & 0 & -N_{1i} & R_{1i} \\ * & * & * & * & * & \Phi_{22i} & 0 & -N_{2i} & R_{2i} \\ * & * & * & * & * & * & -\gamma^2 I & 0 & 0 \\ * & * & * & * & * & * & * & -Q_2 & 0 \\ * & * & * & * & * & * & * & * & -Q_3 \end{bmatrix} < 0,$$

$$k \in \aleph(i), \, (i, j) \in \bar{\Omega} \tag{9.24}$$

$$P_i U_i = I, \; i \in \bar{L}, \; Z_\beta S_\beta = I, \; \beta \in \{1, 2\} \tag{9.25}$$

$$\begin{bmatrix} X_i & M_i \\ * & Z_1 \end{bmatrix} \geq 0, \; i \in \bar{L} \tag{9.26}$$

$$\begin{bmatrix} X_i + Y_i & N_i \\ * & Z_1 + Z_2 \end{bmatrix} \geq 0, \; i \in \bar{L} \tag{9.27}$$

$$\begin{bmatrix} Y_i & R_i \\ * & Z_2 \end{bmatrix} \geq 0, \; i \in \bar{L}, \tag{9.28}$$

where

$$\Phi_{11i} := -P_i + (\tau_2 - \tau_1 + 1)Q_1 + Q_2 + Q_3 + M_{1i} + M_{1i}^T + \tau_2 X_{11i} + (\tau_2 - \tau_1)Y_{11i},$$

$$\Phi_{12i} := N_{1i} - M_{1i} - R_{1i} + M_{2i}^T + \tau_2 X_{12i} + (\tau_2 - \tau_1)Y_{12i},$$

$$\Phi_{22i} := N_{2i} + N_{2i}^T - M_{2i} - M_{2i}^T - R_{2i} - R_{2i}^T + \tau_2 X_{22i} + (\tau_2 - \tau_1)Y_{22i}. \tag{9.29}$$

Proof: Define $e(t) := x(t+1) - x(t)$ and consider the following piecewise Lyapunov–Krasovskii functional candidate,

$$V(t) := V_1(t) + V_2(t) + V_3(t) + V_4(t)$$

$$V_1(t) := x^T(t)P_i x(t), \; z(t) \in S_i, \; i \in \bar{L}$$

$$V_2(t) := \sum_{s=-\tau_2}^{-\tau_1} \sum_{v=t+s}^{t-1} x^T(v)Q_1 x(v) \tag{9.30}$$

$$V_3(t) := \sum_{s=t-\tau_2}^{t-1} x^T(s)Q_2 x(s) + \sum_{s=t-\tau_1}^{t-1} x^T(s)Q_3 x(s)$$

$$V_4(t) := \sum_{s=-\tau_2}^{-1} \sum_{m=t+s}^{t-1} e^T(m)Z_1 e(m) + \sum_{s=-\tau_2}^{-\tau_1-1} \sum_{m=t+s}^{t-1} e^T(m)Z_2 e(m),$$

where $P_i = P_i^T > 0$, $i \in \bar{L}$, $Q_\alpha = Q_\alpha^T \geq 0$, $\alpha \in \{1, 2, 3\}$ and $Z_\beta = Z_\beta^T > 0$, $\beta \in \{1, 2\}$ are Lyapunov matrices to be determined.

With the Lyapunov functional candidate defined in (9.30) and similar to the proof of Theorem 9.1, it suffices to show the inequality (9.13) to prove that the closed-loop system (9.9) is globally exponentially stable with H_∞ performance γ under zero initial conditions for any nonzero $v \in l_2[0, \infty)$.

Define $\Delta V(t) := V(t+1) - V(t)$, and along the trajectory of the system (9.9), one has

$$\Delta V_1(t) = x^T(t+1)P_j x(t+1) - x^T(t)P_i x(t), \; (i, j) \in \bar{\Omega} \tag{9.31}$$

$$\Delta V_2(t) \le (\tau_2 - \tau_1 + 1)x^T(t)Q_1 x(t) - x^T(t - \tau(t))Q_1 x(t - \tau(t)), \tag{9.32}$$

$$\Delta V_3(t) = x^T(t)(Q_2 + Q_3)x(t) - x^T(t - \tau_2)Q_2 x(t - \tau_2) - x^T(t - \tau_1)Q_3 x(t - \tau_1), \tag{9.33}$$

$$\Delta V_4(t) = \sum_{s=-\tau_2}^{-1} \left[e^T(t)Z_1 e(t) - e^T(t+s)Z_1 e(t+s) \right]$$

$$+ \sum_{s=-\tau_2}^{-\tau_1-1} \left[e^T(t)Z_2 e(t) - e^T(t+s)Z_2 e(t+s) \right]$$

$$= e^T(t)[\tau_2 Z_1 + (\tau_2 - \tau_1)Z_2]e(t) - \sum_{m=k-\tau_2}^{k-1} e^T(m)Z_1 e(m) - \sum_{m=k-\tau_2}^{k-\tau_1-1} e^T(m)Z_1 e(m)$$

$$= e^T(t)[\tau_2 Z_1 + (\tau_2 - \tau_1)Z_2]e(t) - \sum_{m=t-\tau_2}^{k-\tau(t)-1} e^T(m)(Z_1 + Z_2)e(m)$$

$$- \sum_{m=t-\tau(t)}^{t-1} e^T(m)Z_1 e(m) - \sum_{m=t-\tau(t)}^{t-\tau_1-1} e^T(m)Z_2 e(m). \tag{9.34}$$

In addition, define $\xi_1(t) := [x^T(t) \, x^T(t - \tau(t))]^T$. Then from the definition of $e(t)$, for any appropriately dimensioned matrices,

$$M_i = \begin{bmatrix} M_{1i} \\ M_{2i} \end{bmatrix}, \ N_i = \begin{bmatrix} N_{1i} \\ N_{2i} \end{bmatrix}, \ R_i = \begin{bmatrix} R_{1i} \\ R_{2i} \end{bmatrix}, \ X_i = X_i^T = \begin{bmatrix} X_{11i} & X_{12i} \\ X_{12i}^T & X_{22i} \end{bmatrix},$$

$$Y_i = Y_i^T = \begin{bmatrix} Y_{11i} & Y_{12i} \\ Y_{12i}^T & Y_{22i} \end{bmatrix}, \ i \in \bar{L}, \tag{9.35}$$

one has

$$0 \equiv 2\xi_1^T(t)M_i \left[x(t) - x(t - \tau(t)) - \sum_{m=t-\tau(t)}^{t-1} e(m) \right], \tag{9.36}$$

$$0 \equiv 2\xi_1^T(t)N_i \left[x\big(t - \tau(t)\big) - x(t - \tau_2) - \sum_{m=t-\tau_2}^{t-\tau(t)-1} e(m) \right], \tag{9.37}$$

$$0 \equiv 2\xi_1^T(t)R_i \left[x(t - \tau_1) - x(t - \tau(t)) - \sum_{m=t-\tau(t)}^{t-\tau_1-1} e(m) \right], \tag{9.38}$$

$$0 \equiv \sum_{m=t-\tau_2}^{t-1} \xi_1^T(t) X_i \xi_1(t) + \sum_{m=t-\tau_2}^{t-\tau_1-1} \xi_1^T(t) Y_i \xi_1(t) - \sum_{m=t-\tau_2}^{t-1} \xi_1^T(t) X_i \xi_1(t) - \sum_{m=t-\tau_2}^{t-\tau_1-1} \xi_1^T(t) Y_i \xi_1(t)$$

$$\equiv \tau_2 \xi_1^T(t) X_i \xi_1(t) + (\tau_2 - \tau_1) \xi_1^T(t) Y_i \xi_1(t) - \sum_{m=t-\tau(t)}^{t-1} \xi_1^T(t) X_i \xi_1(t) \tag{9.39}$$

$$- \sum_{m=t-\tau_2}^{t-\tau(t)-1} \xi_1^T(t)(X_i + Y_i)\xi_1(t) - \sum_{m=t-\tau(t)}^{t-\tau_1-1} \xi_1^T(t) Y_i \xi_1(t).$$

On the other hand, under the conditions (9.26)–(9.28), the following inequalities are also true,

$$0 \leq \sum_{m=t-\tau(t)}^{t-1} \left[\xi_1^T(t) X_i \xi_1(t) + 2\xi_1^T(t) M_i e(m) + e^T(m) Z_1 e(m) \right]$$

$$= \sum_{m=t-\tau(t)}^{t-1} \begin{bmatrix} \xi_1^T(t) \\ e(m) \end{bmatrix}^T \begin{bmatrix} X_i & M_i \\ * & Z_1 \end{bmatrix} \begin{bmatrix} \xi_1(t) \\ e(m) \end{bmatrix}, \tag{9.40}$$

$$0 \leq \sum_{m=t-\tau_2}^{t-\tau(t)-1} \left[\xi_1^T(t)(X_i + Y_i)\xi_1(t) + 2\xi_1^T(t) N_i e(m) + e^T(m)(Z_1 + Z_2) e(m) \right]$$

$$= \sum_{m=t-\tau_2}^{t-\tau(t)-1} \begin{bmatrix} \xi_1(t) \\ e(m) \end{bmatrix}^T \begin{bmatrix} X_i + Y_i & N_i \\ * & Z_1 + Z_2 \end{bmatrix} \begin{bmatrix} \xi_1(t) \\ e(m) \end{bmatrix}, \tag{9.41}$$

$$0 \leq \sum_{m=t-\tau(t)}^{t-\tau_1-1} \left[\xi_1^T(t) Y_i \xi_1(t) + 2\xi_1^T(t) R_i e(m) + e^T(m) Z_2 e(m) \right]$$

$$= \sum_{m=t-\tau(t)}^{t-\tau_1-1} \begin{bmatrix} \xi_1(t) \\ e(m) \end{bmatrix}^T \begin{bmatrix} Y_i & R_i \\ * & Z_2 \end{bmatrix} \begin{bmatrix} \xi_1(k) \\ e(m) \end{bmatrix}. \tag{9.42}$$

Thus, based on the piecewise Lyapunov–Krasovskii functional candidate defined in (9.30), together with consideration of (9.31)–(9.42), it is easy to see that the following inequality implies (9.13),

$$x^T(t+1) P_j x(t+1) + \tau_2 e^T(t) Z_1 e(t) + (\tau_2 - \tau_1) e^T(t) Z_2 e(t) + q^T(t) q(t) + \xi_2^T(t) \phi_i \xi_2(t) < 0, \tag{9.43}$$

where

$$\xi_2(t) := \begin{bmatrix} x^T(t) & x^T(t-\tau(t)) & v^T(t) & x^T(t-\tau_2) & x^T(t-\tau_1) \end{bmatrix}^T,$$

$$
\Phi_i := \begin{bmatrix}
\Phi_{11i} & \Phi_{12i} & 0 & -N_{1i} & R_{1i} \\
* & \Phi_{22i} & 0 & -N_{2i} & R_{2i} \\
* & * & -\gamma^2 I & 0 & 0 \\
* & * & * & -Q_2 & 0 \\
* & * & * & * & -Q_3
\end{bmatrix},
$$

and Φ_{11i}, Φ_{12i}, and Φ_{22i} are defined in (9.29).

Moreover, it follows from (9.9) that

$$
\begin{bmatrix}
x(t+1) \\
\sqrt{\tau_2}\, e(t) \\
\sqrt{\tau_2 - \tau_1}\, e(t) \\
q(t)
\end{bmatrix}
=
\begin{bmatrix}
\bar{A}_i & A_{di} & D_{1i} & 0 & 0 \\
\sqrt{\tau_2}(\bar{A}_i - I) & \sqrt{\tau_2}\, A_{di} & \sqrt{\tau_2}\, D_{1i} & 0 & 0 \\
\sqrt{\tau_2 - \tau_1}(\bar{A}_i - I) & \sqrt{\tau_2 - \tau_1}\, A_{di} & \sqrt{\tau_2 - \tau_1}\, D_{1i} & 0 & 0 \\
\bar{C}_i & C_{di} & D_{2i} & 0 & 0
\end{bmatrix}
\begin{bmatrix}
x(t) \\
x(t - \tau(t)) \\
v(t) \\
x(t - \tau_2) \\
x(t - \tau_1)
\end{bmatrix}.
$$

(9.44)

Then, substituting (9.44) into (9.43), together with consideration of (9.10) and the Schur complement lemma, it is easy to see that (9.24) and (9.25) imply (9.43). The proof is thus completed. □

Remark 9.1

Note that from the proof of Theorem 9.2, the delay-dependent criterion is realized by using a free weighting matrix technique, which enables one to avoid performing any model transformation to the original system and thus no bounding technique is employed to estimate the inner product of the involved crossing terms. Moreover, when treating the time-varying delay and estimating the upper bound of the difference of Lyapunov functional, some useful terms such as $\sum_{m=t-\tau_2}^{t-\tau(t)-1} e^T(m)(\bullet)e(m)$ are fully utilized by introducing some additional terms into the proposed Lyapunov–Krasovskii functional. Thus, compared with the exiting delay-independent and delay-dependent approaches for discrete-time T–S fuzzy systems with delay, these features have the potential to enable one to obtain less conservative results.

Also note that the conditions given in Theorem 9.2 are not convex due to the matrix equality constraints in (9.25). Using the cone complementarity linearization technique, the original nonconvex feasibility problem formulated in Theorem 9.2 can be converted to the following nonlinear minimization problem involving LMI conditions.

$$
\text{Minimize } Tr\left(\sum_{i \in \bar{L}} P_i U_i + \sum_{\beta=1}^{2} Z_\beta S_\beta \right)
$$

subject to (9.24), (9.26)–(9.28) and

$$
P_i > 0, \, U_i > 0, \, \begin{bmatrix} P_i & I \\ I & U_i \end{bmatrix} \geq 0, \, i \in \bar{L} \qquad (9.45)
$$

$$Z_\beta > 0, \, S_\beta > 0, \begin{bmatrix} Z_\beta & I \\ I & S_\beta \end{bmatrix} \geq 0, \, \beta \in \{1, 2\}. \tag{9.46}$$

Then, for given delay bounds τ_1 and τ_2, the suboptimal value of γ can be found by the following algorithm.

Algorithm 9.1

Step 1. Choose a sufficiently large initial $\gamma > 0$ such that there exists a feasible solution to (9.24), (9.26)–(9.28), (9.45), and (9.46). Set $\gamma_0 = \gamma$.

Step 2. Find a feasible set

$$\{P_{i0}, U_{i0}, M_{i0}, N_{i0}, R_{i0}, X_{i0}, Y_{i0}, K_{i0}, i \in \bar{L}, Q_{\alpha 0}, \alpha \in \{1, 2, 3\}, Z_{\beta 0}, S_{\beta 0}, \beta \in \{1, 2\}\}$$

satisfying (9.24), (9.26)–(9.28), (9.45), and (9.46). Set $\sigma = 0$.

Step 3. Solve the following LMI feasibility problem for the variables P_i, U_i, M_i, N_i, R_i, X_i, Y_i, K_i, $i \in \bar{L}$, Q_1, $\alpha \in \{1, 2, 3\}$, Z_β, S_β, $\beta \in \{1, 2\}$,

$$\text{Minimize } Tr\left(\sum_{i \in \bar{L}} (P_{i\sigma} U_i + P_i U_{i\sigma}) + \sum_{\beta=1}^{2} (Z_{\beta\sigma} S_\beta + Z_\beta S_{\beta\sigma}) \right)$$

subject to (9.24), (9.26)–(9.28), (9.45), and (9.46).

Set $P_{i(\sigma+1)} = P_i$, $U_{i(\sigma+1)} = U_i$, $i \in \bar{L}$, $S_{\beta(\sigma+1)} = S_\beta$, $\beta \in \{1, 2\}$.

Step 4. Substitute the controller gains K_i obtained in Step 3 into (9.24) and if the conditions (9.24) and (9.26)–(9.28) are feasible for the variables P_i, M_i, N_i, R_i, X_i, Y_i, $i \in \bar{L}$, Q_α, $\alpha \in \{1, 2, 3\}$, Z_β, $\beta \in \{1, 2\}$, then set $\gamma_0 = \gamma$ and return to Step 2 after decreasing γ to some extent. If the conditions (9.24) and (9.26)–(9.28) are infeasible within the maximum number of iterations allowed, then exit. Otherwise, set $\sigma = \sigma + 1$ and go to step 3.

9.4 CONTROLLER SYNTHESIS BASED ON FUZZY QUADRATIC LYAPUNOV FUNCTIONALS

In this section, we consider the robust H_∞ control problem of the T–S fuzzy model (9.1) based on fuzzy quadratic Lyapunov functionals. Similar to Section 9.3, based on two different Lyapunov–Krasovskii functionals and some matrix inequality convexifying procedures, both delay-independent and delay-dependent approaches are developed in terms of a set of linear matrix inequalities.

Consider the following smooth fuzzy control law for the fuzzy system (9.1),

$$R^l: \quad \text{IF} \quad z_1 \text{ is } F_1^l \text{ and } \dots \text{ and } z_v \text{ is } F_1^l,$$

$$\text{THEN} \quad u(t) = K_l x(t), \tag{9.47}$$

$$l \in L,$$

which can be rewritten as

$$u(t) = \sum_{l=1}^{m} \mu_l K_l x(t) = K(\mu)x(t). \tag{9.48}$$

Under the control law (9.48), the closed-loop system can be obtained as

$$x(t+1) = \bar{A}(\mu)x(t) + A_d(\mu)x(t - \tau(t)) + D_1(\mu)v(t)$$

$$q(t) = \bar{C}(\mu)x(t) + C_d(\mu)x(t - \tau(t)) + D_2(\mu)v(t) \tag{9.49}$$

$$x(t) = \varphi(t), \ -\tau_2 \leq t \leq 0,$$

where

$$\bar{A}(\mu) := A(\mu) + B_1(\mu)K(\mu), \ \bar{C}(\mu) := C(\mu) + B_2(\mu)K(\mu). \tag{9.50}$$

9.4.1 DELAY-INDEPENDENT H_∞ CONTROLLER DESIGN

In this section, based on a fuzzy quadratic Lyapunov functional, the delay-independent H_∞ control problem is studied for the fuzzy system (9.1) with the fuzzy control law (9.48).

Theorem 9.3

The closed-loop system (9.49) is globally exponentially stable with H_∞ performance γ if there exist two sets of matrices $U_i = U_i^T > 0$, $\bar{K}_i$, $i \in L$, matrices $Q_1 = Q_1^T \geq 0$, and G such that the following linear matrix inequalities are satisfied,

$$\bar{\Sigma}_{iil} < 0, \quad i, l \in L \tag{9.51}$$

$$\bar{\Sigma}_{ijl} + \bar{\Sigma}_{jil} < 0, \ 1 \leq i < j \leq m, l \in L, \tag{9.52}$$

where

$$\bar{\Sigma}_{ijl} := \begin{bmatrix} -U_l & 0 & A_i G + B_{1i}\bar{K}_j & A_{di}G & D_{1i} \\ * & -I & C_i G + B_{2i}\bar{K}_j & C_{di}G & D_{2i} \\ * & * & U_i - G - G^T + (\tau_2 - \tau_1 + 1)Q_1 & 0 & 0 \\ * & * & * & -Q_1 & 0 \\ * & * & * & * & -\gamma^2 I \end{bmatrix}. \tag{9.53}$$

Moreover, the controller gains are given by

$$K_i = \bar{K}_i G^{-1}, \quad i \in L. \tag{9.54}$$

Proof: Consider the following fuzzy Lyapunov–Krasovskii functional candidate

$$V(t) := V_1(t) + V_2(t)$$

$$V_1(t) := x^T(t)P(\mu)x(t) = x^T(t)\left(\sum_{i=1}^{m}\mu_i P_i\right)x(t) \tag{9.55}$$

$$V_2(t) := \sum_{s=-\tau_2}^{-\tau_1} \sum_{v=t+s}^{t-1} x^T(v)Q_1 x(v),$$

where $P_i = P_i^T > 0$, $i \in L$, $Q_1 = Q_1^T \geq 0$ are Lyapunov matrices to be determined.
Define $\Delta V(t) := V(t+1) - V(t)$, and along the trajectory of system (9.49), one has

$$\Delta V_1(t) = x^T(t+1)P(\mu^+)x(t+1) - x^T(t)P(\mu)x(t) \tag{9.56}$$

$$\Delta V_2(t) \leq (\tau_2 - \tau_1 + 1)x^T(t)Q_1 x(t) - x^T(t-\tau(t))Q_1 x(t-\tau(t)), \tag{9.57}$$

where $\mu^+ := \mu(z(t+1)) = [\mu_1(z(t+1)) \quad \mu_2(z(t+1)) \quad \cdots \quad \mu_m(z(t+1))]$.
Then, based on (9.55)–(9.57), it is easy to see that the following inequality implies (9.13),

$$\begin{bmatrix} x(t+1) \\ q(t) \end{bmatrix}^T \begin{bmatrix} P(\mu^+) & 0 \\ 0 & I \end{bmatrix} \begin{bmatrix} x(t+1) \\ q(t) \end{bmatrix}$$

$$- \begin{bmatrix} x(t) \\ x(t-\tau(t)) \\ v(t) \end{bmatrix}^T \begin{bmatrix} -P(\mu)+(\tau_2-\tau_1+1)Q_1 & 0 & 0 \\ 0 & -Q_1 & 0 \\ 0 & 0 & -\gamma^2 I \end{bmatrix} \begin{bmatrix} x(t) \\ x(t-\tau(t)) \\ v(t) \end{bmatrix} < 0, \tag{9.58}$$

and thus it suffices to show (9.58) instead of (9.13).
On the other hand, (9.49) can be rewritten as

$$\begin{bmatrix} x(t+1) \\ q(t) \end{bmatrix} = \begin{bmatrix} \bar{A}(\mu) & A_d(\mu) & D_1(\mu) \\ \bar{C}(\mu) & C_d(\mu) & D_2(\mu) \end{bmatrix} \begin{bmatrix} x(t) \\ x(t-\tau(t)) \\ v(t) \end{bmatrix}. \tag{9.59}$$

Then, it follows from (9.58) and (9.59) and applying the Schur complement lemma that the following inequality implies (9.58),

$$\begin{bmatrix} -P(\mu^+) & 0 & P(\mu^+)\bar{A}(\mu) & P(\mu^+)A_d(\mu) & P(\mu^+)D_1(\mu) \\ * & -I & \bar{C}(\mu) & C_d(\mu) & D_2(\mu) \\ * & * & -P(\mu)+(\tau_2-\tau_1+1)Q_1 & 0 & 0 \\ * & * & * & -Q_1 & 0 \\ * & * & * & * & -\gamma^2 I \end{bmatrix} < 0, \tag{9.60}$$

which can be rewritten as

$$\sum_{i=1}^{m}\sum_{j=1}^{m}\sum_{l=1}^{m}\mu_i\mu_j\mu_l^+\Sigma_{ijl} = \sum_{i=1}^{m}\sum_{l=1}^{m}\mu_i^2\mu_l^+\Sigma_{iil} + \sum_{i=1}^{m-1}\sum_{j=i+1}^{m}\sum_{l=1}^{m}\mu_i\mu_j\mu_l^+(\Sigma_{ijl}+\Sigma_{jil})<0, \quad (9.61)$$

where

$$\Sigma_{ijl} := \begin{bmatrix} -P_l & 0 & P_l(A_i+B_{1i}K_j) & P_lA_{di} & P_lD_{1i} \\ * & -I & C_i+B_{2i}K_j & C_{di} & D_{2i} \\ * & * & -P_i+(\tau_2-\tau_1+1)Q_1 & 0 & 0 \\ * & * & * & -Q_i & 0 \\ * & * & * & * & -\gamma^2 I \end{bmatrix}. \quad (9.62)$$

Thus, it is easy to see that the following inequalities imply (9.61)

$$\Sigma_{iil} < 0, \ i, l \in L \quad (9.63)$$

$$\Sigma_{ijl}+\Sigma_{jil} < 0, \ 1 \le i < j \le m, \ l \in L. \quad (9.64)$$

The Lyapunov matrices P_l, $l \in L$ in (9.62) are associated with the system matrices. By defining $U_i := P_i^{-1}$, introducing a slack variable $G \in R^{n \times n}$, and using a similar decoupling procedure as in the proof of Theorem 9.1, it is easy to show that (9.51) and (9.52) imply (9.63) and (9.64), respectively. The proof is thus completed. □

9.4.2   DELAY-DEPENDENT H_∞ CONTROLLER DESIGN

In this section, delay-dependent H_∞ control of the system (9.3) with fuzzy control law (9.48) is considered by constructing a new fuzzy Lyapunov–Krasovskii functional. The result is summarized in the following theorem.

Theorem 9.4
The closed-loop system (9.49) is globally exponentially stable with H_∞ performance γ if there exist matrices $P_i = P_i^T > 0$, $U_i = U_i^T > 0$,

$$M_i = \begin{bmatrix} M_{1i} \\ M_{2i} \end{bmatrix}, \ N_i = \begin{bmatrix} N_{1i} \\ N_{2i} \end{bmatrix}, \ R_i = \begin{bmatrix} R_{1i} \\ R_{2i} \end{bmatrix}, \ X_i = X_i^T = \begin{bmatrix} X_{11i} & X_{12i} \\ X_{12i}^T & X_{22i} \end{bmatrix},$$

$$Y_i = Y_i^T = \begin{bmatrix} Y_{11i} & Y_{12i} \\ Y_{12i}^T & Y_{22i} \end{bmatrix},$$

K_i, $i \in L$, $Q_\alpha = Q_\alpha^T > 0$, $\alpha \in \{1, 2, 3\}$, $Z_\beta = Z_\beta^T > 0$, and $S_\beta = S_\beta^T > 0$, $\beta \in \{1, 2\}$ satisfying

$$\Xi_{iil} < 0, \ i, l \in L \quad (9.65)$$

$$\Xi_{ijl}+\Xi_{jil} < 0, \ 1 \le i < j \le m, \ l \in L \quad (9.66)$$

$$P_iU_i = I, \ i \in L, \ Z_\beta S_\beta = I, \ \beta \in \{1,2\} \quad (9.67)$$

$$\begin{bmatrix} X_i & M_i \\ * & Z_1 \end{bmatrix} \geq 0, i \in L \tag{9.68}$$

$$\begin{bmatrix} X_i + Y_i & N_i \\ * & Z_1 + Z_2 \end{bmatrix} \geq 0, i \in L \tag{9.69}$$

$$\begin{bmatrix} Y_i & R_i \\ * & Z_2 \end{bmatrix} \geq 0, i \in L, \tag{9.70}$$

where

$$\Xi_{ijl} := \begin{bmatrix} -U_l & 0 & 0 & 0 & A_i + B_{1i}K_j & A_{di} & D_{1i} & 0 & 0 \\ * & -S_1 & 0 & 0 & \sqrt{\tau_2}(A_i + B_{1i}K_j - I) & \sqrt{\tau_2}A_{di} & \sqrt{\tau_2}D_{1i} & 0 & 0 \\ * & * & -S_2 & 0 & \sqrt{\tau_2 - \tau_1}(A_i + B_{1i}K_j - I) & \sqrt{\tau_2 - \tau_1}A_{di} & \sqrt{\tau_2 - \tau_1}D_{1i} & 0 & 0 \\ * & * & * & -I & C_i + B_{2i}K_j & C_{di} & D_{2i} & 0 & 0 \\ * & * & * & * & \Theta_{11i} & \Theta_{12i} & 0 & -N_{1i} & R_{1i} \\ * & * & * & * & * & \Theta_{22i} & 0 & -N_{2i} & R_{2i} \\ * & * & * & * & * & * & -\gamma^2 I & 0 & 0 \\ * & * & * & * & * & * & * & -Q_2 & 0 \\ * & * & * & * & * & * & * & * & -Q_3 \end{bmatrix}.$$

$$\Theta_{11i} := -P_i + (\tau_2 - \tau_1 + 1)Q_1 + Q_2 + Q_3 + M_{1i} + M_{1i}^T + \tau_2 X_{11i} + (\tau_2 - \tau_1)Y_{11i},$$

$$\Theta_{12i} := N_{1i} - M_{1i} - R_{1i} + M_{2i}^T + \tau_2 X_{12i} + (\tau_2 - \tau_1)Y_{12i},$$

$$\Theta_{22i} := N_{2i} + N_{2i}^T - M_{2i} - M_{2i}^T - R_{2i} - R_{2i}^T + \tau_2 X_{22i} + (\tau_2 - \tau_1)Y_{22i}. \tag{9.71}$$

Proof: Consider the following fuzzy Lyapunov–Krasovskii functional candidate,

$$V(t) := V_1(t) + V_2(t) + V_3(t) + V_4(t), \tag{9.72}$$

where $V_1(t)$ is defined in (9.55), and $V_2(t)$, $V_3(t)$, and $V_4(t)$ are defined in (9.30).
Define the following matrices

$$M(\mu) = \begin{bmatrix} M_1(\mu) \\ M_2(\mu) \end{bmatrix} = \sum_{i=1}^{m} \mu_i M_i = \sum_{i=1}^{m} \mu_i \begin{bmatrix} M_{1i} \\ M_{2i} \end{bmatrix},$$

$$N(\mu) = \begin{bmatrix} N_1(\mu) \\ N_2(\mu) \end{bmatrix} = \sum_{i=1}^{m} \mu_i N_i = \sum_{i=1}^{m} \mu_i \begin{bmatrix} N_{1i} \\ N_{2i} \end{bmatrix},$$

$$R(\mu) = \begin{bmatrix} R_1(\mu) \\ R_2(\mu) \end{bmatrix} = \sum_{i=1}^{m} \mu_i R_i = \sum_{i=1}^{m} \mu_i \begin{bmatrix} R_{1i} \\ R_{2i} \end{bmatrix},$$

$$X(\mu) = \begin{bmatrix} X_{11}(\mu) & X_{12}^T(\mu) \\ X_{12}(\mu) & X_{22}(\mu) \end{bmatrix} = \sum_{i=1}^m \mu_i X_i = \sum_{i=1}^m \mu_i \begin{bmatrix} X_{11i} & X_{12i}^T \\ X_{12i} & X_{22i} \end{bmatrix},$$

$$Y(\mu) = \begin{bmatrix} Y_{11}(\mu) & Y_{12}^T(\mu) \\ Y_{12}(\mu) & Y_{22}(\mu) \end{bmatrix} = \sum_{i=1}^m \mu_i Y_i = \sum_{i=1}^m \mu_i \begin{bmatrix} Y_{11i} & Y_{12i}^T \\ Y_{12i} & Y_{22i} \end{bmatrix}.$$

$$(9.73)$$

Then following a similar procedure as in the proofs of Theorems 9.2 and 9.3, it is easy to see that the following inequality implies (9.13),

$$x^T(t+1)P(\mu^+)x(t+1) + \tau_2 e^T(t)Z_1 e(t) + (\tau_2 - \tau_1)e^T(t)Z_2 e(t)$$

$$+ q^T(t)q(t) + \xi_2^T(t)\Theta(\mu)\xi_2(t) < 0,$$

$$(9.74)$$

where

$$\xi_2(t) := [x^T(t) \quad x^T(t - \tau(t)) \quad v^T(t) \quad x^T(t-\tau_2) \quad x^T(t-\tau_1)]^T,$$

$$\Theta(\mu) := \begin{bmatrix} \Theta_{11}(\mu) & \Theta_{12}(\mu) & 0 & -N_1(\mu) & R_1(\mu) \\ * & \Theta_{22}(\mu) & 0 & -N_2(\mu) & R_2(\mu) \\ * & * & -\gamma^2 I & 0 & 0 \\ * & * & * & -Q_2 & 0 \\ * & * & * & * & -Q_3 \end{bmatrix},$$

$$\Theta_{11}(\mu) := -P(\mu) + (\tau_2 - \tau_1 + 1)Q_1 + Q_2 + Q_3 + M_1(\mu) + M_1^T(\mu)$$

$$+ \tau_2 X_{11}(\mu) + (\tau_2 - \tau_1)Y_{11}(\mu),$$

$$(9.75)$$

$$\Theta_{12}(\mu) := N_1(\mu) - M_1(\mu) - R_1(\mu) + M_2^T(\mu) + \tau_2 X_{12}(\mu) + (\tau_2 - \tau_1)Y_{12}(\mu),$$

$$\Theta_{22}(\mu) := N_2(\mu) + N_2^T(\mu) - M_2(\mu) - M_2^T(\mu) - R_2(\mu) - R_2^T(\mu)$$

$$+ \tau_2 X_{22}(\mu) + (\tau_2 - \tau_1)Y_{22}(\mu).$$

On the other hand, it follows from (9.49) that

$$\begin{bmatrix} x(t+1) \\ \sqrt{\tau_2}\, e(t) \\ \sqrt{\tau_2 - \tau_1}\, e(t) \\ q(t) \end{bmatrix} = \begin{bmatrix} \bar{A}(\mu) & A_d(\mu) & D_1(\mu) & 0 & 0 \\ \sqrt{\tau_2}(\bar{A}(\mu) - I) & \sqrt{\tau_2} A_d(\mu) & \sqrt{\tau_2} D_1(\mu) & 0 & 0 \\ \sqrt{\tau_2 - \tau_1}(\bar{A}(\mu) - I) & \sqrt{\tau_2 - \tau_1} A_d(\mu) & \sqrt{\tau_2 - \tau_1} D_1(\mu) & 0 & 0 \\ \bar{C}(\mu) & C_d(\mu) & D_2(\mu) & 0 & 0 \end{bmatrix} \begin{bmatrix} x(t) \\ x(t - \tau(t)) \\ v(t) \\ x(t-\tau_2) \\ x(t-\tau_1) \end{bmatrix}.$$

$$(9.76)$$

Then substituting (9.76) into (9.74) and applying the Schur complement lemma together with consideration of a similar bounding inequality as in (9.61), it is easy to see that (9.65)–(9.67) guarantee (9.74). The proof is thus completed. ∎

The conditions given in Theorem 9.4 are not convex due to the matrix equality constraints in (9.67). Similar to Algorithm 9.1, by using a cone complementarity technique, the original nonconvex feasibility problem formulated in Theorem 9.4 can be converted to the following nonlinear minimization problem involving LMI conditions,

$$\text{Minimize } Tr\left(\sum_{i \in L} P_i U_i + \sum_{\beta=1}^{2} Z_\beta S_\beta \right)$$

subject to (9.65), (9.66), (9.68)–(9.70), and

$$P_i > 0,\ U_i > 0,\ \begin{bmatrix} P_i & I \\ I & U_i \end{bmatrix} \geq 0,\ i \in L \tag{9.77}$$

$$Z_\beta > 0,\ S_\beta > 0,\ \begin{bmatrix} Z_\beta & I \\ I & S_\beta \end{bmatrix} \geq 0,\ \beta \in \{1,\ 2\}. \tag{9.78}$$

Then, for given delay bounds τ_1 and τ_2, the suboptimal value of γ can be found by the following algorithm.

Algorithm 9.2
Step 1. Choose a sufficiently large initial $\gamma > 0$ such that there exists a feasible solution to (9.65), (9.66), (9.68)–(9.70), (9.77), and (9.78). Set $\gamma_0 = \gamma$.
Step 2. Find a feasible set

$$\{P_{i0}, U_{i0}, M_{i0}, N_{i0}, R_{i0}, X_{i0}, Y_{i0}, K_{i0}, i \in L, Q_{\alpha 0}, \alpha \in \{1,\ 2,\ 3\}, Z_{\beta 0}, S_{\beta 0}, \beta \in \{1,\ 2\}\}$$

satisfying (9.65), (9.66), (9.68)–(9.70), (9.77), and (9.78). Set $\sigma = 0$.
Step 3. Solve the following LMI feasibility problem for the variables P_i, U_i, M_i, N_i, R_i, X_i, Y_i, K_i, $i \in L$, Q_α, $\alpha \in \{1,\ 2,\ 3\}$, Z_β, S_β, $\beta \in \{1,\ 2\}$,

$$\text{Minimize } Tr\left(\sum_{i \in L} (P_{i\sigma} U_i + P_i U_{i\sigma}) + \sum_{\beta=1}^{2} (Z_{\beta\sigma} S_\beta + Z_\beta S_{\beta\sigma}) \right)$$

subject to (9.65), (9.66), (9.68)–(9.70), (9.77), and (9.78).
Set $P_{i(\sigma+1)} = P_i$, $U_{i(\sigma+1)} = U_i$, $i \in L$, $Z_{\beta(\sigma+1)} = Z_\beta$, $S_{\beta(\sigma+1)} = S_\beta$, $\beta \in \{1,\ 2\}$.

Step 4. Substitute the controller gains K_i obtained in Step 3 into (9.65) and (9.66) and if the conditions (9.65),(9.66), and (9.68)–(9.70) are feasible for the variables P_i, M_i, N_i, R_i, X_i, Y_i, $i \in L$, $Q_{-\alpha}$, $\alpha \in \{1, 2, 3\}$, Z_β, $\beta \in \{1, 2\}$, then set $\gamma_0 = \gamma$ and return to Step 2 after decreasing γ to some extent. If the conditions (9.65), (9.66), and (9.68)–(9.70) are infeasible within the maximum number of iterations allowed, then exit. Otherwise, set $\sigma = \sigma + 1$ and go to step 3.

9.5   AN EXAMPLE

Example 9.1

Consider the following discrete-time time-delay T–S fuzzy system of the form (9.1) with three rules,

R^l: IF $x_1(t)$ is F_1^l

   THEN $x(t+1) = A_l x(t) + A_{dl} x(t - \tau(t)) + B_{1l} u(t) + D_{1l} w(t)$

$$q(t) = C_l x(t) + C_{dl} x(t - \tau(t)) + B_{2l} u(t) + D_{2l} w(t),$$

$l \in L := \{1, 2, 3\},$

where

$$A_1 = \begin{bmatrix} 1.210 & 0.363 \\ 0.242 & 1.210 \end{bmatrix}, A_2 = \begin{bmatrix} 1.210 & 0.363 \\ 0.484 & 1.210 \end{bmatrix}, A_3 = \begin{bmatrix} 1.452 & 0.605 \\ 0.242 & 1.210 \end{bmatrix},$$

$$A_{d1} = A_{d2} = A_{d3} = \begin{bmatrix} -0.1 & 0.05 \\ 0 & 0.1 \end{bmatrix},$$

$$B_{11} = \begin{bmatrix} 0.5 \\ 1.5 \end{bmatrix}, B_{12} = \begin{bmatrix} 0.5 \\ 1 \end{bmatrix}, B_{13} = \begin{bmatrix} 0.6 \\ 1 \end{bmatrix},$$

$$D_{11} = D_{12} = D_{13} = \begin{bmatrix} 0.5 \\ 0.5 \end{bmatrix},$$

$$C_1 = C_3 = [0.3 \quad 0], C_2 = [0.4 \quad 0],$$

$$C_{d1} = C_{d2} = C_{d3} = [0.2 \quad 0.2],$$

$$B_{21} = B_{22} = B_{23} = 0.1,$$

$$D_{21} = D_{22} = D_{23} = 0.$$

The membership functions are shown in Figure 9.1 and the time-varying delay $\tau(t)$ is assumed to satisfy (9.2) with $\tau_1 = 0$ and $\tau_2 = 4$. Note that the open-loop system is unstable.

The objective is to design a suitable state feedback controller for the above fuzzy system such that the resulting closed-loop system is globally exponentially stable with H_∞ disturbance attenuation level γ. Algorithms 9.1 and 9.2 provide a suboptimal disturbance attenuation performance for delay-dependent approaches based on piecewise and fuzzy quadratic Lyapunov functionals, respectively. Also note that the corresponding algorithm for the approach based on common quadratic Lyapunov functionals can be obtained by setting $U_i = U$, $P_i = P$, $l \in \bar{L}$ in Algorithm 9.1, which is subsequently referred to as Algorithm 9.3.

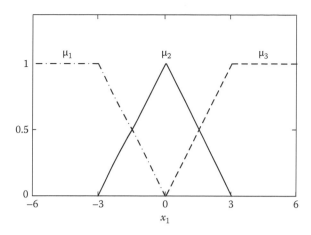

FIGURE 9.1 Membership functions in Example 9.1.

By applying Algorithms 9.1–9.3 with the fixed iteration number of 30, suboptimal disturbance attenuation performance with the corresponding controller gains can be obtained and are summarized in Table 9.1.

It can be observed that for the delay-dependent approaches, both piecewise and fuzzy quadratic Lyapunov functional-based approaches achieve better performance than the common quadratic Lyapunov functional-based approach, and this again has confirmed that the former approaches are less conservative than the latter.

On the other hand, algorithms minimizing the performance index γ for the delay-independent approaches proposed in this chapter can also be developed and are summarized as follows.

Algorithm 9.4: $\displaystyle\min_{U_l, \bar{K}_l, l \in L, G, Q_1} \gamma$, subject to LMIs (9.11)

Algorithm 9.5: $\displaystyle\min_{U_l, \bar{K}_l, l \in L, G, Q_1} \gamma$, subject to LMIs (9.51) and (9.52)

Algorithm 9.6: $\displaystyle\min_{U_l, \bar{K}_l, l \in L, G, Q_1} \gamma$, subject to LMIs (9.11) with $U_l = U$, $l \in \bar{L}$

TABLE 9.1

Minimum H_∞ Performance for Algorithms 9.1–9.3

Algorithm	Minimum γ	Controller Gains		
9.1	1.5146	$K_1 = [-0.8115 \quad -0.7014]$, $K_2 = [-0.8795 \quad -0.7385]$		
		$K_3 = [-1.1837 \quad -0.9633]$, $K_4 = [-1.1212 \quad -1.0060]$		
9.2	1.2484	$K_1 = [-0.8101 \quad -0.6718]$, $K_2 = [-1.1722 \quad -0.8522]$		
		$K_3 = [-1.1241 \quad -0.9728]$		
9.3	1.6942	$K_1 = [-0.7564 \quad -0.6966]$, $K_2 = [-0.8442 \quad -0.7396]$		
		$K_3 = [-1.1361 \quad -0.9655]$, $K_4 = [-1.1101 \quad -1.0244]$		

TABLE 9.2

Minimum H_∞ Performance for Algorithms 9.4–9.6

Algorithm	Minimum γ	Controller Gains		
9.4	2.2807	$K_1 = [-1.7655 \quad -0.7209]$, $K_2 = [-1.6551 \quad -0.7590]$		
		$K_3 = [-2.1524 \quad -0.8878]$, $K_4 = [-1.9271 \quad -1.0092]$		
9.5	1.7576	$K_1 = [-1.8614 \quad -0.7135]$, $K_2 = [-2.1336 \quad -0.7809]$		
		$K_3 = [-2.1246 \quad -1.0024]$		
9.6	2.6943	$K_1 = [-1.6796 \quad -0.7268]$, $K_2 = [-1.6919 \quad -0.7539]$		
		$K_3 = [-2.1076 \quad -0.8968]$, $K_4 = [-2.0256 \quad -1.0125]$		

Algorithms 9.4–9.6 correspond to delay-independent control based on piecewise, fuzzy, and common quadratic Lyapunov functionals, respectively.

By applying these three algorithms to the system, the suboptimal disturbance attenuation performance with the corresponding control gains can be obtained and are summarized in Table 9.2.

It can also be observed that for the delay-independent approaches, both piecewise and fuzzy quadratic Lyapunov functional-based approaches also achieve better performance than the common quadratic Lyapunov functional-based approach. Moreover, it can be observed from both Tables 9.1 and 9.2 that the delay-dependent approaches achieve much better performance than their delay-independent counterparts although the computational cost for the former approaches could be much higher.

9.6 CONCLUSIONS

In this chapter, based on piecewise and fuzzy quadratic Lyapunov–Krasovskii functionals, both delay-independent and delay-dependent approaches to H_∞ state feedback control have been presented for discrete-time T–S fuzzy systems with time-varying state delay. It has been shown that via some matrix inequality convexifying techniques, controller gains can be obtained by solving an optimization problem involving a number of LMIs. A numerical example is presented to demonstrate the performance of the proposed approaches and comparison of those approaches in terms of performance.

10 Fuzzy Model Predictive Control

10.1 INTRODUCTION

Model predictive control (MPC), also known as receding horizon control, is widely used in industrial processes, especially in processes with slow dynamics, for it is able to deal with physical constraints and multivariable systems efficiently in an online optimal way. Typically, based on an explicit plant model and information about input and output constraints, MPC solves a constrained optimization problem online at each time instant and implements the first element of the optimal control profile resulting from the optimization procedure. The process is repeated at the next time instant with the updated new state. One of the most appealing features of MPC is that the transient control performance of model predictive control systems can be adequately addressed in terms of certain optimal performance cost. In this chapter, two approaches to model predictive controller synthesis of T–S fuzzy systems are presented.

The rest of the chapter is organized as follows. Problem formation of model predictive control for T–S fuzzy systems is introduced in Section 10.2. Two min–max fuzzy model predictive control schemes are presented in Section 10.3. Several simulation examples are used to demonstrate the performance of the proposed schemes in Section 10.4, followed by some concluding remarks in Section 10.5.

10.2 PROBLEM FORMULATION

In this section, some preliminary knowledge of model predictive control of uncertain linear time-varying systems is introduced inasmuch as T–S fuzzy systems can be treated as uncertain linear time-varying systems in general. For easy reference and convenience, T–S fuzzy models are rewritten here as follows.

$$R^l: \quad \text{IF} \qquad z_1 \text{ is } F_1^l \text{ and } \dots z_v \text{ is } F_v^l$$

$$\text{THEN} \quad x(t+1) = A_l x(t) + B_l u(t) \tag{10.1}$$

$$y(t) = C_l x(t)$$

$$l \in L := \{1, 2, \dots, m\},$$

or in a compact form,

$$x(t+1) = A(\mu)x(t) + B(\mu)u(t)$$
$$y(t) = C(\mu)x(t), \tag{10.2}$$

where

$$A(\mu) = \sum_{l=1}^{m} \mu_l A_l, \; B(\mu) = \sum_{l=1}^{m} \mu_l B_l, \; C(\mu) = \sum_{l=1}^{m} \mu_l C_l. \tag{10.3}$$

We now review an approach to quasi-min–max infinite horizon model predictive control (Lu and Arkun, 2000, 2002), which is an extension of the well-known min–max infinite horizon MPC (Kothare, Balakrishnan, and Morari, 1996; Ozkan, Kothare, and Georgakis, 2000; Wan and Kothare, 2003). Consider an uncertain linear time-varying (LTV) system

$$x(t+1) = A(t)x(t) + B(t)u(t), \quad [A(t) \; B(t)] \in \Xi, \tag{10.4}$$

where $x(t) \in \Re^n$ is the state vector, and $u(t) \in \Re^g$ is the control input subject to $u_r(t) \le u_{r,\max}$, $r = 1, 2, \ldots, g$. Ξ is the polytope $Co\{[A_1 \; B_1], \ldots, [A_m \; B_m]\}$, where Co denotes the convex hull, and $[A_l \; B_l]$, $l = 1, 2, \ldots, m$ are vertices of the convex hull. Any $[A \; B]$ within the convex set Ξ is a linear combination of the vertices; that is, $A = \sum_{l=1}^{m} \alpha_l A_l$, $B = \sum_{l=1}^{m} \alpha_l B_l$ with $\sum_{l=1}^{m} \alpha_l = 1, 0 \le \alpha_l \le 1$.

The system (10.4) can be rewritten in predictive form as

$$x(t+i+1|t) = A(t+i|t)x(t+i|t) + B(t+i|t)u(t+i|t), \tag{10.5}$$

where $x(t+i|t), i \ge 1$ is the predicted state at time $t + i$ based on the current state $x(t)$, denoted as $x(t|t)$.

The objective of model predictive control is to minimize the worst-case infinite horizon quadratic objective function,

$$\min_{u(t|t), u(t+i|t) = F(t)x(t+i|t)} \; \max_{[A(t+i|t) \; B(t+i|t)] \in \Xi, i \ge 0} \; J_0^{\infty}(t) \tag{10.6}$$

subject to

$$|u_r(t+i|t)| \le u_{r,\max}, \quad r = 1, 2, \ldots, g, i \ge 0, \tag{10.7}$$

where

$$J_0^{\infty}(t) = \sum_{i=0}^{\infty} [x(t+i|t)^T Q x(t+i|t) + u(t+i|t)^T R u(t+i|t)]$$

$$= \sum_{i=0}^{H-1} [x(t+i|t)^T Q x(t+i|t) + u(t+i|t)^T R u(t+i|t)] + J_H^{\infty}(t) \tag{10.8}$$

with $Q > 0, R > 0, H = 1$.

Note from (10.6) that at each time instant t, both $u(t|t)$ and state feedback laws $u(t+i|t) = K(t)x(t+i|t), i \geq 1$ are used to minimize the worst-case value of $J_0^\infty(t)$. Following the approach given in Kothare, Balakrishnan, and Morari (1996), it is easy to derive an upper bound on $J_H^\infty(t)$. At time t, define a quadratic function $V(x(t+i|t)) = x(t+i|t)^T P(t)x(t+i|t)$ with $P(t)$ being a positive definite matrix at time t. For any $[A(t+i|t) \; B(t+i|t)] \in \Xi, i \geq 1$, suppose $V(x(t))$ satisfies the following robust stability constraint,

$$V(x(t+i+1|t)) - V(x(t+i|t))$$
$$\leq -[x(t+i|t)^T Q x(t+i|t) + u(t+i|t)^T R u(t+i|t)]. \tag{10.9}$$

Summing (10.9) from $i = 1$ to ∞ and letting $x(\infty|t) = 0$ or $V(x(\infty|t)) = 0$, it follows that

$$\max_{[A(t+i|t) \; B(t+i|t)] \in \Xi, i \geq 1} J_1^\infty(t) \leq V(x(t+1|t)) = x(t+1|t)^T P(t)x(t+1|t). \tag{10.10}$$

So an upper bound for $J_0^\infty(t)$ can be derived from (10.8) and (10.10); that is,

$$J_0^\infty(t) \leq x(t|t)^T Q x(t|t) + u(t|t)^T R u(t|t) + x(t+1|t)^T P(t)x(t+1|t).$$

Obviously, its minimization with $P(t) > 0$ is equivalent to

$$\min_{\gamma, u(t|t), P(t)} \gamma \tag{10.11}$$

subject to

$$x(t|t)^T Q x(t|t) + u(t|t)^T R u(t|t)$$
$$+ [A(t|t)x(t|t) + B(t|t)u(t|t)]^T P(t)[A(t|t)x(t|t) + B(t|t)u(t|t)] \leq \gamma, \tag{10.12}$$

where $A(t|t) = A(t), B(t|t) = B(t)$.

By using Schur complements and defining $X(t) = \gamma P(t)^{-1} > 0$, condition (10.12) can be expressed equivalently as the following LMI,

$$\begin{bmatrix} 1 & * & * & * \\ A(t|t)x(t|t) + B(t|t)u(t|t) & X(t) & * & * \\ Q^{1/2}x(t|t) & O & \gamma I & * \\ R^{1/2}u(t|t) & O & O & \gamma I \end{bmatrix} \geq 0, \tag{10.13}$$

where the asterisk, *, in a matrix stands for the corresponding terms of a symmetric matrix as in the other chapters. The robust stability condition (10.9) is satisfied (Kothare, Balakrishnan, and Morari, 1996) if for each vertex of Ξ,

$$
\begin{bmatrix}
X(t) & * & * & * \\
A_j X(t) + B_j Y(t) & X(t) & * & * \\
Q^{1/2} X(t) & O & \gamma I & * \\
R^{1/2} Y(t) & O & O & \gamma I
\end{bmatrix} \geq 0, \, j = 1, 2, \ldots, m,
\tag{10.14}
$$

where $X(t) = \gamma P(t)^{-1}$, $Y(t) = K(t)X(t)$.

The input constraints (10.7) can be expressed as follows,

$$
|u_r(t\,|\,t)| \leq u_{r,\max}, \, r = 1, 2, \ldots, g,
\tag{10.15}
$$

$$
|u_r(t+i\,|\,t)| \leq u_{r,\max}, \, r = 1, 2, \ldots, g, \, i \geq 1.
\tag{10.16}
$$

The constraint (10.16) is satisfied if there exists a symmetric matrix U such that

$$
\begin{bmatrix}
U & Y(t) \\
Y(t)^T & X(t)
\end{bmatrix} \geq 0, \quad \text{with} \quad U_{rr} \leq u_{r,\max}^2, \, r = 1, 2, \ldots, g,
\tag{10.17}
$$

where U_{rr} is the diagonal element of the matrix U (Lu and Arkun, 2000).

Summarizing the above arguments one can obtain the following theorem (Lu and Arkun, 2000).

Theorem 10.1

For system (10.4), at sampling time t, the control action $u(t\,|\,t)$ and state feedback control law $u(t+i\,|\,t) = K(t)x(t+i\,|\,t), i \geq 1$ minimize the worst-case MPC objective function $J_0^\infty(t)$ with input constraints (10.7), if there exist $u(t\,|\,t)$, $X(t) > 0$, and $Y(t)$ such that the following minimization problem,

$$
\min_{\gamma, u(t|t), X, Y, U} \gamma \text{ subject to (10.13)–10.15) and (10.17),}
$$

is feasible, and $K(t) = Y(t)X(t)^{-1}$. Moreover, the resulting closed-loop model predictive control system is asymptotically stable.

Remark 10.1 (Lu and Arkun, 2000)

A feasible solution of the optimization in Theorem 10.1 at time t implies that the optimization is also feasible for all time greater than t.

With the current system state known exactly at every sampling time, the cost $x(t\,|\,t)^T Qx(t\,|\,t) + u(t\,|\,t)^T Ru(t\,|\,t)$ at the current time can be computed without any uncertainty. Given the state feedback gain matrix $K(t)$ and the stability constraint (10.9), the infinite horizon objective function $J_{\tilde{H}}^{\infty}(t)$ beyond the finite prediction horizon H is represented by an endpoint state penalty, which is the main idea of the quasi-min–max MPC with infinite horizon cost in Theorem 10.1 (Lu and Arkun, 2000, 2002).

10.3 FUZZY MODEL PREDICTIVE CONTROL APPROACHES

In this section, two fuzzy model predictive controller design methods based on common and piecewise quadratic Lyapunov functions, respectively, are presented using the min–max optimization of worst-case infinite horizon performance cost. In the first approach, a smooth state feedback law is employed. In the other one, a switching state feedback control law is utilized.

10.3.1 Fuzzy Min–Max MPC Based on Common Quadratic Lyapunov Functions

Consider the following smooth fuzzy state feedback control law,

$$R^l: \quad \text{IF} \quad\quad z_1 \text{ is } F_1^l \text{ and } \dots z_v \text{ is } F_v^l$$

$$\text{THEN} \quad u(t) = K_l\, x(t) \tag{10.18}$$

$$l \in L := \{1, 2, \dots, m\},$$

which can be represented by

$$u(t) = \sum_{l=1}^{m} \mu_l(z(t)) K_l x(t) = K(\mu) x(t). \tag{10.19}$$

With the following common quadratic Lyapunov function defined,

$$V(x(t+i\mid t)) = x(t+i\mid t)^T P(t) x(t+i\mid t), \tag{10.20}$$

one readily has the following theorem.

Theorem 10.2

For the system (10.1), the control action $u(t\mid t)$ and state feedback control law $u(t+i\mid t) = \sum_{l=1}^{m} \mu_l(x(t+i\mid t)) K_l(t) x(t+i\mid t), i \geq 1$ minimize the worst-case MPC objective function $J_0^{\infty}(t)$, if there exist decision variables $u(t\mid t)$, $X(t) > 0$, and $Y_l(t)$ such that the solution to the following convex objective minimization problem is feasible,

$$\min_{\gamma, u(t\mid t), X, Y_l} \gamma \tag{10.21}$$

subject to the following LMIs,

$$
\begin{bmatrix}
1 & * & * & * \\
A_{t|t}x(t|t)+B_{t|t}u(t|t) & X(t) & * & * \\
Q^{1/2}x(t|t) & O & \gamma \cdot I & * \\
R^{1/2}u(t|t) & O & O & \gamma \cdot I
\end{bmatrix} \geq 0, \tag{10.22}
$$

$$
\begin{bmatrix}
X(t) & * & * & * \\
A_j X(t)+B_j Y_j(t) & X(t) & * & * \\
Q^{1/2}X(t) & O & \gamma \cdot I & * \\
R^{1/2}Y_j(t) & O & O & \gamma \cdot I
\end{bmatrix} > 0, \; j \in L, \tag{10.23}
$$

$$
\begin{bmatrix}
X(t) & * & * \\
[A_j X(t)+A_l X(t)+B_j Y_l(t)+B_l Y_j(t)]/2 & X(t) & * \\
Q^{1/2}X(t) & O & \gamma \cdot I
\end{bmatrix} \geq 0, 1 \leq j \leq l \leq m, \tag{10.24}
$$

where $A_{t|t} = \sum_{j=1}^{m} \mu_j(z(t))A_j$, $B_{t|t} = \sum_{j=1}^{m} \mu_j(z(t))B_j$, $X(t) = \gamma P(t)^{-1}$, $Y_j(t) = K_j(t)X(t)$. Moreover, the resulting closed-loop predictive control system is asymptotically stable.

Proof: Rewrite the T–S fuzzy model (10.1), or equivalently (10.2), and the fuzzy state feedback law (10.19) in the predictor form, respectively, as follows,

$$
x(t+i+1|t) = \sum_{j=1}^{m} \mu_j(z(t+i|t))[A_j x(t+i|t)+B_j u(t+i|t)], \tag{10.25}
$$

$$
u(t+i|t) = \sum_{l=1}^{m} \mu_l(z(t+i|t))K_l x(t+i|t). \tag{10.26}
$$

Substituting (10.26) into (10.25), one can get

$$
x(t+i+1|t) = \sum_{j=1}^{m}\sum_{l=1}^{m} \mu_j \mu_l (A_j + B_j K_l)x(t+i|t), \tag{10.27}
$$

where $\mu_j = \mu_j(z(t+i|t))$, $\mu_l = \mu_l(z(t+i|t))$.

Because the system state at the sampling time t, $(t|t)$, is known, one can then obtain

$$x(t+1|t) = A_{t|t}x(t|t) + B_{t|t}u(t|t), \qquad (10.28)$$

where $A_{t|t} = \sum_{j=1}^{m} \mu_j(z(t))A_j$, $B_{t|t} = \sum_{j=1}^{m} \mu_j(z(t))B_j$. It then follows that the quasi-min–max inequality (10.12) can be expressed equivalently by the LMI (10.22), which is similar to (10.13).

Now we consider the robust stability condition. It follows from (10.27) that the robust stability constraint (10.9) is equivalent to

$$\left[\sum_{j=1}^{m}\sum_{l=1}^{m}\mu_j\mu_l(A_j + B_jK_l)x(t+i|t)\right]^T P\left[\sum_{k=1}^{m}\sum_{n=1}^{m}\mu_k\mu_n(A_k + B_kK_n)x(t+i|t)\right]$$

$$+ x(t+i|t)^T Px(t+i|t) + x(t+i|t)^T Qx(t+i|t)$$

$$+ \left[\sum_{j=1}^{m}\mu_jK_jx(t+i|t)\right]^T R\left[\sum_{l=1}^{m}\mu_lK_lx(t+i|t)\right] \le 0.$$

Let $G_{jl} = A_j + B_jK_l$; the above inequality becomes

$$\sum_{j=1}^{m}\sum_{l=1}^{m}\sum_{k=1}^{m}\sum_{n=1}^{m}\mu_j\mu_l\mu_k\mu_n x(t+i|t)^T (G_{jl}^T PG_{kn} - P)x(t+i|t) + x(t+i|t)^T Qx(t+i|t)$$

$$+ \sum_{j=1}^{m}\sum_{l=1}^{m}\mu_j\mu_l x(t+i|t)^T K_j^T RK_l x(t+i|t) \le 0.$$

It follows from $2X^T RY \le X^T RX + Y^T RY$ and Lemma 2 in Guan and Chen (2004) that the above inequality is satisfied for all $i \ge 1$ if

$$\sum_{j=1}^{m}\sum_{l=1}^{m}\mu_j\mu_l x(t+i|t)^T (G_{jl}^T PG_{jl} - P)x(t+i|t) + x(t+i|t)^T Qx(t+i|t)$$

$$+ \sum_{j=1}^{m}\mu_l x(t+i|t)^T K_j^T RK_j x(t+i|t) \le 0,$$

which is equivalent to

$$\sum_{j=1}^{m}\sum_{l=1}^{m}\mu_j\mu_l x(t+i|t)^T (G_{jl}^T PG_{jl} - P + Q + K_j^T RK_j)x(t+i|t) \le 0. \qquad (10.29)$$

Furthermore, (10.29) can be rewritten as

$$\sum_{j=1}^{m} \mu_j^2 x(t+i\mid t)^T \left(G_{jj}^T P G_{jj} - P + Q + K_j^T R K_j \right) x(t+i\mid t) +$$

$$2\sum_{j=1}^{m}\sum_{j<l}^{m} \mu_j \mu_l x(t+i\mid t)^T \left[\left(\frac{G_{jl}+G_{lj}}{2} \right)^T P \left(\frac{G_{jl}+G_{lj}}{2} \right) - P + Q \right] x(t+i\mid t) \le 0.$$

$$(10.30)$$

So one can obtain the following sufficient condition for stability,

$$G_{jj}^T P G_{jj} - P + Q + K_j^T R K_j < 0,\ j \in L \tag{10.31}$$

$$\left(\frac{G_{jl}+G_{lj}}{2} \right)^T P \left(\frac{G_{jl}+G_{lj}}{2} \right) - P + Q \le 0,\ 1 \le j \le l \le m. \tag{10.32}$$

By defining $X(t) = \gamma P(t)^{-1}$, $Y_j(t) = K_j(t)X(t)$, and using Schur complements, it is easy to transform (10.31) and (10.32) into the LMIs as in (10.23) and (10.24), respectively. It thus follows that the closed-loop system is asymptotically stable if (10.23) and (10.24) are satisfied. Then via Theorem 10.1 one can conclude that the solution to the minimization problem (10.21) minimizes the worst-case MPC objective function and the closed-loop system is asymptotically stable. The proof is thus completed. ❑

10.3.2 Fuzzy Min–Max MPC Based on Piecewise Quadratic Lyapunov Functions

Suppose the premise variable space is partitioned into a number of closed polyhedral regions, denoted $\{S_l\}_{l \in L}$, where L is the index set of regions. In this case similar to the previous chapters, a set Ω that represents all possible transitions from one region to another region is defined as

$$\Omega := \{(l,j)\mid z(t) \in S_l, z(t+1) \in S_j, \forall l,j \in L, l \ne j\}.$$

Consider the following piecewise quadratic Lyapunov function candidate,

$$V(x(t+i\mid t)) = x(t+i\mid t)^T P_j(t)x(t+i\mid t),\ z(t) \in S_j. \tag{10.33}$$

Then one has the following result.

Theorem 10.3

For the system (10.1), the control action $u(t|t)$, and the state feedback control law $u(t+i|t) = K_s(t)x(t+i|t), i \geq 1$ minimize the worst-case MPC objective function $J_0^{\infty}(t)$, if there exist decision variables $u(t|t)$, $X_s(t) > 0$, and $Y_s(t)$ such that the following convex objective minimization problem is feasible,

$$\min_{\gamma, u(t|t), X_s, Y_s} \gamma \tag{10.34}$$

subject to the following LMIs,

$$\begin{bmatrix} 1 & * & * & * \\ A_{t|t}x(t|t) + B_{t|t}u(t|t) & X_j(t) & * & * \\ Q^{1/2}x(t|t) & 0 & \gamma \cdot I & * \\ R^{1/2}u(t|t) & 0 & 0 & \gamma \cdot I \end{bmatrix} \geq 0, j \in L, \tag{10.35}$$

$$\begin{bmatrix} X_s(t) & * & * & * \\ A_j X_s(t) + B_j Y_s(t) & X_k(t) & * & * \\ Q^{1/2}X_s(t) & 0 & \gamma \cdot I & * \\ R^{1/2}Y_s(t) & 0 & 0 & \gamma \cdot I \end{bmatrix} > 0, j \in L_s, s, k \in \Omega, \tag{10.36}$$

where $A_{t|t} = \sum_{j=1}^{m} \mu_j(z(t))A_j, B_{t|t} = \sum_{j=1}^{m} \mu_j(z(t))B_j$, $X_k(t) = \gamma P_k(t)^{-1}$, $Y_s = K_s(t)X_s(t)$, $s, k \in \Omega$, and L_s is the index set of firing fuzzy rules in the region S_s. Moreover, the resulting closed-loop model predictive control system is asymptotically stable.

Proof: For the predicted system state $x(t+i|t) (i \geq 1)$, without loss of generality, suppose it belongs to some region, $x(t+i|t) \in S_s$, then the T–S fuzzy model (10.2) or equivalently (10.25), can be rewritten as follows in the region S_s,

$$x(t+i+1|t) = \sum_{j=1}^{\bar{m}} \mu_j(z(t+i|t))[A_j x(t+i|t) + B_j u(t+i|t)], \tag{10.37}$$

where $\bar{m}$ is the number of firing fuzzy rules in the region S_s. In this approach, the state feedback control law is a switching one, dependent on the region S_s instead of fuzzy state feedback law (10.19). In this case the closed-loop fuzzy control system becomes

$$x(t+i+1|t) = \sum_{j=1}^{\bar{m}} \mu_j(x)(A_j + B_j K_s)x(t+i|t), z(t) \in S_s. \tag{10.38}$$

Similar to Theorem 10.2, it is easy see that the quasi-min–max inequality (10.12) can be expressed equivalently by the LMI (10.35). Without loss of generality, suppose that the state will transit from arbitrary region S_s to another region S_k at time t, $s, k \in \Omega$. Then with (10.38), the robust stability constraint (10.9) becomes

$$\left[\sum_{j=1}^{\bar{m}} \mu_j (A_j + B_j K_s) x(t+i|t) \right]^T P_k \left[\sum_{l=1}^{\bar{m}} \mu_l (A_l + B_l K_s) x(t+i|t) \right]$$

$$+ x(t+i|t)^T P_s x(t+i|t) + x(t+i|t)^T Q x(t+i|t)$$

$$+ [K_s x(t+i|t)]^T R[K_s x(t+i|t)] \leq 0.$$

Letting $G_{js} = A_j + B_j K_s$, the above inequality can be rewritten as

$$\sum_{j=1}^{\bar{m}} \sum_{l=1}^{\bar{m}} \mu_j \mu_l x(t+i|t)^T \left(G_{js}^T P_k G_{ls} - P_s \right) x(t+i|t)$$

$$+ x(t+i|t)^T Q x(t+i|t) + x(t+i|t)^T K_s^T R K_s x(t+i|t) \leq 0.$$

It follows from $2X^T RY \leq X^T RX + Y^T RY$ and Lemma 2 in Guan and Chen (2004) that the above inequality is satisfied for all $i \geq 1$ if

$$\sum_{j=1}^{\bar{m}} \mu_j x(t+i|t)^T (G_{js}^T P_k G_{js} - P_s) x(t+i|t)$$

$$+ x(t+i|t)^T Q x(t+i|t) + x(t+i|t)^T K_s^T R K_s x(t+i|t) \leq 0. \tag{10.39}$$

So one can obtain the following sufficient condition for stability,

$$G_{js}^T P_k G_{js} - P_s + Q + K_s^T R K_s < 0, \ j \in L_s, s, k \in \Omega. \tag{10.40}$$

By defining $X_k(t) = \gamma P_k(t)^{-1}$, $Y_s(t) = K_s(t) X_s(t)$, and using Schur complements, it is easy to transform (10.40) into (10.36). It thus follows that the closed-loop system is asymptotically stable if (10.36) is satisfied. Then via Theorem 10.1 one can conclude that the solution to the minimization problem (10.34) minimizes the worst-case MPC objective function and the closed-loop system is asymptotically stable. The proof is thus completed. ❑

10.3.3 Constrained Fuzzy MPC

To deal with the input constraints with respect to the proposed two fuzzy MPC approaches, the following properties of stable invariant ellipsoids are introduced. This follows from a similar idea in Lu and Arkun (2000, 2002).

Corollary 10.1

Consider the closed-loop control system composed of the system model (10.1), $u(t|t)$, and the state feedback control law $u(t+i|t) = \sum_{l=1}^{m} \mu_l(z(t+i|t))K_l(t)x(t+i|t), i \geq 1$, which are obtained by applying the fuzzy MPC approach defined in Theorem 10.2 to the current system state $x(t)$. Then the subset $\Im = \{x \in \Re^n | x^T X^{-1} x \leq 1, X > 0\}$ of the state space $\Re^n$ is an asymptotically stable invariant ellipsoid.

Proof: If there exist $X > 0$, γ, $Y_j = K_j X$, and such that (10.23) and (10.24) hold, and if (10.12) holds, then

$$x(t+1|t)^T Px(t+1|t) \leq \gamma \quad \text{or} \quad x(t+1|t)^T X^{-1} x(t+1|t) \leq 1.$$

From the inequality (10.9), one can easily get

$$V(x(t+i+1|t)) - V(x(t+i|t)) \leq 0, i \geq 1,$$

or

$$x(t+i+1|t)^T Px(t+i+1|t) \leq x(t+i|t)^T Px(t+i|t) \leq \gamma.$$

That is,

$$x(t+i+1|t)^T X^{-1}(t)x(t+i+1|t) \leq x(t+i|t)^T X^{-1}(t)x(t+i|t) \leq 1, i \geq 1.$$

Thus $x(t+i|t) \in \Im, i \geq 1$ and $x(t+i|t) \to 0$ as $i \to \infty$, establishing that $\Im$ is an asymptotically stable invariant ellipsoid. ❑

Corollary 10.2

Consider the closed-loop control system composed of the system model (10.1), $u(t|t)$, and the state feedback control law $u(t+i|t) = K_s(t)x(t+i|t), i \geq 1$, which are obtained by applying the fuzzy MPC approach defined in Theorem 10.3 to the current system state $x(t)$. Then the subset $\Im = \{x \in \Re^n | x^T X_s^{-1} x \leq 1, X_s > 0\}$ of the state space $\Re^n$ is an asymptotically stable invariant ellipsoid.

Proof: Similar to the proof of Corollary 10.1, it is therefore omitted. ❑

Once Corollary 10.1 and Corollary 10.2 are established, the constraints on $\{u(t+i|t), i > 0\}$ can be cast into LMIs for the proposed fuzzy MPC approaches. In the first approach, one has that

$$\max_{i>0} |u_r(t+i|t)|^2 = \max_{i>0} |\left(Y_j X^{-1} x(t+i|t)\right)_r|^2$$

$$\leq \max_{z \in \Im} \left|\left(Y_j X^{-1} z\right)_r\right|^2 \leq \left\|\left(Y_j X^{-1/2}\right)_r\right\|_2^2 \max_{z \in \Im} \left\|\left(X^{-1/2} z\right)_r\right\|_2^2 \leq \left(Y_j X^{-1} Y_j^T\right)_{rr}$$

$$r = 1, 2, \ldots, g, j \in L.$$

Thus the input constraints (10.16),

$$|u_r(t+i|t)| \le u_{r,\max}, i > 0, r = 1,2,\ldots,g,$$

can be achieved if there exists a matrix U such that the following LMIs are feasible,

$$\begin{bmatrix} U & Y_j(t) \\ Y_j(t)^T & X(t) \end{bmatrix} \ge 0, \quad \text{and} \quad U_{rr} \le u_{r,\max}^2, r = 1,2,\ldots,g, j \in L, \quad (10.41)$$

where U_{rr} is the diagonal element of the matrix U.

Therefore the input constraints (10.7) in the first fuzzy MPC approach (Theorem 10.2) are satisfied if

$$|u_r(t|t)| \le u_{r,\max}, r = 1,2,\ldots,g, \quad (10.42)$$

and if there exists a matrix U such that LMIs in (10.41) are feasible.

Similarly for the second approach, one has that

$$\max_{i>0} |u_r(t+i|t)|^2 = \max_{i>0} |(Y_s X_s^{-1} x(t+i|t))_r|^2$$

$$\le \max_{z \in \mathfrak{I}} |(Y_s X_s^{-1} z)_r|^2 \le \|(Y_s X_s^{-1/2})_r\|_2^2 \max_{z \in \mathfrak{I}} \|(X_s^{-1/2} z)_r\|_2^2 \le (Y_s X_s^{-1} Y_s^T)_{rr}$$

$$r = 1,2,\ldots,g, s \in L.$$

Therefore the input constraints (10.7) in the second fuzzy MPC approach (Theorem 10.3) are satisfied if

$$|u_r(t|t)| \le u_{r,\max}, r = 1,2,\ldots,g, \quad (10.43)$$

and if there exists a matrix U such that the following LMIs are feasible,

$$\begin{bmatrix} U & Y_s(t) \\ Y_s(t)^T & X_s(t) \end{bmatrix} \ge 0, \quad \text{and} \quad U_{rr} \le u_{r,\max}^2, r = 1,2,\ldots,g, s \in L, \quad (10.44)$$

where U_{rr} is the diagonal element of the matrix U.

In addition, the speed of the closed-loop control system response can be manipulated by specifying a desired decay rate on the state as follows,

$$x(t+i+1|t)^T Px(t+i+1|t) \le \alpha^2 x(t+i|t)^T Px(t+i|t), \quad i \ge 0, \quad (10.45)$$

for the first approach, and

$$x(t+i+1|t)^T P_s x(t+i+1|t) \leq \alpha^2 x(t+i|t)^T P_k x(t+i|t), \quad i \geq 0, \quad (10.46)$$

for the second approach, respectively.

Following the proof of Theorem 10.2, it can be shown that the requirement (10.45) becomes the following LMIs,

$$\begin{bmatrix} \alpha^2 X(t) & * \\ A_j X(t) + B_j Y_j(t) & X(t) \end{bmatrix} > 0, j \in L \quad (10.47)$$

$$\begin{bmatrix} \alpha^2 X(t) & * \\ [A_j X(t) + A_l X(t) + B_j Y_l(t) + B_l Y_j(t)]/2 & X(t) \end{bmatrix} \geq 0, 1 \leq j < l < m. \quad (10.48)$$

Similarly, following the proof of Theorem 10.3, it can be shown that the requirement (10.46) becomes the following LMI,

$$\begin{bmatrix} \alpha^2 X_s(t) & X_s(t) A_j^T + Y_s(t)^T B_j^T \\ A_j X_s(t) + B_j Y_s(t) & X_k(t) \end{bmatrix} > 0, j \in L_s, s, k \in \Omega. \quad (10.49)$$

Remark 10.2

If the receding horizon implementation is not employed in Theorems 10.2 and 10.3, the decision variables would just depend on the initial state of the system, and $X(t)$, and $Y_l(t)$ in Theorem 10.2 and $X_s(t)$ and $Y_s(t)$ in Theorem 10.3 would become constant. In this case those decision variables can be obtained offline before implementation of real-time control. And these controllers are referred to as the offline forms of the fuzzy model predictive control laws.

10.4   SIMULATION EXAMPLES

In this section, two examples, including the highly nonlinear benchmark model of a continuous stirred tank reactor (CSTR; Henson and Seborg, 1997), are presented to show the performance of the proposed fuzzy model predictive control approaches and also the advantages of the fuzzy MPC approach based on piecewise quadratic Lyapunov functions over the fuzzy MPC approach based on common quadratic Lyapunov functions. For simplicity, the following abbreviations are adopted to denote the control approaches discussed in the previous sections.

- **FMPCCLF**: Fuzzy Model Predictive Control based on Common Lyapunov Function (CLF) in Theorem 10.2 with input constraints as in (10.41) and (10.42)

- **FMPCPLF**: Fuzzy Model Predictive Control based on Piecewise Lyapunov Function (PLF) in Theorem 10.3 with input constraints as in (10.43) and (10.44)
- **PDCFC**: Parallel Distributed Compensation Fuzzy Control with input constraints (Tanaka and Wang, 2001)

Example 10.1

Consider stabilization of a discrete-time T–S fuzzy system

$$R^l: \quad \text{IF} \qquad x_2(t) \text{ is } M^l(t)$$

$$\text{THEN} \qquad x(t+1) = A_l x(t) + B_l u(t), \qquad (10.50)$$

$$l = 1, 2, 3, 4,$$

with the membership functions given by

$$M^1(t) = \begin{cases} \dfrac{-x_2(t)+1}{4}, & -3 \le x_2(t) < 1 \\ 1, & x_2(t) < -3 \end{cases}$$

$$M^2(t) = \begin{cases} \dfrac{x_2(t)+3}{4}, & -3 \le x_2(t) < 1 \\ 0, & x_2(t) < -3 \end{cases}$$

$$M^3(t) = \begin{cases} \dfrac{-x_2(t)+5}{4}, & 1 < x_2(t) \le 5 \\ 0, & x_2(t) > 5 \end{cases}$$

$$M^4(t) = \begin{cases} \dfrac{x_2(t)-1}{4}, & 1 < x_2(t) \le 5 \\ 1, & x_2(t) > 5, \end{cases}$$

which are shown in Figure 10.1.

The system matrices are given by

$$A_1 = \begin{bmatrix} 1 & 2 \\ 0 & 1 \end{bmatrix}, B_1 = \begin{bmatrix} 0.5 \\ -1 \end{bmatrix}, A_2 = \begin{bmatrix} 1 & 1.5 \\ 0 & 1 \end{bmatrix}, B_2 = \begin{bmatrix} 1 \\ -0.5 \end{bmatrix}$$

$$A_3 = \begin{bmatrix} 1 & 0.5 \\ 0 & 1 \end{bmatrix}, B_3 = \begin{bmatrix} 0.5 \\ 1 \end{bmatrix}, A_4 = \begin{bmatrix} 1 & 2.5 \\ 0 & 1 \end{bmatrix}, B_4 = \begin{bmatrix} 1 \\ 0.5 \end{bmatrix}.$$

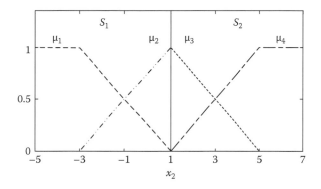

FIGURE 10.1 Membership functions and space partitions of Example 10.1.

FMPCCLF and FMPCPLF approaches are applied to this T–S fuzzy model with the initial state $x(0) = [-3\ 3]^T$, $Q = 0.04 \cdot I_2$, and $R = 0.01$. There is no feasible solution for MPC based on common quadratic Lyapunov functions whereas the approach based on piecewise quadratic Lyapunov functions not only leads to feasible solutions but also achieves good performance even with implementation of the offline form of the proposed controllers (Remark 10.2). The closed-loop responses are shown in Figure 10.2.

In offline FMPCPLF, the state space is partitioned into two regions as shown in Figure 10.1, and the corresponding two positive definite matrices and switching state feedback gains are given, respectively, by

$$P_1 = \begin{bmatrix} 0.4272 & 1.2927 \\ 1.2927 & 6.3928 \end{bmatrix} \square\, P_2 = \begin{bmatrix} 0.3163 & 0.3160 \\ 0.3160 & 2.2223 \end{bmatrix} \square$$

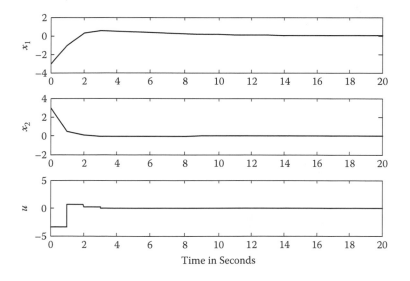

FIGURE 10.2 Closed-loop responses of Example 10.1.

and

$$K_1 = [0.1588 \quad 1.8673], \; K_2 = [-0.2264 \; -1.3582].$$

This example clearly demonstrates the advantages of the approach based on piecewise quadratic Lyapunov functions over the approach based on common quadratic Lyapunov functions.

Example 10.2

Consider a highly nonlinear continuous stirred tank reactor model (Henson and Seborg, 1997). With constant liquid volume, the CSTR for an exothermic irreversible reaction $A \rightarrow B$ is described by the following dynamic model based on a component balance for reactant A and on an energy balance,

$$\dot{C}_A = \frac{q}{V}(C_{Af} - C_A) - k_0 \exp\left(-\frac{E}{RT}\right) C_A$$

$$\dot{T} = \frac{q}{V}(T_f - T) + \frac{(-\Delta H)}{\rho C_p} k_0 \exp\left(-\frac{E}{RT}\right) C_A + \frac{UA}{V\rho C_p}(T_c - T), \qquad (10.50)$$

where C_A is the concentration of A in the reactor, T is the reactor temperature, and T_c is the temperature of the coolant stream. The objective is to control T by manipulating T_c. Table 10.1 contains nominal operating conditions, which correspond to an unstable steady state. The open-loop response in Figure 10.3 reveals that the reactor exhibits highly nonlinear behavior in this operating regime.

The equation (10.50) can be expressed in the form of nonlinear state space model $\dot{x} = f(x) + g(x)u$ by defining the state vector as $x = [C_A \; T]^T$, and the manipulated input as $u = T_c$. At steady input $u = 300$, the CSTR system (10.50) has the following three equilibrium states

$$x_s^1 = [0.875, 324.4]^T, \quad x_s^2 = [0.5, 350]^T, \quad x_s^3 = [0.2, 370.65]^T.$$

TABLE 10.1

Nominal Operating Conditions for CSTR

Variable	Value	Variable	Value
q	100 L/min	E/R	8750 K
C_{AF}	1 mol/L	k_0	7.2×10^{10} min^{-1}
T_f	350 K	UA	5×10^4 J/min K
V	100 L	T_c	300 K
ρ	1000 g/L	C_A	0.5 mol/L
C_p	0.239 J/g K	T	350 K
$(-\Delta H)$	5×10^4 J/mol		

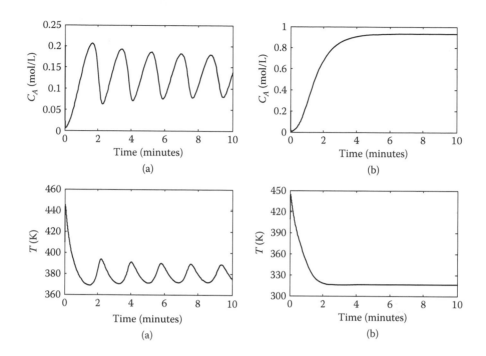

FIGURE 10.3 Open-loop response for (a) +6K and (b) −6K step changes in T_c.

Using the approximation based modeling method described in Teixeira and Zak (1999), and choosing the membership functions as

$$\text{``low":}\quad M^1(t) = \begin{cases} 1 & \text{if } x_2(t) \leq 324.4 \\ 1 - \dfrac{x_2(t) - 324.4}{350 - 324.4} & \text{if } 324.4 < x_2(t) < 350 \\ 0 & \text{if } x_2(t) \geq 350 \end{cases}$$

$$\text{``middle":}\quad M^2(t) = \begin{cases} 1 - M^1(t) & \text{if } x_2(t) \leq 350 \\ 1 - M^3(t) & \text{if } x_2(t) \geq 350 \end{cases}$$

$$\text{``high":}\quad M^3(t) = \begin{cases} 0 & \text{if } x_2(t) \leq 350 \\ \dfrac{x_2(t) - 350}{370.65 - 350} & \text{if } 350 < x_2(t) < 370.65 \\ 1 & \text{if } x_2(t) \geq 370.65, \end{cases}$$

which are shown in Figure 10.4, one can easily obtain the following T–S fuzzy model (10.51) of the nonlinear CSTR (10.50)

Rule 1: IF $x_2(t)$ is low

$$\text{THEN}\quad \delta\dot{x}(t) = \begin{bmatrix} -1.1390 & 0.0031 \\ 29.0858 & -2.0131 \end{bmatrix}\delta x(t) + \begin{bmatrix} 0 \\ 2.0921 \end{bmatrix}\delta u(t)$$

Rule 2: IF $x_2(t)$ is middle

$$\text{THEN}\quad \delta\dot{x}(t) = \begin{bmatrix} -1.9999 & 0.0029 \\ 209.1709 & -2.0920 \end{bmatrix}\delta x(t) + \begin{bmatrix} 0 \\ 2.0921 \end{bmatrix}\delta u(t) \quad (10.51)$$

Rule 3: IF $x_2(t)$ is high

$$\text{THEN}\quad \delta\dot{x}(t) = \begin{bmatrix} -5.0259 & 0.0027 \\ 842.2159 & -2.1477 \end{bmatrix}\delta x(t) + \begin{bmatrix} 0 \\ 2.0921 \end{bmatrix}\delta u(t),$$

where $\delta x(t) = x(t) - x_d$, $\delta u(t) = u(t) - u_d$, and (x_d, u_d) is an arbitrary specified operating point, which is a stationary point of the nonlinear system (10.50). In the following simulation, the above continuous time fuzzy dynamic model is converted to a discrete time one under the sampling time $T_s = 0.05$ minutes. Choose $Q = 0.1I$, $R = 0.0005$, $T_c \in [200, 400]$. Suppose the desired operating reference involves several stepwise changes from $[0.05\ 398.1]^T$ to $[0.16\ 374.8]^T$, to $[0.33\ 360]^T$, to $[0.67\ 340.6]^T$, and finally to $[0.5\ 350]^T$. With a space partition of the second kind, the partition of the premise space is shown in Figure 10.4.

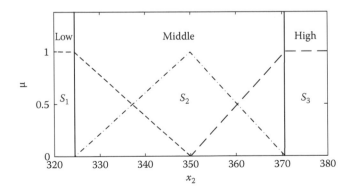

FIGURE 10.4 Membership functions and space partitions of Example 10.2.

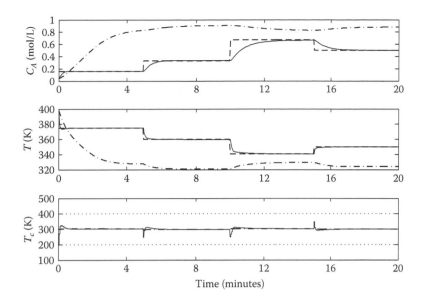

FIGURE 10.5   Closed-loop control performance (solid: FMPCCLF; dotted-dashed: PDCFC; dashed: reference; dotted: input constraints).

At first, the PDCFC and FMPCCLF are applied to the CSTR plant and the simulation results are shown as in Figures 10.5 and 10.6.

Then the FMPCPLF is applied to the CSTR plant. The simulation results are shown in Figures 10.7 and 10.8. Furthermore, the offline approach of the FMPCPLF in Remark 10.2 is also applied to the same benchmark CSTR plant, and the results are also shown in Figure 10.7 for easy comparison.

Simulation results show that under any initial condition the closed-loop state response follows the specified reference very well. It is also shown that fuzzy MPC based on piecewise quadratic Lyapunov functions achieves much better control performance than the approach based on common quadratic Laypunov functions under the same tuning parameters and input constraints, and that the online FMPCPLF achieves better performance than its offline counterpart.

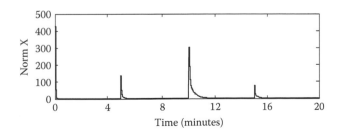

FIGURE 10.6   Profile of common $X(t)'$ norm in FMPCCLF.

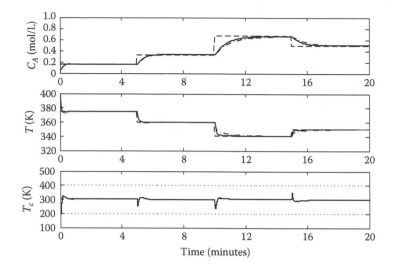

FIGURE 10.7 Closed-loop control performance for FMPCPLF (solid: online FMPCPLF; dotted-dashed: offline FMPCPLF; dashed: reference; dotted: input constraints).

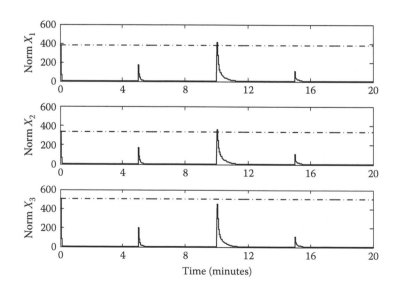

FIGURE 10.8 Profile of piecewise $Xs(t)'s$ norm in FMPCPLF.

10.5 CONCLUSIONS

This chapter has presented two approaches to the design of stable fuzzy model predictive controllers for T–S fuzzy systems based on common and piecewise quadratic Lyapunov functions, respectively. Both the issues of closed-loop asymptotic stability and suboptimal transient control performance can be addressed by solving a convex optimization problem. It has been demonstrated that the fuzzy model predictive control approach based on piecewise quadratic Lyapunov functions is able to achieve better control performance than that based on common quadratic Lyapunov functions.

11 Robust Filtering of T–S Fuzzy Systems

11.1 INTRODUCTION

In Chapters 5–10, stabilization and robust H_∞ control of various T–S fuzzy systems via state or output feedback have been studied. These systems include normal T–S fuzzy systems, uncertain T–S fuzzy systems, and T–S fuzzy systems with time delay. In addition, model predictive control of T–S fuzzy systems has also been studied. In this chapter a number of results on robust filtering of T–S fuzzy systems are presented.

The rest of the chapter is organized as follows. Section 11.2 introduces the problem formulation. Sections 11.3, 11.4, and 11.5 present robust H_∞ and generalized H_2 filter design methods based on common, piecewise, and fuzzy quadratic Lyapunov functions, respectively. Two simulation examples, including a chaotic system, are used to illustrate the procedure and the performance of the proposed approaches in Section 11.6, followed by conclusions in Section 11.7.

11.2 PROBLEM FORMULATION

Consider the following T–S fuzzy model,

$$R^l : \quad \text{IF} \quad z_1 \text{ is } F_1^l, \text{AND}, \dots z_v \text{ is } F_v^l$$

$$\text{THEN} \quad x(t+1) = A_l x(t) + B_l v(t) \tag{11.1}$$

$$y(t) = C_l x(t) + D_l v(t)$$

$$q(t) = Hx(t)$$

$$l \in L := \{1, 2, \dots, m\},$$

where most variables are the same as those defined in the previous chapters; that is, R^l denotes the lth fuzzy inference rule, m the number of inference rules, F_j^l ($j = 1, 2, \dots, v$) the fuzzy sets, $x(t) \in \mathfrak{R}^n$ the state vector, $v(t) \in \mathfrak{R}^s$ the noise vector that belongs to $l_2[0, \infty)$, $y(t) \in \mathfrak{R}^p$ the output vector, $q(t)$ a linear combination of the state variables to be estimated, $z(t) := [z_1(t), z_2(t), \dots, z_v(t)]$ some measurable variables of the system, and (A_l, B_l, C_l, D_l, H) the matrices of the lth subsystem.

Following a similar procedure to that in previous chapters, that is, by using a singleton fuzzifier, product fuzzy inference, and center-average defuzzifier, the following global T–S fuzzy dynamic model can be obtained,

$$x(t+1) = A(\mu)x(t) + B(\mu)v(t)$$

$$y(t) = C(\mu)x(t) + D(\mu)v(t)$$

(11.2)

$$q(t) = Hx(t),$$

where

$$A(\mu) = \sum_{l=1}^{m} \mu_l A_l, \quad B(\mu) = \sum_{l=1}^{m} \mu_l B_l, \quad C(\mu) = \sum_{l=1}^{m} \mu_l C_l, \quad D(\mu) = \sum_{l=1}^{m} \mu_l D_l, \quad (11.3)$$

$\mu_l(z(t))$ is the normalized membership function, and

$$\sum_{l=1}^{m} \mu_l = 1.$$

(11.4)

Both H_∞ filtering and generalized H_2 filtering of the T–S fuzzy systems are considered in this chapter, and the following assumption is adopted.

Assumption 11.1

The state variables of the system (11.1) are bounded for every initial condition and all admissible disturbances.

The first objective of this chapter is to design a suitable filter for the systems (11.2) for the estimation of the variable q such that the filter error system is globally exponentially stable with a guaranteed performance in the H_∞ sense. That is, given a prescribed level of disturbance attenuation $\gamma > 0$, find a suitable filter such that the filter error system is globally exponentially stable and the induced l_2-norm of the operator from v to the filtering error, $\hat{q} - q$, is less than γ under zero initial conditions

$$\| \hat{q} - q \|_2 < \gamma \| v \|_2$$

(11.5)

for all nonzero $v \in l_2[0,\infty)$, where $\hat{q}$ represents the estimate of the output q. In this case, the filter error dynamics is said to be globally exponentially stable with H_∞ performance γ.

The second objective of this chapter is to design a suitable filter for the systems (11.2) for estimation of the variable q such that the filter error system is globally exponentially stable with a guaranteed performance in the generalized H_2 sense. That is, given a prescribed level of noise attenuation $\sigma > 0$, find a suitable filter such that the filter error system is globally exponentially stable and the generalized H_2-norm of the error system defined as

$$\Xi := \sup \left\{ \| \hat{q}(T_f) - q(T_f) \| \mid \tilde{x}(0) = 0, T_f \geq 0, \sum_{t=0}^{T_f} v^T v \leq 1 \right\} \qquad (11.6)$$

is less than σ, where $\tilde{x}(t) = \hat{x}(t) - x(t)$. In this case, the filter error dynamics is said to be globally exponentially stable with generalized H_2 performance σ.

11.3 FILTER DESIGN BASED ON COMMON QUADRATIC LYAPUNOV FUNCTIONS

11.3.1 H_∞ FILTER DESIGN

In this section, we first present a robust H_∞ filter design approach based on a common Lyapunov function for the T–S fuzzy dynamic systems introduced in Section 11.2.

We choose the following fuzzy filter as

$$R^l : \quad \text{IF} \quad z_1 \text{ is } F_1^l, \text{ AND}, \dots z_v \text{ is } F_v^l$$

$$\text{THEN} \quad \hat{x}(t+1) = A_l \hat{x}(t) + G_l(\hat{y}(t) - y(t)) \qquad (11.7)$$

$$\hat{y}(t) = C_l \hat{x}(t)$$

$$\hat{q}(t) = H \hat{x}(t)$$

$$l \in L := \{1, 2, \dots, m\},$$

which can be rewritten as

$$\hat{x}(t+1) = A(\mu)\hat{x}(t) + G(\mu)(\hat{y}(t) - y(t))$$

$$\hat{y}(t) = C(\mu)\hat{x}(t) \qquad (11.8)$$

$$\hat{q}(t) = H \hat{x}(t),$$

where $G(\mu) = \sum_{l=1}^{m} \mu_l G_l$ and other variables are as defined in (11.3).

Define the filter errors $\tilde{q}(t) = \hat{q}(t) - q(t)$ and $\tilde{x}(t) = \hat{x}(t) - x(t)$. Then the filter error dynamic system is obtained by combining the filter (11.8) and the system (11.2), and can be described by the equation,

$$\tilde{x}(t+1) = A_c(\mu)\tilde{x}(t) + B_c(\mu)v(t)$$

$$\tilde{q}(t) = H \tilde{x}(t), \qquad (11.9)$$

where

$$A_c(\mu) = A(\mu) + G(\mu)C(\mu) = \sum_{j=1}^{m} \sum_{l=1}^{m} \mu_l \mu_j (A_l + G_l C_j), \tag{11.10}$$

$$B_c(\mu) = -(B(\mu) + G(\mu)D(\mu)) = -\sum_{j=1}^{m} \sum_{l=1}^{m} \mu_l \mu_j (B_l + G_l D_j).$$

Then one readily has the following result.

Theorem 11.1

Given a constant $\gamma > 0$, the filter error system (11.9) is globally exponentially stable with H_∞ performance γ if there exist a positive definite matrix P, a matrix R, and a set of matrices $Q_l, l \in L$, such that the following LMIs are satisfied,

$$\begin{bmatrix} P - R - R^T & RA_l + Q_l C_j & -RB_l - Q_l D_j \\ A_l^T R^T + C_j^T Q_l^T & -P + H^T H & 0 \\ -B_l^T R^T - D_j^T Q_l^T & 0 & -\gamma^2 I \end{bmatrix} < 0, \quad l, j \in L. \tag{11.11}$$

Moreover, the filter gains are given by

$$G_l = R^{-1} Q_l, \quad l \in L. \tag{11.12}$$

Proof: It follows from Lemma 6.1 that if one can show (6.5) and (6.6) rewritten as follows with different notations of matrices,

$$\gamma^2 I - B_c^T P B_c > 0, \tag{11.13}$$

$$A_c^T P A_c - P + A_c^T P B_c \left(\gamma^2 I - B_c^T P B_c \right)^{-1} B_c^T P A_c + H^T H < 0, \tag{11.14}$$

then the claimed results follow directly. By the Schur complement, (11.13) and (11.14) are equivalent to

$$\begin{bmatrix} A_c^T P A_c - P + H^T H & A_c^T P B_c \\ B_c^T P A_c & B_c^T P B_c - \gamma^2 I \end{bmatrix} < 0, \tag{11.15}$$

which can be rewritten as

$$\begin{bmatrix} A_c^T \\ B_c^T \end{bmatrix} P [A_c \quad B_c] - \begin{bmatrix} P - H^T H & 0 \\ 0 & \gamma^2 I \end{bmatrix} < 0. \tag{11.16}$$

In addition, performing congruence transformation to the inequality (11.17) by

$$W := \begin{bmatrix} [A_c \quad B_c] \\ I \end{bmatrix},$$

that is, multiplying (11.17) by W^T on the left and by W on the right, leads to exactly (11.16).

$$
\begin{bmatrix}
P - R - R^T & R[A_c \ \ B_c] \\
\begin{bmatrix} A_c^T \\ B_c^T \end{bmatrix} R^T & \begin{bmatrix} -P + H^T H & 0 \\ 0 & -\gamma^2 I \end{bmatrix}
\end{bmatrix} < 0.
\tag{11.17}
$$

Thus if one can show (11.17) then the inequalities (11.13) and (11.14) can be established. Note that (11.17) can be rewritten as

$$
\sum_{j=1}^{m} \sum_{l=1}^{m} \mu_l \mu_j
\begin{bmatrix}
P - R - R^T & R[A_l + G_l C_j \ \ B_l - G_l D_j] \\
\begin{bmatrix} A_l^T + C_j^T G_l^T \\ -B_l^T - D_j^T G_l^T \end{bmatrix} R^T & \begin{bmatrix} -P + H^T H & 0 \\ 0 & -\gamma^2 I \end{bmatrix}
\end{bmatrix} < 0.
$$

$$\tag{11.18}$$

Considering (11.14), (11.18) becomes

$$
\sum_{j=1}^{m} \sum_{l=1}^{m} \mu_l \mu_j
\begin{bmatrix}
P - R - R^T & [RA_l + Q_l C_j \ \ -RB_l - Q_l D_j] \\
\begin{bmatrix} A_l^T R^T + C_j^T Q_l^T \\ -B_l^T R^T - D_j^T Q_l^T \end{bmatrix} & \begin{bmatrix} -P + H^T H & 0 \\ 0 & -\gamma^2 I \end{bmatrix}
\end{bmatrix} < 0.
\tag{11.19}
$$

It then follows from (11.19) that LMIs in (11.11) imply (11.19). Thus it has been shown that (11.11) implies (11.13) and (11.14), and the proof is completed by invoking Lemma 6.1. ❑

The performance index γ described in the above theorem can also be optimized by the following convex optimization algorithm.

Algorithm 11.1: $\min_{P,R,Q_l} \gamma^2$, subject to LMIs (11.11).

11.3.2 GENERALIZED H_2 FILTER DESIGN

In this section, the filter design based on the second objective, that is, the generalized H_2 performance, is discussed. Reconsider the system model in (11.2), the fuzzy filter in (11.8), and the resulting filter error dynamic system in (11.9).

The following lemma is introduced.

Lemma 11.1

Given a constant $\sigma > 0$, the filter error system (11.9) is globally exponentially stable with generalized H_2 performance σ if there exists a positive definite matrix P such that the following matrix inequalities are satisfied,

$$P - \frac{1}{\sigma^2} H^T H > 0, \tag{11.20}$$

$$I - B_c^T P B_c > 0, \tag{11.21}$$

$$A_c^T P A_c - P + A_c^T P B_c \left(I - B_c^T P B_c \right)^{-1} B_c^T P A_c < 0. \tag{11.22}$$

Proof: It is easily seen that (11.21) and (11.22) imply the following inequality,

$$A_c^T P A_c - P < 0, \tag{11.23}$$

thus it follows that the filter error dynamic system is globally exponentially stable. Now we show the disturbance attenuation performance. Consider the Lyapunov function candidate,

$$V(\tilde{x}(t)) = \tilde{x}(t)^T P \tilde{x}(t). \tag{11.24}$$

Then its difference along the solution of the filter error system (11.9) can be described as

$$\Delta V(t) := V(t+1) - V(t)$$

$$= \tilde{x}(t+1)^T P \tilde{x}(t+1) - \tilde{x}(t)^T P \tilde{x}(t)$$

$$= \tilde{x}(t)^T \left(A_c^T P A_c - P \right) \tilde{x}(t) + v(t)^T B_c^T P A_c \tilde{x}(t) + \tilde{x}(t)^T A_c^T P B_c v(t) + v(t)^T B_c^T P B_c v(t)$$

$$\leq \tilde{x}(t)^T \left[-A_c^T P B_c (I - B_c^T P B_c)^{-1} B_c^T P A_c \right] \tilde{x}(t)$$

$$\quad + v(t)^T B_c^T P A_c \tilde{x}(t) + \tilde{x}(t)^T A_c^T P B_c v(t) + v(t)^T B_c^T P B_c v(t)$$

$$= v(t)^T v(t) - \varphi(t)^T \Pi \varphi(t)$$

$$\leq v(t)^T v(t), \tag{11.25}$$

where $\Pi = I - B_c^T P B_c$, $\varphi(t) = v(t) - \Pi^{-1} B_c^T P A_c \tilde{x}$, and (11.22) has been used. Then it follows from (11.25) that

$$V(\tilde{x}(T_f)) - V(\tilde{x}(0)) \leq \sum_{t=0}^{T_f} v(t)^T v(t);$$

that is, with $\tilde{x}(0) = 0$,

$$V(\tilde{x}(T_f)) \leq \sum_{t=0}^{T_f} v(t)^T v(t). \tag{11.26}$$

In addition, it follows from (11.20) that

$$\frac{1}{\sigma^2} \tilde{x}(T_f)^T H^T H \tilde{x}(T_f) < \tilde{x}(T_f)^T P_l \tilde{x}(T_f); \tag{11.27}$$

that is,

$$\|\tilde{q}(T_f)\|^2 < \sigma^2 V(\tilde{x}(T_f)). \tag{11.28}$$

Then combining (11.28) and (11.26) leads to

$$\| \tilde{q}(T_f) \|^2 < \sigma^2 \sum_{t=0}^{T_f} v(t)^T v(t), \tag{11.29}$$

which implies that the generalized H_2 norm of the filter error system is less than σ, and the proof is thus completed. $\square$

Now one readily has the following result.

Theorem 11.2
Given a constant $\sigma > 0$, the filter error system (11.9) is globally exponentially stable with generalized H_2 performance σ, if there exist a positive definite matrix P, a matrix R, and a set of matrices $Q_l, l \in L$, such that the following LMIs are satisfied,

$$\begin{bmatrix} P & H^T \\ H & \sigma^2 I \end{bmatrix} > 0, \tag{11.30}$$

$$\begin{bmatrix} P - R - R^T & RA_l + Q_l C_j & -RB_l - Q_l D_j \\ A_l^T R^T + C_j^T Q_l^T & -P & 0 \\ -B_l^T R^T - D_j^T Q_l^T & 0 & -I \end{bmatrix} < 0, \quad l, j \in L. \tag{11.31}$$

Moreover, the filter gains are given by

$$G_l = R^{-1} Q_l, \quad l \in L. \tag{11.32}$$

Proof: The proof follows directly from Lemma 11.1 and a similar procedure as in the proof of Theorem 11.1 and is thus omitted. $\square$

Similarly, the performance index σ described in the theorem can also be optimized by the following convex optimization algorithm.

Algorithm 11.2: $\min_{P,R,Q_l} \sigma^2$, subject to LMIs (11.30) and (11.31).

11.4 FILTER DESIGN BASED ON PIECEWISE QUADRATIC LYAPUNOV FUNCTIONS

Consider a space partition of the second kind and the following piecewise fuzzy system,

$$x(t+1) = \sum_{k \in \aleph(l)} \mu_k(z(t))[A_k x(t) + B_k v(t)]$$

$$y(t) = \sum_{k \in \aleph(l)} \mu_k(z(t))[C_k x(t) + D_k v(t)] \qquad (11.33)$$

$$q(t) = Hx(t)$$

$$z(t) \in S_l, l \in \bar{L}.$$

The corresponding fuzzy filter can be designed as

$$\hat{x}(t+1) = \sum_{k \in \aleph(l)} \mu_k(z(t))[A_k \hat{x}(t) + B_k v(t) + G_l(\hat{y}(t) - y(t))]$$

$$\hat{y}(t) = \sum_{k \in \aleph(l)} \mu_k(z(t))C_k \hat{x}(t) \qquad (11.34)$$

$$\hat{q}(t) = H\hat{x}(t)$$

$$z(t) \in S_l, l \in \bar{L}.$$

With the same definition of the filter errors $\tilde{x}(t) = \hat{x}(t) - x(t)$ and $\tilde{q}(t) = \hat{q}(t) - q(t)$, the fuzzy filter error dynamic equation can be described as

$$\tilde{x}(t+1) = \sum_{k \in \aleph(l)} \mu_k(z(t))[A_{clk}\tilde{x}(t) + B_{clk}v(t)]$$

$$\tilde{q}(t) = H\tilde{x}(t) \qquad z(t) \in S_l, l \in \bar{L}, \quad (11.35)$$

where

$$A_{clk} = A_k + G_l C_k, \, B_{clk} = -(B_k + G_l D_k). \qquad (11.36)$$

Define the set $\bar{\Omega}$ as in (4.32) representing all possible trajectory transitions among regions.

11.4.1 H_∞ FILTER DESIGN

In this section, a robust H_∞ filter design approach is presented based on a piecewise quadratic Lyapunov function.

Theorem 11.3

Given a constant $\gamma > 0$, the filter error system (11.35) is globally exponentially stable with H_∞ performance γ if there exist a set of positive definite matrices P_l, $l \in L$, and two sets of matrices R_l, Q_l, $l \in L$, such that the following LMIs are satisfied,

$$
\begin{bmatrix}
P_l - R_l - R_l^T & R_l A_k + Q_l C_k & -R_l B_k - Q_l D_k \\
A_k^T R_l^T + C_k^T Q_l^T & -P_l + H^T H & 0 \\
-B_k^T R_l^T - D_k^T Q_l^T & 0 & -\gamma^2 I
\end{bmatrix} < 0, \quad l \in \bar{L}, k \in \aleph(l), \quad (11.37)
$$

$$
\begin{bmatrix}
P_j - R_l - R_l^T & R_l A_k + Q_l C_k & -R_l B_k - Q_l D_k \\
A_k^T R_l^T + C_k^T Q_l^T & -P_l + H^T H & 0 \\
-B_k^T R_l^T - D_k^T Q_l^T & 0 & -\gamma^2 I
\end{bmatrix} < 0, \quad (l, j) \in \bar{\Omega}. \qquad (11.38)
$$

Moreover, the filter gains are given by

$$
G_l = R_l^{-1} Q_l, \quad l \in \bar{L}. \tag{11.39}
$$

Proof: It follows from Lemma 6.4 that if one can show (6.47)–(6.50) for the filter error system described in (11.35), then the claimed results follow directly. Note that (11.37) is a special case of (11.38) with $j = l$, so one only needs to prove (11.38), and (11.37) follows directly. By the Schur complement, (6.49) and (6.50) in this case are equivalent to

$$
\begin{bmatrix}
A_{clk}^T P_j A_{clk} - P_l + H^T H & A_{clk}^T P_j B_{clk} \\
B_{clk}^T P_j A_{clk} & B_{clk}^T P_j B_{clk} - \gamma^2 I
\end{bmatrix} < 0, \tag{11.40}
$$

which can be rewritten as

$$
\begin{bmatrix} A_{clk}^T \\ B_{clk}^T \end{bmatrix} P_j [A_{clk} \quad B_{clk}] - \begin{bmatrix} P_l - H^T H & 0 \\ 0 & \gamma^2 I \end{bmatrix} < 0. \tag{11.41}
$$

On the other hand, performing congruence transformation on the following inequality (11.42) by

$$
W := \begin{bmatrix} [A_{clk} \quad B_{clk}] \\ I \end{bmatrix},
$$

that is, multiplying (11.42) by W^T on the left and by W on the right, leads to exactly (11.41),

$$
\begin{bmatrix}
P_j - R_l - R_l^T & R_l[A_{clk} \quad B_{clk}] \\
\begin{bmatrix} A_{clk}^T \\ B_{clk}^T \end{bmatrix} R_l^T & \begin{bmatrix} -P_l + H^T H & 0 \\ 0 & -\gamma^2 I \end{bmatrix}
\end{bmatrix} < 0.
\tag{11.42}
$$

Thus if one can show (11.42) then the inequalities (6.49) and (6.50) can be established. Note that with (11.36) and (11.39), (11.42) can be expressed as

$$
\begin{bmatrix}
P_j - R_l - R_l^T & R_l A_k + Q_l C_k & -R_l B_k - Q_l D_k \\
A_k^T R_l^T + C_k^T Q_l^T & -P_l + H^T H & 0 \\
-B_k^T R_l^T - D_k^T Q_l^T & 0 & -\gamma^2 I
\end{bmatrix} < 0,
\tag{11.43}
$$

which is exactly (11.38). Therefore, one has shown that (11.38) implies both (6.49) and (6.50). Similarly it can be shown that (11.37) implies both (6.47) and (6.48), and the proof is thus completed by invoking Lemma 6.4. ❑

Similarly, the performance index γ described in the above theorem can also be optimized by the following convex optimization algorithm.

Algorithm 11.3: $\min\limits_{P_l, R_l, Q_l} \gamma^2$, subject to LMIs (11.37) and (11.38).

Remark 11.3

The results obtained in the theorem are made possible by following the technique of introducing extra slack variables R_l as in de Oliveira, Bernussou, and Geromel (1999). It should also be noted that the matrices R_l are not required to be symmetric.

11.4.2 GENERALIZED H_2 FILTER DESIGN

In this section, the filter design based on generalized H_2 performance is studied. Reconsider the system model in (11.33), the fuzzy filter in (11.34), and the resulting filter error dynamic system in (11.35).

The following lemma is useful for subsequent development.

Lemma 11.3

Given a constant $\sigma > 0$, the filter error system (11.35) is globally exponentially stable with generalized H_2 performance σ, if there exists a set of positive definite matrices $P_l, l \in L$ such that the following matrix inequalities are satisfied,

$$
P_l - \frac{1}{\sigma^2} H^T H > 0, l \in \bar{L},
\tag{11.44}
$$

$$
I - B_{clk}^T P_l B_{clk} > 0, l \in \bar{L}, k \in \aleph(l),
\tag{11.45}
$$

$$A_{clk}^T P_l A_{clk} - P_l + A_{clk}^T P_l B_{clk} \left(I - B_{clk}^T P_l B_{clk} \right)^{-1} B_{clk}^T P_l A_{clk} < 0, \, l \in \bar{L}, \, k \in \aleph(l), \quad (11.46)$$

$$I - B_{clk}^T P_j B_{clk} > 0, \quad l, \, j \in \bar{\Omega}, \, k \in \aleph(l), \quad (11.47)$$

$$A_{clk}^T P_j A_{clk} - P_l + A_{clk}^T P_j B_{clk} \left(I - B_{clk}^T P_j B_{clk} \right)^{-1} B_{clk}^T P_j A_{clk} < 0, \quad l, \, j \in \bar{\Omega}, \, k \in \aleph(l). \quad (11.48)$$

Proof: It is easily seen that (11.45)–(11.48) imply the following inequalities,

$$A_{clk}^T P_l A_{clk} - P_l < 0, \, l \in \bar{L}, \, k \in \aleph(l), \quad (11.49)$$

$$A_{clk}^T P_j A_{clk} - P_l < 0, \, l, \, j \in \bar{\Omega}, \, k \in \aleph(l), \quad (11.50)$$

thus it follows from Theorem 4.3 and its proof that the filter error dynamic system is globally exponentially stable. Then by considering the following Lyapunov function,

$$V(\tilde{x}) = \tilde{x}^T P_l \tilde{x}, \, z(t) \in S_l, \quad (11.51)$$

the proof of the disturbance attenuation performance can be established by following similar arguments as in the proof of Lemma 11.1, and it is thus omitted. ❑

Then one readily obtains the following result.

Theorem 11.4

Given a constant $\sigma > 0$, the filter error system (11.35) is globally exponentially stable with generalized H_2 performance σ if there exist a set of positive definite matrices $P_l, l \in \bar{L}$, and two sets of matrices $R_l, Q_l, l \in \bar{L}$, such that the following LMIs are satisfied,

$$\begin{bmatrix} P_l & H^T \\ H & \sigma^2 I \end{bmatrix} > 0, \quad l \in \bar{L}, \quad (11.52)$$

$$\begin{bmatrix} P_l - R_l - R_l^T & R_l A_k + Q_l C_k & -R_l B_k - Q_l D_k \\ A_k^T R_l^T + C_k^T Q_l^T & -P_l & 0 \\ -B_k^T R_l^T - D_k^T Q_l^T & 0 & -I \end{bmatrix} < 0, \quad l \in \bar{L}, \, k \in \aleph(l), \quad (11.53)$$

$$\begin{bmatrix} P_j - R_l - R_l^T & R_l A_k + Q_l C_k & -R_l B_k - Q_l D_k \\ A_k^T R_l^T + C_k^T Q_l^T & -P_l & 0 \\ -B_k^T R_l^T - D_k^T Q_l^T & 0 & -I \end{bmatrix} < 0, \quad l, \, j \in \bar{\Omega}, \, k \in \aleph(l). \quad (11.54)$$

Moreover, the filter gains are given by

$$G_l = R_l^{-1} Q_l, \quad l \in \bar{L}. \quad (11.55)$$

Proof: The proof follows directly from Lemma 11.2 and a similar procedure as in the proof of Theorem 11.3 and is thus omitted. ❑

Similarly, the performance index σ described in the theorem can also be optimized by the following convex optimization algorithm.

Algorithm 11.4: $\min\limits_{P_l, R_l, Q_l} \sigma^2$, subject to LMIs (11.52)–(11.54).

11.5 FILTER DESIGN BASED ON FUZZY QUADRATIC LYAPUNOV FUNCTIONS

11.5.1 H_∞ FILTER DESIGN

In this section, a robust H_∞ filter design approach is presented based on a fuzzy quadratic Lyapunov function.

Reconsider the T–S fuzzy model (11.2) and the fuzzy filter (11.7) or (11.8). The filter error dynamic system is as described in (11.9), which can also be expressed as

$$\tilde{x}(t+1) = \sum_{j=1}^{m} \sum_{l=1}^{m} \mu_l \mu_j [A_{clj} \tilde{x}(t) + B_{clj} v(t)]$$

(11.56)

$$\tilde{q}(t) = H\tilde{x}(t),$$

where

$$A_{clj} = (A_l + G_l C_j), \quad B_{clj} = -(B_l + G_l D_j).$$

(11.57)

Then one readily has the following result.

Theorem 11.5

Given a constant $\gamma > 0$, the filter error system (11.9) or equivalently (11.56) is globally exponentially stable with H_∞ performance γ if there exist a set of positive definite matrices P_l, $l \in L$, and two sets of matrices R_l, Q_l, $l \in L$, such that the following LMIs are satisfied,

$$\begin{bmatrix} P_i - R_l - R_l^T & R_l A_i + Q_l C_j & -R_l B_l - Q_l D_j \\ A_i^T R_l^T + C_j^T Q_l^T & -P_j + H^T H & 0 \\ -B_l^T R_l^T - D_j^T Q_l^T & 0 & -\gamma^2 I \end{bmatrix} < 0, \quad i, l, j \in L. \quad (11.58)$$

Moreover, the filter gains are given by

$$G_l = R_l^{-1} Q_l, \quad l \in L.$$

(11.59)

Proof: It follows from Lemma 6.5 that if one can show (6.5) and (6.6) rewritten with different notations of matrices as

$$\gamma^2 I - B_{clj}^T P_i B_{clj} < 0, \quad i, l, j \in L,$$

(11.60)

$$A_{clj}^T P_i A_{clj} - P_j + A_{clj}^T P_i B_{clj} \left(\gamma^2 I - B_{clj}^T P_i B_{clj} \right)^{-1} B_{clj}^T P_i A_{clj} + H^T H < 0, \quad i, l, j \in L, \quad (11.61)$$

then the claimed results follow directly. By the Schur complement, (11.60) and (11.61) are equivalent to

$$
\begin{bmatrix}
A_{clj}^T P_i A_{clj} - P_j + H^T H & A_{clj}^T P_i B_{clj} \\
B_{clj}^T P_i A_{clj} & B_{clj}^T P_i B_{clj} - \gamma^2 I
\end{bmatrix} < 0, \tag{11.62}
$$

which can be rewritten as

$$
\begin{bmatrix} A_{clj}^T \\ B_{clj}^T \end{bmatrix} P_i [A_{clj} \quad B_{clj}] - \begin{bmatrix} P_j - H^T H & 0 \\ 0 & \gamma^2 I \end{bmatrix} < 0. \tag{11.63}
$$

On the other hand, performing congruence transformation to the following inequality (11.64) by

$$
W := \begin{bmatrix} [A_{clj} \quad B_{clj}] \\ I \end{bmatrix},
$$

that is, multiplying (11.64) by W^T on the left and by W on the right, leads exactly to (11.63).

$$
\begin{bmatrix}
P_i - R_l - R_l^T & R_l[A_{clj} \quad B_{clj}] \\
\begin{bmatrix} A_{clj}^T \\ B_{clj}^T \end{bmatrix} R_l^T & \begin{bmatrix} -P_j + H^T H & 0 \\ 0 & -\gamma^2 I \end{bmatrix}
\end{bmatrix} < 0. \tag{11.64}
$$

Thus if one can show (11.64) then the inequalities (11.60) and (11.61) can be established. Note that via (11.57), (11.64) can be rewritten as

$$
\begin{bmatrix}
P_i - R_l - R_l^T & R_l[A_l + G_l C_j \quad -B_l - G_l D_j] \\
\begin{bmatrix} A_l^T + C_j^T G_l^T \\ -B_l^T - D_j^T G_l^T \end{bmatrix} R_l^T & \begin{bmatrix} -P_j + H^T H & 0 \\ 0 & -\gamma^2 I \end{bmatrix}
\end{bmatrix} < 0. \tag{11.65}
$$

Considering (11.59), (11.65) becomes

$$
\begin{bmatrix}
P_i - R_l - R_l^T & [R_l A_l + Q_l C_j \quad -R_l B_l - Q_l D_j] \\
\begin{bmatrix} A_l^T R_l^T + C_j^T Q_l^T \\ -B_l^T R_l^T - D_j^T Q_l^T \end{bmatrix} & \begin{bmatrix} -P_j + H^T H & 0 \\ 0 & -\gamma^2 I \end{bmatrix}
\end{bmatrix} < 0, \tag{11.66}
$$

which is exactly (11.58). It then follows that LMIs in (11.58) imply (11.64). Therefore one has shown that (11.58) implies (11.60) and (11.61), and the proof is thus completed by invoking Lemma 6.5. ❑

Similarly, the performance index γ described in the above theorem can also be optimized by the following convex optimization algorithm.

Algorithm 11.5: $\min\limits_{P_l,R_l,Q_l} \gamma^2$, subject to LMIs (11.58).

11.5.2 GENERALIZED H_2 FILTER DESIGN

Reconsider the system model in (11.2), the fuzzy filter in (11.8), and the resulting filter error dynamic system in (11.9) or (11.56).

The following lemma is introduced.

Lemma 11.3

Given a constant $\sigma > 0$, the filter error system (11.9) or equivalently (11.56) is globally exponentially stable with generalized H_2 performance σ if there exists a set of positive definite matrices $P_l, l \in L$ such that the following matrix inequalities are satisfied,

$$P_i - \frac{1}{\sigma^2} H^T H > 0, \quad i \in L, \tag{11.67}$$

$$I - B_{clj}^T P_i B_{clj} > 0, \quad i, l, j \in L, \tag{11.68}$$

$$A_{clj}^T P_i A_{clj} - P_j + A_{clj}^T P_i B_{clj} \left(I - B_{clj}^T P_i B_{clj} \right)^{-1} B_{clj}^T P_i A_{clj} < 0, \quad i, l, j \in L. \tag{11.69}$$

Proof: It is easily seen that (11.68) and (11.69) imply the following inequality,

$$A_{clj}^T P_i A_{clj} - P_j < 0, \tag{11.70}$$

thus it follows from Theorem 4.4 that the filter error dynamic system is globally exponentially stable.

Then by considering the Lyapunov function,

$$V(x) = \sum_{l=1}^{m} \mu_l(z) x^T P_l x, \tag{11.71}$$

the proof of the disturbance attenuation performance can be readily established by following similar arguments to those in the proof of Lemma 11.1, and it is thus omitted. $\square$

Now one readily obtains the following result.

Theorem 11.6

Given a constant $\sigma > 0$, the filter error system (11.9) or equivalently (11.56) is globally exponentially stable with generalized H_2 performance σ if there exist a set of positive definite matrices $P_l, l \in L$, and two sets of matrices $R_l, Q_l, l \in L$, such that the following LMIs are satisfied,

$$\begin{bmatrix} P_i & H^T \\ H & \sigma^2 I \end{bmatrix} > 0, \quad i \in L, \tag{11.72}$$

$$\begin{bmatrix} P_i - R_l - R_l^T & R_l A_l + Q_l C_j & -R_l B_l - Q_l D_j \\ A_l^T R_l^T + C_j^T Q_l^T & -P_j & 0 \\ -B_l^T R_l^T - D_j^T Q_l^T & 0 & -I \end{bmatrix} < 0, \quad i, l, j \in L. \tag{11.73}$$

Moreover, the filter gains are given by

$$G_l = R_l^{-1} Q_l, \quad l \in L. \tag{11.74}$$

Proof: The proof follows directly from Lemma 11.3 and a similar procedure to that in the proof of Theorem 11.5 and is thus omitted. ❑

Similarly, the performance index σ described in the theorem can also be optimized by the following convex optimization algorithm.

Algorithm 11.6: $\min_{P_i, R_l, Q_l} \sigma^2$, subject to LMIs (11.72) and (11.73).

11.6   SIMULATION EXAMPLES

Two numerical examples are presented to illustrate the application of the proposed filter design approaches and their performance.

Example 11.1
Consider the modified chaotic Henon mapping model without external disturbance

$$\begin{cases} x_1(t+1) = -x_1^2(t) - 2x_f x_1(t) + 0.3 x_2(t) \\ x_2(t+1) = x_1(t) \end{cases}, \tag{11.75}$$

where $x_f = 0.8839$. It is noted that the modified Henon model is obtained from the error model with respect to one of the fixed points of the original Henon model. The system dynamic appears in a chaotic manner as shown in Figure 11.1 with initial condition $x(0) = [0.1 \quad 0]^T$.

Consider the modified chaotic Henon model with external disturbance which can be represented by the following T–S fuzzy model,

$$R^l : \quad \text{IF} \quad x_1(t) \text{ is } F_1^l$$
$$\text{THEN} \quad x(t+1) = A_l x(t) + B_l v(t)$$
$$y(t) = C_l x(t) + D_l v(t)$$
$$q(t) = Hx(t), \tag{11.76}$$
$$l = 1, 2, 3,$$

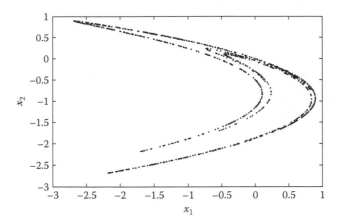

FIGURE 11.1 Chaotic behavior of the modified Henon mapping model.

where

$$A_1 = \begin{bmatrix} -d & 0.3 \\ 1 & 0 \end{bmatrix}, A_2 = \begin{bmatrix} 0 & 0.3 \\ 1 & 0 \end{bmatrix} A_3 = \begin{bmatrix} d & 0.3 \\ 1 & 0 \end{bmatrix}, B_1 = B_3 = \begin{bmatrix} 1 \\ 0 \end{bmatrix}, B_2 = \begin{bmatrix} 0 \\ 0 \end{bmatrix},$$

$$C_1 = C_2 = C_3 = [2 \quad 0], \quad D_1 = D_3 = 1, \quad D_2 = 0, \quad H = [1 \quad 3],$$

d is a constant, the disturbance is given by $v(t) = e^{-0.5t} \sin(0.01\pi t)$, and the membership functions are given by

$$\mu_1(x_1(t)) = \frac{x_1(t) + 2x_f}{d}, \quad x_1(t) \in [-2x_f, d - 2x_f]$$

$$\mu_2(x_1(t)) = \begin{cases} 1 - \dfrac{x_1(t) + 2x_f}{d}, & x_1(t) \in [-2x_f, d - 2x_f] \\ 1 + \dfrac{x_1(t) + 2x_f}{d}, & x_1(t) \in [-d - 2x_f, -2x_f] \end{cases}$$

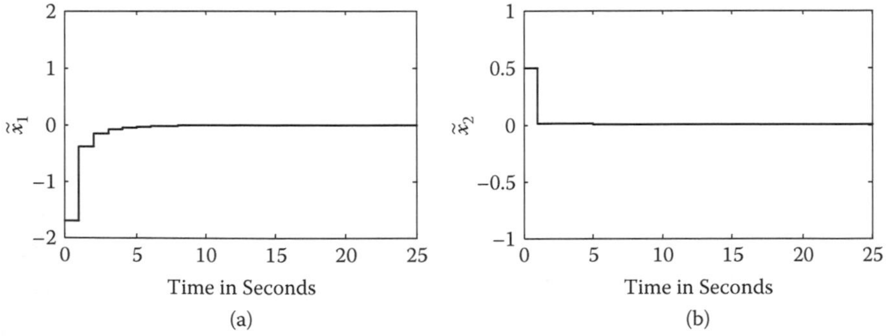

FIGURE 11.2 Responses of H_∞ filter based on common Lyapunov function.

$$\mu_3(x_1(t)) = -\frac{x_1(t) + 2x_f}{d}, \quad x_1(t) \in [-d - 2x_f, -2x_f].$$

It is chosen that $d = 1$ in this study. Without external disturbance (11.76) represents the model (11.75) exactly when $x_1(t) \in [-d - 2x_f, d - 2x_f]$.

Case A: Filtering Based on Common Quadratic Lyapunov Functions

By applying Algorithm 11.1, the robust H_∞ filtering results are obtained as

$$P = \begin{bmatrix} 1.8487 & 3.8585 \\ 3.8585 & 9.8686 \end{bmatrix}, \quad R = \begin{bmatrix} 1.8487 & 3.8585 \\ 3.8585 & 9.8686 \end{bmatrix},$$

$$G_1 = \begin{bmatrix} 0.6483 \\ -0.5000 \end{bmatrix}, \quad G_2 = \begin{bmatrix} 0.1483 \\ -0.5000 \end{bmatrix}, \quad G_3 = \begin{bmatrix} -0.3518 \\ -0.5000 \end{bmatrix},$$

with the optimal H_∞ performance index $\gamma_{\min} = 1.5122$. Satisfactory filter error responses with the above solutions and initial conditions $x(0) = [0.2 \quad 0]^T$ and $\hat{x}(0) = [-1.5 \quad 0.5]^T$ are shown in Figure 11.2, and also in the subsequent figures, where (a) represents the error response $\tilde{x}_1$ and (b) the error response $\tilde{x}_2$.

By applying Algorithm 11.2, the robust H_2 filtering results are obtained as follows,

$$P = \begin{bmatrix} 0.9813 & 1.5698 \\ 1.5698 & 4.7206 \end{bmatrix}, \quad R = \begin{bmatrix} 0.9813 & 1.5698 \\ 1.5698 & 4.7206 \end{bmatrix},$$

$$G_1 = \begin{bmatrix} 0.3836 \\ -0.3961 \end{bmatrix}, \quad G_2 = \begin{bmatrix} 0.0447 \\ -0.4485 \end{bmatrix}, \quad G_3 = \begin{bmatrix} -0.5217 \\ -0.4553 \end{bmatrix},$$

with the optimal H_2 performance index $\sigma_{\min} = 1.3808$. Satisfactory filter error responses with the above solutions and the same initial conditions are shown in Figure 11.3.

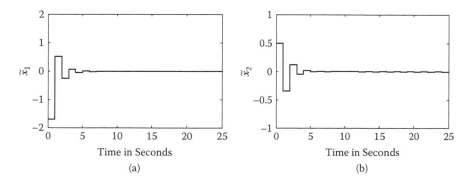

Time in Seconds
(a)

Time in Seconds
(b)

FIGURE 11.3 Responses of H_2 filter based on common Lyapunov function.

Case B: Filtering Based on Piecewise Quadratic Lyapunov Functions

By applying Algorithm 11.3, the H_∞ filtering results are obtained as follows,

$$P_1 = \begin{bmatrix} 1.6114 & 3.3627 \\ 3.3627 & 10.2523 \end{bmatrix}, \quad P_2 = \begin{bmatrix} 1.6114 & 3.3628 \\ 3.3628 & 10.2524 \end{bmatrix},$$

$$R_1 = \begin{bmatrix} 1.6114 & 3.3630 \\ 3.3630 & 10.2537 \end{bmatrix}, \quad R_2 = \begin{bmatrix} 1.6114 & 3.3627 \\ 3.3628 & 10.2524 \end{bmatrix},$$

$$G_1 = \begin{bmatrix} 0.4241 \\ -0.5626 \end{bmatrix}, \quad G_2 = \begin{bmatrix} -0.0759 \\ -0.5626 \end{bmatrix},$$

with the optimal H_∞ performance index $\gamma_{min} = 1.0666$. Satisfactory filter error responses with the above solutions and the same initial conditions are shown in Figure 11.4.

By applying Algorithm 11.4, the generalized H_2 filtering results are obtained as

$$P_1 = \begin{bmatrix} 1.6167 & 3.3990 \\ 3.3990 & 9.9624 \end{bmatrix}, \quad P_2 = \begin{bmatrix} 1.6168 & 3.3990 \\ 3.3990 & 9.9623 \end{bmatrix},$$

$$R_1 = \begin{bmatrix} 1.6168 & 3.3992 \\ 3.3992 & 9.9629 \end{bmatrix}, \quad R_2 = \begin{bmatrix} 1.6168 & 3.3991 \\ 3.3991 & 9.9623 \end{bmatrix},$$

$$G_1 = \begin{bmatrix} 0.4368 \\ -0.5645 \end{bmatrix}, \quad G_2 = \begin{bmatrix} -0.0632 \\ -0.5645 \end{bmatrix},$$

with the optimal generalized H_2 performance $\sigma_{min} = 0.9511$. Satisfactory filter error responses with the above solutions and the same initial conditions are shown in Figure 11.5.

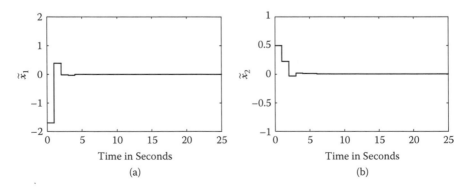

FIGURE 11.4 Responses of H_∞ filter based on piecewise Lyapunov function.

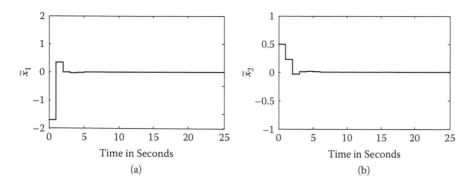

FIGURE 11.5   Responses of H_2 filter based on piecewise Lyapunov function.

Case C: Filtering Based on Fuzzy Quadratic Lyapunov Functions

By applying Algorithm 11.5, the robust H_∞ filtering results are obtained as

$$P_1 = \begin{bmatrix} 1.8649 & 3.8158 \\ 3.8158 & 9.8294 \end{bmatrix}, P_2 = \begin{bmatrix} 1.6796 & 3.4497 \\ 3.4497 & 9.9222 \end{bmatrix}, P_3 = \begin{bmatrix} 1.8649 & 3.8158 \\ 3.8158 & 9.8294 \end{bmatrix},$$

$$R_1 = \begin{bmatrix} 1.8655 & 3.8319 \\ 3.9438 & 14.6294 \end{bmatrix}, R_2 = \begin{bmatrix} 9.2063 & 2.7206 \\ 2.7205 & 9.9929 \end{bmatrix}, R_3 = \begin{bmatrix} 1.8669 & 3.8038 \\ 3.7716 & 16.9138 \end{bmatrix},$$

$$G_1 = \begin{bmatrix} 0.6410 \\ -0.5298 \end{bmatrix}, G_2 = \begin{bmatrix} 0.1465 \\ -0.4966 \end{bmatrix}, G_3 = \begin{bmatrix} -0.4599 \\ -0.4812 \end{bmatrix},$$

with the optimal H_∞ performance index $\gamma_{min} = 1.5106$. Satisfactory filter error responses with the above solutions and the same initial conditions are shown in Figure 11.6.

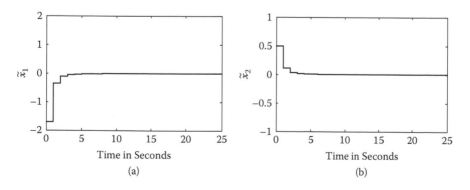

FIGURE 11.6   Responses of H_∞ filter based on fuzzy Lyapunov function.

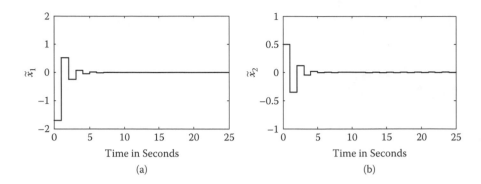

FIGURE 11.7 Responses of H_2 filter based on fuzzy Lyapunov function.

By applying Algorithm 11.6, the robust H_2 filtering results are obtained as follows,

$$P_1 = \begin{bmatrix} 0.9813 & 1.5699 \\ 1.5699 & 4.7206 \end{bmatrix}, P_2 = \begin{bmatrix} 0.9236 & 1.6326 \\ 1.6326 & 4.7295 \end{bmatrix}, P_3 = \begin{bmatrix} 0.9813 & 1.5699 \\ 1.5699 & 4.7206 \end{bmatrix},$$

$$R_1 = \begin{bmatrix} 0.9813 & 1.5699 \\ 1.5699 & 4.7208 \end{bmatrix}, R_2 = \begin{bmatrix} 52.2949 & 5.2232 \\ 5.2218 & 4.9806 \end{bmatrix}, R_3 = \begin{bmatrix} 0.9813 & 1.5699 \\ 1.6015 & 7.2912 \end{bmatrix}$$

$$G_1 = \begin{bmatrix} 0.3834 \\ -0.3948 \end{bmatrix}, G_2 = \begin{bmatrix} 0.0455 \\ -0.4487 \end{bmatrix}, G_3 = \begin{bmatrix} -0.5229 \\ -0.4533 \end{bmatrix},$$

with the optimal H_2 performance index $\sigma_{min} = 1.3808$. Satisfactory filter error responses with the above solutions and the same initial conditions are shown in Figure 11.7.

For easy comparison, the filtering performances under different filter design approaches are summarized in Table 11.1.

It can be observed that the filtering approaches based on piecewise quadratic Lyapunov functions or fuzzy quadratic Lyapunov functions achieve better performance than the approaches based on common quadratic Laypunov functions, whereas

TABLE 11.1

Comparison of H_∞ and H_2 Filtering of Example 11.1

Approach	H_∞ Performance	H_2 Performance
CQLF based filter	$\gamma_{min} = 1.5122$	$\sigma_{min} = 1.3808$
PQLF based filter	$\gamma_{min} = 1.0666$	$\sigma_{min} = 0.9511$
FQLF based filter	$\gamma_{min} = 1.5106$	$\sigma_{min} = 1.3808$

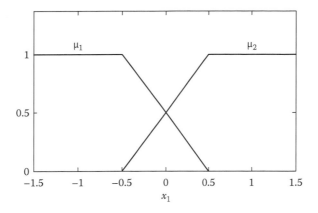

FIGURE 11.8 Membership functions in Example 11.2.

the approaches based on piecewise quadratic Lyapunov functions appear to achieve better performance than those based on fuzzy quadratic Lyapunov functions.

In order to further show the advantages of the approaches based on piecewise or fuzzy quadratic Lyapunov functions over the approaches based on common quadratic Lyapunov functions, the following numerical example is also adopted.

Example 11.2

Consider a fuzzy discrete time system in the form (11.1) with system matrices given by

$$A_1 = \begin{bmatrix} 0.95 & 0.38 \\ -0.48 & 0.89 \end{bmatrix}, \quad A_2 = \begin{bmatrix} 0.95 & 0.48 \\ -0.09 & 0.89 \end{bmatrix}, \quad B_1 = \begin{bmatrix} 0 \\ 0.1 \end{bmatrix}, \quad B_2 = \begin{bmatrix} 0.1 \\ 0 \end{bmatrix},$$

$$C_1 = [1 \quad 0], \quad C_2 = [0 \quad -1], \quad D_1 = 0.1, \quad D_2 = -0.1, \quad H = [1 \quad 0].$$

The membership functions are chosen as shown in Figure 11.8. Applying Algorithms 11.1–11.6, respectively, the achievable minimum performance indices for the H_∞ and H_2 filters can be obtained and have been summarized in Table 11.2.

TABLE 11.2

Comparison of H_∞ and H_2 Filtering of Example 11.2

Approach	H_∞ Performance	H_2 Performance
CQLF based filter	Infeasible	Infeasible
PQLF based filter	$\gamma_{min} = 2.3982$	$\sigma_{min} = 0.5515$
FQLF based filter	$\gamma_{min} = 37.8655$	$\sigma_{min} = 2.9746$

The results clearly demonstrate that the approaches based on piecewise or fuzzy quadratic Lyapunov functions achieve much better performance than those based on common quadratic Lyapunov functions.

11.7 CONCLUSIONS

This chapter has presented a number of approaches to robust H_∞ and generalized H_2 filtering for T–S fuzzy systems based on common, piecewise, and fuzzy quadratic Lyapunov functions, respectively. It has been shown that the filter gains can be obtained by solving a number of linear matrix inequalities. Two examples have also been presented to demonstrate the performance of the proposed approaches, and the advantages of the approaches based on piecewise or fuzzy quadratic Lyapunov functions over those based on common quadratic Lyapunov functions.

12 Adaptive Control of T–S Fuzzy Systems

12.1 INTRODUCTION

In the previous chapters, a number of control design approaches to T–S fuzzy systems have been presented under a common assumption that the parameters of T–S fuzzy systems are known a priori. However, if the parameters of T–S fuzzy systems are not known a priori or if the parameters of T–S fuzzy systems are time-varying, one would not be able to apply those approaches. Instead, some kind of adaptive mechanism has to be adopted. In this chapter, two approaches to adaptive control of T–S fuzzy systems are presented. The basic idea is to design adaptive controllers in such a way that the global closed-loop adaptive control system is guaranteed to be stable.

The rest of the chapter is organized as follows. Section 12.2 introduces the T–S fuzzy system model in polynomial form. Section 12.3 presents two approaches to adaptive control design, and stability analysis of the resulting closed-loop adaptive control systems is also provided. One simulation example of chaos control is presented in Section 12.4, followed by concluding remarks in Section 12.5.

12.2 PROBLEM FORMULATION

Consider the following single-input single-output T–S fuzzy system in polynomial form,

$$R^i: \quad \text{IF} \quad z_1 \text{ is } F_1^i \text{ AND } \dots z_v \text{ is } F_v^i$$

$$\text{THEN} \quad y(t+1) = a_{i1}y(t) + \dots + a_{in}y(t-n+1) + b_{i1}u(t) + \dots + b_{in}u(t-n+1)$$

$$\text{(12.1)}$$

$$i \in L := \{1, 2, \dots, m\},$$

where R^i denotes the ith fuzzy inference rule, m the number of inference rules, $F_j^i (j = 1, 2, \dots, v)$ fuzzy sets, $u(t)$ the input variable, $y(t)$ the system output, $a_{i1}, \dots, a_{in}, b_{i1}, \dots, b_{in}$ are coefficients of the ith subsystem, and $z(t) := [z_1, z_2, \dots, z_v]$ are some measurable system variables.

By applying a standard fuzzy inference method, that is, by using a singleton fuzzifier, product fuzzy inference, and center-average defuzzifier, the following fuzzy global dynamic model is obtained,

$$y(t+1) = \sum_{i=1}^{m} \mu_i [a_{i1}y(t) + \dots + a_{in}y(t-n+1) + b_{i1}u(t) + \dots + b_{in}u(t-n+1)], \quad \text{(12.2)}$$

which can be rewritten as

$$y(t+1) = \sum_{i=1}^{m} \mu_i a_{i1} y(t) + \cdots + \sum_{i=1}^{m} \mu_i a_{in} y(t-n+1) + \sum_{i=1}^{m} \mu_i b_{i1} u(t) \cdots + \sum_{i=1}^{m} \mu_i b_{in} u(t-n+1),$$

(12.3)

where $\mu_i(z)$ is the normalized membership function satisfying

$$\mu_i = \frac{\xi_i(z)}{\sum_{j=1}^{m} \xi_j(z)}, \xi_j(z) = \prod_{j=1}^{v} F_j^i(z_j), \mu_i \geq 0, \sum_{i=1}^{m} \mu_i = 1,$$

(12.4)

and $F_j^i(z_j)$ is the membership grade of z_j in the fuzzy set F_j^i.

The following assumptions are made.

Assumption 12.1

The system in (12.3) is minimum phase in the sense that the zero dynamics of the system defined in (12.5),

$$0 = \sum_{i=1}^{m} \mu_i b_{i1} u(t) \cdots + \sum_{i=1}^{m} \mu_i b_{in} u(t-n+1),$$

(12.5)

are globally exponentially stable.

Assumption 12.2

$$\sum_{i=1}^{m} \mu_i b_{i1} \neq 0.$$

For simplicity in subsequent development, it is further assumed that the sign of $b_i, i = 1, 2, \ldots, m$ is known and positive, and moreover, $\underline{b}_i \leq b_i \leq \overline{b}_i, i = 1, 2, \ldots m$, with known lower bounds $\underline{b}_i$ and upper bounds $\overline{b}_i$.

Remark 12.1

Assumption 12.1 is standard for model reference adaptive control. It is similar to the assumption of minimum phase in linear systems in the sense that all zeros of the system are within the unit circle. Assumption 12.2 is needed to avoid the singularity of control design. If some more advanced adaptive control techniques are used, this assumption can be removed. In other words, this assumption is not essential in many cases.

In this chapter, we address the design of adaptive control for the T–S fuzzy system (12.3) when the coefficients of the fuzzy system are unknown a priori. To simplify the presentation, it is assumed that the membership functions have been given a priori based on an expert's knowledge or some other available information.

The objective is to find an adaptive control law such that the output of the closed-loop adaptive control system tracks a given bounded reference signal with guaranteed stability.

12.3 ADAPTIVE CONTROL SYSTEM DESIGN

In this section, approaches to adaptive control design for the T–S fuzzy systems are presented, and the stability of the resulting closed-loop adaptive control systems is studied.

12.3.1 ADAPTATION ALGORITHM

Define

$$\phi_i(t-1) = [\mu_i y(t-1) \quad \cdots \quad \mu_i y(t-n) \quad \mu_i u(t-1) \quad \cdots \quad \mu_i u(t-n)]^T,$$

$$\theta_i = [a_{i1} \quad \cdots \quad a_{in} \quad b_{i1} \quad \cdots \quad b_{in}]^T, \qquad (12.6)$$

$$\phi(t-1) = [\phi_1^T(t-1) \quad \phi_2^T(t-1) \quad \cdots \quad \phi_m^T(t-1)]^T,$$

$$\theta = \begin{bmatrix} \theta_1^T & \theta_2^T & \cdots & \theta_m^T \end{bmatrix}^T.$$

Then, the T–S fuzzy system (12.3) can be rewritten as

$$y(t) = \theta^T \phi(t-1). \qquad (12.7)$$

This is in a regression form, with θ being the system parameter vector and ϕ being the regression vector. It should be noted that the system (12.7) is in general nonlinear but is linear with respect to its unknown parameters. Therefore, all the parameter adaptation algorithms developed for linear systems can be employed for the estimation of these unknown parameters in (12.1) (Goodwin and Sin, 1984; Narendra and Annaswamy, 1989). Here, the following simple projection algorithm for parameter estimation is adopted,

$$\hat{\theta}(t) = \hat{\theta}(t-1) + \frac{\phi(t-1)e_1(t)}{1 + \phi(t-1)^T \phi(t-1)}, \qquad (12.8)$$

where $e_1(t) := y(t) - \hat{\theta}(t-1)^T \phi(t-1)$ and $\hat{\theta}(t)$ is the estimate of the parameter θ. With this parameter update law, the following convergence result has been proven (Goodwin and Sin, 1984; Middleton et al., 1988).

Lemma 12.1

The parameter update law (12.8), when applied to the T–S fuzzy model (12.7) in the regression form, has the following properties.

$E1$: $\hat{\theta}(t)$ is continuous and bounded.

$E2$: $\|\hat{\theta}(t) - \hat{\theta}(t-1)\| \in l_2$.

$E3$: $\bar{e}_1(t) := \dfrac{e_1(t)}{[1 + \phi(t-1)^T \phi(t-1)]^{1/2}} \in l_2$.

The certainty equivalence principle is applied to the design of adaptive controllers. In doing so, first consider the control design for the case where the parameters of the T–S fuzzy model are known a priori.

12.3.2 CONTROLLER DESIGN WITH KNOWN PARAMETERS

Define

$$g(t) = \sum_{i=1}^{m} \mu_i b_{i1}, \tag{12.9}$$

and it follows from Assumption 12.2 that $g(t) > 0$, $\forall t$.

Then, one can design the following fuzzy control law,

R^i : IF z_1 is F_1^i AND $\ldots$ z_v is F_v^i

THEN $u_i(t) = \dfrac{1}{g(t)}[-a_{i1}y(t) - a_{i2}y(t-1) - \cdots - a_{in}y(t-n+1) - b_{i2}u(t-1) - \cdots$

$$- b_{in}u(t-n+1) + y_m(t+1) - k_1 e(t) - k_2 e(t-1) - \cdots - k_n e(t-n+1)] \tag{12.10}$$

$i \in L := \{1, 2, \ldots, m\}$.

The global control law can be obtained as

$$u(t) = \sum_{i=1}^{m} \mu_i u_i(t)$$

$$= \dfrac{1}{g(t)}[-h(t) + y_m(t+1) - k_1 e(t) - k_2 e(t-1) - \cdots - k_n e(t-n+1)], \tag{12.11}$$

where the tracking error

$$e = y - y_m,$$

with y_m being the reference signal to be tracked,

$$h(t) = \sum_{i=1}^{m} \mu_i a_{i1} y(t) + \sum_{i=1}^{m} \mu_i a_{i2} y(t-1) + \cdots + \sum_{i=1}^{m} \mu_i a_{in} y(t-n+1)$$

$$+ \sum_{i=1}^{m} \mu_i b_{i2} u(t-1) + \cdots + \sum_{i=1}^{m} \mu_i b_{in} u(t-n+1), \tag{12.12}$$

and $\{k_i\}$ are coefficients of a stable polynomial defined by

$$\Psi(z) = z^n + k_1 z^{n-1} + \cdots + k_{n-1} z + k_n,$$

which specifies the desired output tracking error dynamics.

Substituting the control law (12.11) into the T–S fuzzy model (12.3) leads to the following closed-loop control system,

$$e(t+1) + k_1 e(t) + \cdots + k_{n-1} e(t-n+2) + k_n e(t-n+1) = 0. \tag{12.13}$$

It can be easily seen that (12.13) guarantees the tracking error approaches zero as time goes to infinity, where the error dynamics are determined by the coefficients $\{k_i\}$. The boundedness of all the signals in the closed-loop control system is guaranteed by Equation (12.13) and the assumption of the system being minimum phase, that is, Assumption 12.2.

12.3.3 ADAPTIVE CONTROL SYSTEM DESIGN

Based on the certainty equivalence principle, one may choose the following local adaptive control law,

R^i : IF $\qquad\qquad z_1$ is F_1^i AND ... z_v is F_v^i

THEN $u_i(t) = \dfrac{1}{\hat{g}(t)}[-\hat{a}_{i1}y(t) - \hat{a}_{i2}y(t-1) - \cdots - \hat{a}_{in}y(t-n+1) - \hat{b}_{i2}u(t-1) - \cdots$

$\qquad\qquad - \hat{b}_{in}u(t-n+1) + y_m(t+1) - k_1 e(t) - k_2 e(t-1) - \cdots - k_n e(t-n+1)]$

$i \in L := \{1, 2, \ldots, m\},$

$\qquad\qquad\qquad\qquad\qquad\qquad\qquad\qquad\qquad\qquad\qquad\qquad\qquad\qquad\qquad (12.14)$

where

$$\hat{g}(t) = \sum_{i=1}^{m} \mu_i \hat{b}_{i1}.$$

Then the global control law can be obtained as

$$u(t) = \sum_{i=1}^{m} \mu_i u_i(t)$$

$$= \frac{1}{\hat{g}(t)}[-\hat{h}(t) + y_m(t+1) - k_1 e(t) - k_2 e(t-1) - \cdots - k_n e(t-n+1)], \tag{12.15}$$

where

$$\hat{h}(t) = \sum_{i=1}^{m} \mu_i \hat{a}_{i1} y(t) + \sum_{i=1}^{m} \mu_i \hat{a}_{i2} y(t-1) + \cdots + \sum_{i=1}^{m} \mu_i \hat{a}_{in} y(t-n+1)$$

$$\sum_{i=1}^{m} \mu_i \hat{b}_{i2} u(t-1) + \cdots + \sum_{i=1}^{m} \mu_i \hat{b}_{in} u(t-n+1). \tag{12.16}$$

Remark 12.2

In order to guarantee the nonsingularity of the adaptive control law, $\hat{g}(t) \neq 0$ has to be ensured. Under Assumption 12.2, there are a number of methods that can be used to achieve this, such as the projection method (Goodwin and Sin, 1984; Middleton et al., 1988; Feng, 1999). In addition, there are many other advanced adaptive control approaches that are able to avoid the singularity of the adaptive control law but do not need Assumption 12.2.

Substituting the control law (12.15) into the T–S fuzzy system (12.3) leads to the following closed-loop system,

$$
\begin{aligned}
&e(t+1) + k_1 e(t) + \cdots + k_{n-1} e(t-n+2) + k_n e(t-n+1) \\
&= h(t) - \hat{h}(t) + [g(t) - \hat{g}(t)]u(t).
\end{aligned}
\tag{12.17}
$$

Define

$$
x_e(t) = [e(t-n+1) \quad e(t-n+2) \quad \cdots e(t)]^T.
\tag{12.18}
$$

Then, the closed-loop system (12.17) can be expressed, in a state space form, as

$$
x_e(t+1) = A_c x_e(t) + B_c \{h(t) - \hat{h}(t) + [g(t) - \hat{g}(t)]u(t)\}.
\tag{12.19}
$$

That is,

$$
\begin{aligned}
x_e(t+1) &= A_c x_e(t) + B_c \{\theta^T \phi(t) - \hat{\theta}(t)^T \phi(t)\} \\
&= A_c x_e(t) - B_c \tilde{\theta}(t)^T \phi(t) \\
&= A_c x_e(t) + B_c e_1(t+1),
\end{aligned}
\tag{12.20}
$$

where

$$
\tilde{\theta} := \hat{\theta} - \theta,
$$

$$
A_c = \begin{bmatrix} 0 & 1 & 0 & \cdots & 0 \\ 0 & 0 & 1 & & 0 \\ \vdots & \vdots & 0 & \ddots & 0 \\ 0 & 0 & \cdots & 0 & 1 \\ -k_n & -k_{n-1} & \cdots & -k_2 & -k_1 \end{bmatrix}, \quad B_c = \begin{bmatrix} 0 \\ 0 \\ \vdots \\ 0 \\ 1 \end{bmatrix}.
$$

The matrix A_c has all its eigenvalues located inside the unit circle of the z-plane, and thus it is Schur stable. One then has the following stability result on the resulting closed-loop adaptive control system.

Theorem 12.1

For the T–S fuzzy system (12.3), if the adaptive control law is chosen as described in (12.14), or equivalently (12.15), and if the parameter adaptation law is chosen as in (12.8) such as $\hat{g}(t) \neq 0$ (Remark 12.2), then the closed-loop adaptive control system is stable in the sense that all the signals in the closed-loop system are bounded. Furthermore, the output tracking error will approach zero as time goes to infinity.

Proof: Consider the closed-loop control system (12.20). Its autonomous system, that is, $x_e(t+1) = A_c x_e(t)$, is globally uniformly exponentially stable. Hence, there exist positive constants d_1, d_2 and $0 < \lambda < 1$ such that

$$\| x_e(t) \| \leq d_1 + d_2 \sum_{\tau=0}^{t-1} (1-\lambda)^{(t-\tau)} | e_1(\tau) | . \tag{12.21}$$

Using the Schwarz inequality, (12.21) leads to

$$\| x_e(t) \| \leq d_1 + d_2 \left[\sum_{\tau=0}^{t-1} (1-\lambda)^{(t-\tau)} \right] \left[\sum_{\tau=0}^{t-1} | e_1(\tau) | \right]$$

$$\leq d_1 + (d_2/\lambda) \left[\sum_{\tau=0}^{t-1} | e_1(\tau) |^2 \right]^{1/2} .$$

Squaring both sides of the above inequality yields

$$\| x_e(t) \|^2 \leq 2d_1^2 + \frac{2d_2^2}{\lambda^2} \left[\sum_{\tau=0}^{t-1} | e_1(\tau) |^2 \right]$$

$$\leq 2d_1^2 + \frac{2d_2^2}{\lambda^2} \left[\sum_{\tau=0}^{t-1} (1 + \phi(\tau-1)^T \phi(\tau-1)) | \bar{e}_1(\tau) |^2 \right]$$

$$\leq 2d_1^2 + \frac{2d_2^2}{\lambda^2} \left[\sum_{\tau=0}^{t-1} | \bar{e}_1(\tau) |^2 \right] + \frac{2d_2^2}{\lambda^2} \left[\sum_{\tau=0}^{t-1} \| \phi(\tau-1) \|^2 | \bar{e}_1(\tau) |^2 \right]$$

$$\leq d_3 + \frac{2d_2^2}{\lambda^2} \left[\sum_{\tau=0}^{t-1} \| \phi(\tau-1) \|^2 | \bar{e}_1(\tau) |^2 \right], \tag{12.22}$$

where

$$d_3 = 2d_1^2 + \sum_{\tau=0}^{\infty} | \bar{e}_1(\tau) |^2,$$

which is bounded due to property E3 of the estimator (Lemma 12.1).

The following result can be established based on the definitions of $\phi(t-1)$ and $x_e(t)$, as well as the minimum phase Assumption 12.1,

$$\| \phi(t-1) \| \leq \rho_1 \| x_e(t) \| + \rho_2, \tag{12.23}$$

for some positive constants ρ_1 and ρ_2. Then it follows from (12.22) and (12.23) that there exist two positive constants d_4, d_5 such that

$$\| x_e(t) \|^2 \le d_4 + d_5 \sum_{\tau=0}^{t-1} \| x_e(\tau) \|^2 | \overline{e}_1(\tau) |^2. \tag{12.24}$$

Because $\overline{e}_1 \in l_2$, it follows from (12.24) and the Bellman–Gronwall lemma (Middleton et al., 1988; Feng, 1999) that x_e is bounded. It implies that e, ϕ, e_1, and thus y, u are all bounded. Furthermore, it implies that $e_1 \to 0$ as time goes to infinity. Moreover, it follows from (12.20) that $x_e \to 0$, so that $e \to 0$ as time goes to infinity, and the proof is thus completed. ❏

12.3.4   ROBUST ADAPTIVE CONTROL

The T–S fuzzy systems discussed in the previous sections are assumed to be ideal; that is, the underlying systems are modeled precisely by the assumed T–S fuzzy systems. In practice, however, there always exists some discrepancy between the underlying system and its mathematical model, and moreover, the physical systems are always subject to some uncertainties such as unmodeled dynamics or external disturbances. As is known for typical adaptive control systems, even a small uncertainty may lead to instability of the closed-loop adaptive control system. To avoid such problems, various types of robust adaptation and robust adaptive control algorithms have been developed. The techniques include dead zones, relative dead zones, and σ-modification, among many others.

Here, the relative dead zones algorithm is employed to achieve robustness of the adaptive control algorithm with respect to uncertainties. Consider the following uncertain T–S fuzzy system,

$$R^i: \quad \text{IF} \quad z_1 \text{ is } F_1^i \text{ AND } \dots z_v \text{ is } F_v^i$$

$$\text{THEN} \quad y(t+1) = a_{i1} y(t) + a_{i2} y(t-1) + \cdots + a_{in} y(t-n+1)$$

$$+ b_{i1} u(t) + \cdots + b_{in} u(t-n+1) + \eta_i(t) \tag{12.25}$$

$$i \in L := \{1, 2, \dots, m\},$$

where η_i represents the unmodeled dynamics and unknown but bounded disturbances.

By using standard fuzzy system techniques, one can easily obtain the following fuzzy global model,

$$y(t+1) = \sum_{i=1}^m \mu_i a_{i1} y(t) + \sum_{i=1}^m \mu_i a_{i2} y(t-1) + \cdots + \sum_{i=1}^m \mu_i a_{in} y(t-n+1)$$

$$+ \sum_{i=1}^m \mu_i b_{i1} u(t) + \cdots + \sum_{i=1}^m \mu_i b_{in} u(t-n+1) + \sum_{i=1}^m \mu_i \eta_i(t). \tag{12.26}$$

The following assumption is made on the uncertainties of the T–S fuzzy system (12.26).

Assumption 12.3

There exists an upper bounding function $\gamma(t)$ for the uncertainties $\eta(t) = \sum_{i=1}^{m} \mu_i \eta_i(t)$; that is,

$$|\eta(t)| \le \gamma(t), \tag{12.27}$$

where $\gamma(t)$ is assumed to satisfy

$$\gamma(t) = c_1 + c_2 \sup_{0 \le \tau \le t} \|\phi(\tau)\| \tag{12.28}$$

for two unknown but small constants c_1 and c_2, and ϕ is defined in (12.6).

With the same definition of the regression vector and the system parameter vector as in (12.6), and the same adaptive control law (12.14), or equivalently (12.15), one obtains the following closed-loop control system,

$$e(t+1) + k_1 e(t) + \cdots + k_{n-1} e(t-n+2) + k_n e(t-n+1)$$
$$= h(t) - \hat{h}(t) + [g(t) - \hat{g}(t)]u(t) + \eta(t), \tag{12.29}$$

or, in the state space form,

$$x_e(t+1) = A_c x_e(t) - B_c[\tilde{\theta}(t)^T \phi(t) - \eta(t)] = A_c x_e(t) + B_c e_1(t+1). \tag{12.30}$$

In this case, the following modified projection algorithm with relative dead zones can be used for robust parameter estimation,

$$\hat{\theta}(t) = \hat{\theta}(t-1) + a(t) \frac{\phi(t-1)e_1(t)}{1 + \sup_{0 \le \tau \le t} \phi(\tau-1)^T \phi(\tau-1)}, \tag{12.31}$$

where the term $a(t)$ is a dead zone function, which ensures that the parameter estimator is not disrupted by small errors. The dead zone is defined as follows,

$$a(t) = \begin{cases} 0 & \text{if } |e_1| \le \xi \hat{\gamma}(t) \\ \alpha f(\xi \hat{\gamma}(t), e_1) & \text{otherwise} \end{cases}, \tag{12.32}$$

where

$$f(g,e) = \begin{cases} e - g & \text{if } e > g \\ 0 & \text{if } |e| \le g \\ e + g & \text{if } e < -g \end{cases},$$

$0 < \alpha < 1, \xi = \frac{\xi_0}{1-\alpha}, \xi_0 > 1$, and $\hat{\gamma}(t)$ is calculated by

$$\hat{\gamma}(t) = [\hat{c}_2 \quad \hat{c}_1] \begin{bmatrix} \sup_{0 \leq \tau \leq t} \|\phi(\tau - 1)\| \\ 1 \end{bmatrix}$$

$$\hat{c}_2(t) = \hat{c}_2(t-1) + \frac{a(t)\beta}{1 + \sup_{0 \leq \tau \leq t} \phi(\tau-1)^T \phi(\tau-1)} \sup_{0 \leq \tau \leq t} \|\phi(\tau-1)\|$$

$$\hat{c}_1(t) = \hat{c}_1(t-1) + \frac{a(t)\beta}{1 + \sup_{0 \leq \tau \leq t} \phi(\tau-1)^T \phi(\tau-1)}$$

(12.33)

with $\beta > 0$ being an update rate parameter, and $\hat{c}_1(0) = \hat{c}_2(0) = 0$.

The values of $\hat{c}_1$ and $\hat{c}_2$, generated by this estimation scheme, are always nonnegative and nondecreasing.

The following convergence properties for the above robust parameter estimation algorithm can be established (Feng, 1999).

Lemma 12.2

The parameter update law (12.31)–(12.33), when applied to the T–S fuzzy model (12.26), has the following properties.

E1: $\hat{\theta}(t)$ is continuous and bounded.

E2: $\hat{c}_1$ and $\hat{c}_2$ are continuous and bounded, and $\hat{c}_1$ and $\hat{c}_2$ converge to constants, say $\bar{c}_1$ and $\bar{c}_2$, respectively.

E3: $\hat{\theta}(t) \in l2$.

E4: $\tilde{f}(t) := \frac{f(\xi\hat{\gamma}(t), e_1)}{[1 + \sup_{0 \leq \tau \leq t} \phi(\tau-1)^T \phi(\tau-1)]^{1/2}} \in l_2$.

Proof: Define a Lyapunov function candidate,

$$V(t) = \frac{1}{2} \left[\tilde{\theta}(t)^T \tilde{\theta}(t) + \beta^{-1} \tilde{c}_1(t)^2 + \beta^{-1} \tilde{c}_2(t)^2 \right],$$

(12.34)

where $\tilde{\theta} := \hat{\theta} - \theta$, $\tilde{c}_1 = \hat{c}_1 - c_1$, and $\tilde{c}_2 = \hat{c}_2 - c_2$. Then, its difference along the solution of (12.26) can be expressed as

$V(t+1) - V(t)$

$$= \frac{a(t)}{1 + \sup_{0 \leq \tau \leq t} \phi(\tau-1)^T \phi(\tau-1)} \left[2e_1(t)\eta(t) - 2e_1(t)^2 + \frac{\phi(t-1)^T \phi(t-1)}{1 + \sup_{0 \leq \tau \leq t} \phi(\tau-1)^T \phi(\tau-1)} a(t)e_1(t)^2 \right]$$

$$+ 2c_1(t) \frac{a(t)|e_1(t)|}{1 + \sup_{0 \leq \tau \leq t} \phi(\tau-1)^T \phi(\tau-1)} + \frac{a(t)^2 e_1(t)^2}{\left[1 + \sup_{0 \leq \tau \leq t} \phi(\tau-1)^T \phi(\tau-1) \right]^2}$$

$$+ 2c_2(t) \frac{a(t)\,|e_1(t)|}{1 + \sup_{0 \le \tau \le t} \phi(\tau-1)^T \phi(\tau-1)} \sup_{0 \le \tau \le t} \| \phi(\tau-1) \|$$

$$+ \frac{a(t)^2 e_1(t)^2}{\left[1 + \sup_{0 \le \tau \le t} \phi(\tau-1)^T \phi(\tau-1) \right]^2} \left[\sup_{0 \le \tau \le t} \| \phi(\tau-1) \| \right]^2$$

$$= \frac{a(t)}{1 + \sup_{0 \le \tau \le t} \phi(\tau-1)^T \phi(\tau-1)} \left\{ 2e_1(t)\eta(t) - 2e_1(t)^2 + \frac{\phi(t-1)^T \phi(t-1)}{1 + \sup_{0 \le \tau \le t} \phi(\tau-1)^T \phi(\tau-1)} a(t) e_1(t)^2 \right.$$

$$+ 2\hat{\gamma}(t)\,|e_1(t)| - 2\gamma(t)\,|e_1(t)| + \frac{a(t) e_1(t)^2}{1 + \sup_{0 \le \tau \le t} \phi(\tau-1)^T \phi(\tau-1)}$$

$$\left. + \frac{a(t) e_1(t)^2}{1 + \sup_{0 \le \tau \le t} \phi(\tau-1)^T \phi(\tau-1)} \left[\sup_{0 \le \tau \le t} \| \phi(\tau-1) \| \right]^2 \right\}$$

$$\le \frac{a(t)}{1 + \sup_{0 \le \tau \le t} \phi(\tau-1)^T \phi(\tau-1)} \left\{ 2\hat{\gamma}(t)\,|e_1(t)| - 2e_1(t)^2 + 2\alpha e_1(t)^2 \right\}$$

$$= - \frac{a(t)}{1 + \sup_{0 \le \tau \le t} \phi(\tau-1)^T \phi(\tau-1)} \left\{ 2(1-\alpha) e_1(t)^2 - 2\hat{\gamma}(t)\,|e_1(t)| \right\}$$

$$\le - \frac{2a(t)}{1 + \sup_{0 \le \tau \le t} \phi(\tau-1)^T \phi(\tau-1)} \left\{ (1-\alpha) e_1(t)^2 - \frac{|e_1(t)|}{\xi}\,|e_1(t)| \right\}$$

$$= - \frac{(1-\alpha)\xi - 1}{\xi} \frac{2a(t) e_1(t)^2}{1 + \sup_{0 \le \tau \le t} \phi(\tau-1)^T \phi(\tau-1)}$$

$$\le - \frac{(1-\alpha)\xi - 1}{\xi} \frac{2\alpha f^2}{1 + \sup_{0 \le \tau \le t} \phi(\tau-1)^T \phi(\tau-1)}$$

$$\le -2\alpha \frac{\xi_0 - 1}{\xi} \frac{f^2}{1 + \sup_{0 \le \tau \le t} \phi(\tau-1)^T \phi(\tau-1)}, \tag{12.35}$$

where the fact that $a(t) e_1^2 = \alpha f e_1 \ge \alpha f^2$, obtained directly from the properties of the function $f(\cdot, \cdot)$, has been used. Then following the same arguments as in Middleton et al. (1988), the convergence properties for the above modified projection parameter estimator are established, and the proof is thus completed. ❑

Then one has the following robust adaptive control result.

Theorem 12.2

For the T–S fuzzy model (12.26), if the adaptive control law is chosen as in (12.14), or equivalently (12.15), the parameter adaptation law is chosen as in (12.31)–(12.33), and if $\bar{c}_2 < \sigma / (2 d_2^2 \xi^2 \rho_1^2)$ with the parameters defined in the subsequent proof, then the closed-loop adaptive control system will be stable in the sense that all the signals in the system are bounded. Furthermore, the steady-state output tracking performance satisfies

$$|e(t)| \leq \rho,$$

for some positive constant ρ.

Proof: Consider the closed-loop control system (12.30). Notice that its autonomous system, that is, $x_e(t+1) = A_c x_e(t)$, is globally uniformly exponentially stable. Hence, there exist positive constants d_1, d_2, and $0 < \lambda < 1$ such that

$$\|x_e(t)\| \leq d_1 + d_2 \sum_{\tau=0}^{t} (1-\lambda)^{t-\tau} |e_1(\tau)|. \tag{12.36}$$

From the definition of the dead-zone function f, one has

$$|e_1| \leq |f| + |\xi\hat{\gamma}|. \tag{12.37}$$

It follows from (12.36) and (12.37) that

$$\|x_e(t)\| \leq d_1 + d_2 \sum_{\tau=0}^{t} (1-\lambda)^{t-\tau} (|f(\tau)| + \xi\hat{\gamma}(\tau))$$

$$= \leq d_1 + d_2 \sum_{\tau=0}^{t} (1-\lambda)^{t-\tau} |f(\tau)| + \frac{d_2\xi}{\lambda} \sup_{0 \leq \tau \leq t} \hat{\gamma}(\tau). \tag{12.38}$$

Denote some suitable constants $d_3, d_4, \ldots, d_9$ in the subsequent equations. Via the Schwarz inequality, (12.38) yields

$$\|x_e(t)\| \leq d_1 + d_3 \left[\sum_{\tau=0}^{t} f(\tau)^2 \right]^{1/2} + \frac{d_2\xi}{\lambda} \sup_{0 \leq \tau \leq t} \{\hat{\gamma}(\tau)\}$$

$$\leq d_4 + d_3 \left[\sum_{\tau=0}^{t} f(\tau)^2 \right]^{1/2} + \frac{d_2\xi}{\lambda} \sup_{0 \leq \tau \leq t} \hat{c}_2(\tau) \sup_{0 \leq \tau \leq t} \{\phi(\tau-1)\}. \tag{12.39}$$

Because the right-hand side of (12.39) is monotonically nondecreasing, one has

$$\sup_{0 \leq \tau \leq t} \|x_e(\tau)\| \leq d_4 + d_3 \left[\sum_{\tau=0}^{t} f(\tau)^2 \right]^{1/2} + \frac{d_2\xi}{\lambda} \sup_{0 \leq \tau \leq t} \hat{c}_2(\tau) \sup_{0 \leq \tau \leq t} \{\phi(\tau-1)\}$$

$$\leq d_4 + d_3 \left[\sum_{\tau=0}^{t} f(\tau)^2 \right]^{1/2} + \frac{d_2 \xi}{\lambda} \bar{c}_2 \sup_{0 \leq \tau \leq t} \{\phi(\tau-1)\}$$

$$\leq d_5 + d_3 \left[\sum_{\tau=0}^{t} f(\tau)^2 \right]^{1/2} + \frac{d_2 \rho_1 \xi}{\lambda} \bar{c}_2 \sup_{0 \leq \tau \leq t} \{x_e(\tau)\},$$

where (12.23) has been used in the last step.

Thus, as long as $\bar{c}_2 < \lambda/(d_2 \rho_1 \xi)$, one can show that

$$\sup_{0 \leq \tau \leq t} \| x_e(\tau) \| \leq d_6 + d_7 \left[\sum_{\tau=0}^{t} f(\tau)^2 \right]^{1/2}.$$

Squaring both sides of the above equation, one obtains

$$\sup_{0 \leq \tau \leq t} \| x_e(\tau) \|^2 \leq 2d_6^2 + 2d_7^2 \left[\sum_{\tau=0}^{t} f(\tau)^2 \right]$$

$$\leq 2d_6^2 + 2d_7^2 \left[\sum_{\tau=0}^{t-1} (1 + \sup_{0 \leq \tau \leq t} \phi(\tau-1)^T \phi(\tau-1)) | \tilde{f}(\tau) |^2 \right]$$

$$\leq 2d_6^2 + 2d_7^2 \sum_{\tau=0}^{t} | \tilde{f}(\tau) |^2 + 2d_7^2 \sum_{\tau=0}^{t-1} [\sup_{0 \leq \tau \leq t} \phi(\tau-1)^T \phi(\tau-1)] | \tilde{f}(\tau) |^2. \quad (12.40)$$

It then follows from (12.23) and (12.40) that there exist some constants d_8, d_9 such that

$$\sup_{0 \leq \tau \leq t} \| x_e(\tau) \|^2 \leq d_8 + d_9 \sum_{\tau=0}^{t} \sup_{0 \leq \tau \leq t} \| x_e(\tau) \|^2 | \tilde{f}(\tau) |^2. \quad (12.41)$$

Because $\tilde{f} \in l_2$, it follows from the Bellman–Gronwall lemma (Middleton et al., 1988; Feng, 1999) that $x_e(t)$ is bounded. It implies that ϕ, e, and thus y, u are all bounded.

Furthermore, because ϕ is bounded, $\tilde{f} \in l_2$ implies that f tends to zero as time goes to infinity (Middleton et al., 1988; Feng, 1999). It finally follows from (12.29) and (12.37) that when time goes to infinity,

$$| e(t+1) + k_1 e(t) + \cdots + k_{n-1} e(t-n+2) + k_n e(t-n+1) | \leq | e_1 | \leq | f | + \xi \hat{\gamma}.$$

Then it follows directly from the above inequality and the fact that $f \to 0$ as time goes to infinity, that when time goes to infinity, there exists a constant δ such that

$$| e(t) | \leq \delta \xi \hat{\gamma}(t).$$

Moreover, because ϕ is bounded, which implies that $\hat{\gamma}(t)$ is also bounded, there exists a constant ρ such that

$$|e(t)| \leq \rho.$$

The proof is thus completed. ❏

Remark 12.3

It can be observed that the above stability results closely depend on the parameter $\bar{c}_2$ which is generated from the adaptation algorithm with relative dead zones. Three cases can be considered for this parameter.

1. If $\bar{c}_2 = c_2$, then the tolerable unmodeled dynamics of the adaptive control algorithm with relative dead zones is the same as that of the ordinary dead-zone case with a known c_2.
2. If $\bar{c}_2 < c_2$, then the tolerable unmodeled dynamics is larger than that of the ordinary dead-zone case.
3. If $\bar{c}_2 > c_2$, then the tolerable unmodeled dynamics is smaller than that of the ordinary dead-zone case. In view of the estimation equation for the parameter, to obtain a smaller $\bar{c}_2$, that is, to tolerate a larger unmodeled dynamics, a smaller update rate parameter α is preferred.

Remark 12.4

If the parameter c_2 is known a priori, the above algorithm can be easily modified to accommodate this knowledge by the well-known projection technique, so that the estimated $\hat{c}_2$ is always less than or equal to the known parameter value. In such a case, the algorithm presented here is better than the ordinary dead-zone algorithm with a fixed c_2, in the sense that it can tolerate larger unmodeled dynamics and possibly achieve a better tracking performance.

If only bounded disturbances are present, one has the following result.

Corollary 12.1

With the same conditions as in Theorem 12.2, except that there is no unmodeled dynamics, that is, $c_2 = 0$, the adaptive control system with the following simpler update law is stable in the same sense as that defined in Theorem 12.2,

$$\hat{\theta}(t) = \hat{\theta}(t-1) + a(t)\frac{\phi(t-1)e_1(t)}{1+\phi(t-1)^T\phi(t-1)}$$

$$\hat{c}_1(t) = \hat{c}_1(t-1) + \frac{a(t)\,\beta\,|e_1(t)|}{1+\phi(t-1)^T\phi(t-1)}, \quad \beta > 0. \tag{12.42}$$

Furthermore, the steady-state output tracking performance satisfies

$$|e(t)| \leq \rho,$$

for some positive constant ρ.

Proof: It follows directly from the proof of Theorem 12.2. ❏

Remark 12.5
The relative dead zone is introduced in the adaptation law (12.31)–(12.33) to guarantee that the estimated parameter would not drift to infinity. However, it should be noted that in this case, the output tracking error does not approach zero as time goes to infinity, due to the existence of the dead zone. Instead, the tracking error would be bounded by some positive constant.

12.4   A SIMULATION EXAMPLE

This section presents a simulation example to show an application of the proposed adaptive control algorithms.

Example 12.1
Consider the following chaotic Henon model,

$$y(t) = -y(t-1)^2 + 0.3y(t-2) + 1.4 + u(t-1). \tag{12.43}$$

By choosing the following membership functions,

$$\mu_1(y(t-1)) = 0.5\left(1 - \frac{y(t-1)}{d}\right), \quad \mu_2(y(t-1)) = 0.5\left(1 + \frac{y(t-1)}{d}\right), \tag{12.44}$$

where $d = 30$ is a constant large enough to cover the range of the output, it can be shown that the Henon model can be expressed as a T–S fuzzy system in the following form,

$$R^1 : \text{IF} \qquad y(t-1) \text{ is } \mu_1$$

$$\text{THEN} \quad y(t) = a_{11}y(t-1) + a_{12}y(t-2) + b_1 u^*(t-1) \tag{12.45}$$

$$R^2 : \text{IF} \qquad y(t-1) \text{ is } \mu_2$$

$$\text{THEN} \quad y(t) = a_{21}y(t-1) + a_{22}y(t-2) + b_2 u^*(t-1),$$

where $u^* = u + 1.4$, $a_{11} = -a_{21} = d$, $a_{12} = a_{22} = 0.3$, $b_1 = b_2 = 1$.

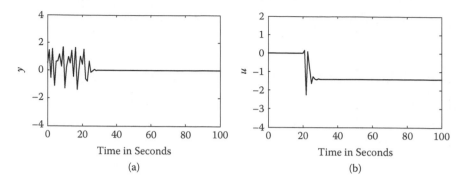

FIGURE 12.1 Responses with adaptive control.

In simulation, we only consider the regulation response of the adaptive control system. Suppose that the parameters of the Henon model are unknown a priori. The initial parameter is chosen to be

$$\hat{\theta}(0) := [a_{11}(0) \quad a_{12}(0) \quad b_1(0) \quad a_{21}(0) \quad a_{22}(0) \quad b_2(0)]^T = [d \quad 1 \quad 1 \quad -d \quad 1 \quad 1]^T.$$

With the initial condition $[y(0) \quad y(-1)] = [0.5 \quad 1]^T$, and the chosen parameters $k_1 = 0.5$, $k_2 = 0.05$, the response is recorded in Figure 12.1, where the controller is switched on at $t = 20$ samples.

In order to show the adaptation capability of the proposed adaptive control system, the following case is also considered. That is, the coefficient $a_{12} = a_{22}$ is changed from 0.3 to 0.1 at the instant $t = 20$ samples. The simulation result is shown in Figure 12.2.

It can be seen that the adaptive control algorithm can cope with variation of the plant dynamics quite well. For comparison, the response of the nonadaptive control system with the same initial conditions and the same parameter variation is shown in Figure 12.3, where a significantly worse performance can be observed.

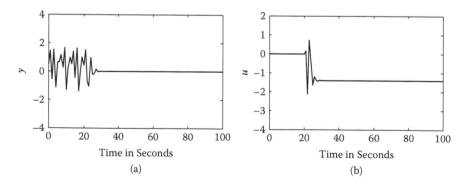

FIGURE 12.2 Responses for adaptive control with parameter change.

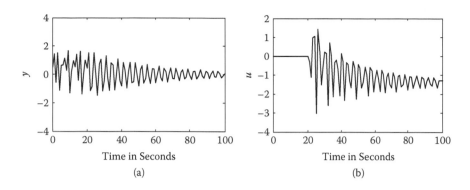

FIGURE 12.3 Responses for nonadaptive control with parameter change.

12.5 CONCLUSIONS

This chapter has presented two approaches to adaptive control of discrete-time T–S fuzzy systems. The basic idea of the approaches is to design a local linear adaptive controller for each local region and then construct a global fuzzy adaptive controller from those local adaptive controllers in such a way that the stability of the closed-loop adaptive control system can be guaranteed. It is shown that the output tracking error can be guaranteed to converge to zero under an ideal case or to a bounded value under a nonideal case (uncertain case). A simulation example has been presented to demonstrate the application and performance of the proposed approaches.

Appendix
Several Useful Lemmas

Lemma A.1 (Xie, 1996)

Suppose that $\Delta(t)$ is given by (8.2) with matrices $M = M^T$, and S and N of appropriate dimensions; the inequality

$$M + S\Delta(t)N + N^T\Delta^T(t)S^T < 0$$

holds if and only if for some scalar $\varepsilon > 0$,

$$M + [\varepsilon^{-1}N^T \quad \varepsilon S] \begin{bmatrix} I & -J \\ -J^T & I \end{bmatrix}^{-1} [\varepsilon^{-1}N^T \quad \varepsilon S]^T < 0.$$

Lemma A.2 (Schur Complement Lemma; Skelton, Iwasaki, and Grigoria-dis, 1998)

For matrices A, B, and C with compatible dimensions, the following two inequalities

$$A > 0 \quad \text{and} \quad C - B^T A^{-1} B > 0$$

are equivalent to the inequality

$$\begin{bmatrix} A & B \\ B^T & C \end{bmatrix} > 0.$$

Lemma A.3 (Dualization Lemma; El Ghaoui and Niculescu, 2000)

Let $P \in \mathfrak{R}^{n \times n}$ be a nonsingular symmetric matrix, and let U, V be two complementary subspaces whose sum equals $\mathfrak{R}^n$; that is, $U \oplus V = \mathfrak{R}^n$. Then

$$x^T P x < 0, \ \forall x \in U \backslash \{0\} \quad \text{and} \quad x^T P x \geq 0, \ \forall x \in V$$

is equivalent to

$$x^T P^{-1} x > 0, \ \forall x \in U^\perp \backslash \{0\} \quad \text{and} \quad x^T P^{-1} x \leq 0, \ \forall x \in V^\perp.$$

Lemma A.4 (Finsler's Lemma; Skelton, Iwasaki, and Grigoriadis, 1998)

Let $x \in \mathfrak{R}^n$, $P = P^T \in \mathfrak{R}^{n \times n}$, $H \in \mathfrak{R}^{m \times n}$ such that $rank(H) = r < n$. The following statements are equivalent,

1. $x^T P x < 0, \ \forall H x = 0, \ x \neq 0.$
2. $\exists N \in \mathfrak{R}^{n \times m} : P + NH + H^T N^T < 0.$
3. $(H^\perp)^T P (H^\perp) < 0.$

References

Abdelazim, T., and O.P. Malik (2003). An adaptive power system stabilizer using on-line self-learning fuzzy systems, *Proc. IEEE Power Engineering Society General Meeting*, Toronto, pp.1715–1720.

Akar, M., and U. Ozguner (2000). Decentralized techniques for the analysis and control of Takagi–Sugeno fuzzy systems, *IEEE Trans. Fuzzy Syst.*, 8(6): 691–704.

Andersen, H.C., A. Lotfi, and A.C. Tsoi (1997). A new approach to adaptive fuzzy control: The controller output error method, *IEEE Trans. Syst. Man Cybern.*, 27(4): 686–691.

Aoul, Y.H., A. Nafaa, D. Negru, and A. Mehaoua (2004). FAFC: Fast adaptive fuzzy AQM controller for TCP/IP networks, *Proc. IEEE Global Telecom. Conf.*, Dallas, pp. 1319–1323.

Assawinchaichote, W., S.K. Nguang, and P. Shi (2004). H_∞ output feedback control design for uncertain fuzzy singularly perturbed systems: An LMI approach, *Automatica*, 40(12): 2147–2152.

Babuska, R. (1998). *Fuzzy Modeling for Control*, Kluwer Academic: Boston.

Bai, Y., H.Q. Zhuang, and Z.S. Roth (2005). Fuzzy logic control to suppress noises and coupling effects in a laser tracking system, *IEEE Trans. Control Syst. Technol.*, 13(1): 113–121.

Barrero, F., A. Gonzalez, A. Torralba, E. Galvan, and L.G. Franquelo (2002). Speed control of induction motors using a novel fuzzy sliding-mode structure, *IEEE Trans. Fuzzy Syst.*, 10(3): 375–383.

Baturone, I., F.J. Moreno-Velo, S. Sanchez-Solano, and A. Ollero (2004). Automatic design of fuzzy controllers for car-like autonomous robots, *IEEE Trans. Fuzzy Syst.*, 12(4): 447–465.

Bellman, R.E., and L.A. Zadeh (1970). Decision making in a fuzzy environment, *Manage. Sci.*, 17: 141–164.

Bergsten, P., R. Palm, and D. Driankov (2002). Observers for Takagi–Sugeno fuzzy systems, *IEEE Trans. Syst. Man Cybern. B*, 32(1): 114–121.

Berstecher, R.G., R. Palm, and H.D. Unbehauen (2001). An adaptive fuzzy sliding-mode controller, *IEEE Trans. Indust. Electron.*, 48(1): 18–31.

Bezdek, J.C., J.M. Keller, R. Krishnapuram, and N.R. Pal (1999). *Fuzzy Models and Algorithms for Pattern Recognition and Image Processing*, Kluwer Academic: Boston.

Bingul, Z., G.E. Cook, and A.M. Strauss (2000). Application of fuzzy logic to spatial thermal control in fusion welding, *IEEE Trans. Indust. Appl.*, 36(6): 1523–1530.

Bonissone, P.P., V. Badami, K.H. Chiang, P.S. Khedkar, K.W. Marcelle, and M.J. Schutten (1995). Industrial applications of fuzzy logic at General Electric, *Proc. IEEE*, 38: 450–465.

Boroushaki, M., M.B. Ghofrani, C. Lucas, and M.J. Yazdanpanah (2003). Identification and control of a nuclear reactor core (VVER) using recurrent neural networks and fuzzy systems, *IEEE Trans. Nuclear Sci.*, 50(1): 159–174.

Boukezzoula, R., S. Galichet, and L. Foulloy (2004). Observer-based fuzzy adaptive control for a class of nonlinear systems: Real-time implementation for a robot wrist, *IEEE Trans. Control Syst. Technol.*, 12(3): 340–351.

Boyd, S., L. El Ghaoui, E. Feron, and V. Balakrishnan (1994). *Linear Matrix Inequalities in Systems and Control Theory*, SIAM: Philadelphia.

Braae, M., and D.A. Rutherford (1979). Theoretical and linguistic aspects of the fuzzy logic controller, *Automatica*, 15: 553–577.

Calcev, G., R. Gorez, and M. De Neyer (1998). Passivity approach to fuzzy control systems, *Automatica*, 34(3): 339–344.

Campello, R.J., L.A.C. Meleiro, and W.C. Amaral (2004). Control of a bioprocess using ortho-normal basis function fuzzy models, *Proc. IEEE Int. Conf. on Fuzzy Systems*, Budapest, Hungary, pp. 801–806.

Campos, J., and F.L. Lewis (1999). Deadzone compensation in discrete time using adaptive fuzzy logic, *IEEE Trans. Fuzzy Syst.*, 7(6): 697–707.

Cao, S.G. and G. Feng (1995). Modeling of complex control systems, *Proc. of the IFAC Symposium on Nonlinear Control System Design*, California, pp. 934–939.

Cao, S.G., N.W. Rees, and G. Feng (1995). Analysis and design of fuzzy control systems using dynamic fuzzy global models, *Fuzzy Sets Syst.*, 75: 47–62.

Cao, S.G., N.W. Rees, and G. Feng (1996a). Stability analysis of fuzzy control systems, *IEEE Trans. Syst. Man Cybern. B*, 26(1): 201–204.

Cao, S.G., N.W. Rees, and G. Feng (1996b). Stability analysis and design for a class of con-tinuous-time fuzzy control systems, *Int. J. Control*, 64: 1069–1087.

Cao, S.G., N.W. Rees, and G. Feng (1996c). H_∞ Control of nonlinear continuous-time systems based on dynamic fuzzy models, *Int. J. Syst. Sci.*, 27: 821–830.

Cao, S.G., N.W. Rees, and G. Feng (1997a). Analysis and design for a class of complex control systems – Part I: Fuzzy modeling and identification, *Automatica*, 33: 1017–1028.

Cao, S.G., N.W. Rees, and G. Feng (1997b). Analysis and design for a class of complex con-trol systems – Part II: Fuzzy controller design, *Automatica*, 33: 1029–1039.

Cao, S.G., N.W. Rees, and G. Feng (1999). Analysis and design of fuzzy control systems using dynamic fuzzy state space models, *IEEE Trans. Fuzzy Syst.*, 7(2): 192–200.

Cao, S.G., N.W. Rees, and G. Feng (2001). Universal fuzzy controllers for a class of nonlinear systems, *Fuzzy Sets Syst.*, 122(1): 117–123.

Cao, Y.Y., and P.M. Frank (2000a). Analysis and synthesis of nonlinear time-delay systems via fuzzy control approach, *IEEE Trans. Fuzzy Syst.*, 8(2): 200–211.

Cao, Y.Y., and P.M. Frank (2000b). Robust H-infinity disturbance attenuation for a class of uncertain discrete-time fuzzy systems, *IEEE Trans. Fuzzy Syst.*, 8(4): 406–415.

Cao, Y.Y., and Z.L. Lin (2003). Robust stability analysis and fuzzy-scheduling control for non-linear systems subject to actuator saturation, *IEEE Trans. Fuzzy Syst.*, 11(1): 57–67.

Castillo-Toledo, B., and J.A. Meda-Campana (2004). The fuzzy discrete-time robust regula-tion problem: An LMI approach, *IEEE Trans. Fuzzy Syst.*, 12(3): 360–367.

Chadli, M., D. Maquin, and J. Ragot (2004). Stabilisation of Takagi-Sugeno models with maximum convergence rate, *Proc. IEEE Int. Conf. on Fuzzy Systems*, Budapest, pp. 1323–1326.

Chakraborty, S., K. Pal, and N.R. Pal (2002). A neuro-fuzzy framework for inferencing, *Neural Netw.*, 15: 247–261.

Chang, Y.C. (2001). Adaptive fuzzy-based tracking control for nonlinear SISO systems via VSS and H_∞ approaches, *IEEE Trans. Fuzzy Syst.*, 9(2): 278–292.

Chang, Y.C., and B.S. Chen (2005). Intelligent robust tracking controls for holonomic and nonholonomic mechanical systems using only position measurements, *IEEE Trans. Fuzzy Syst.*, 13(4): 491–507.

Chen, B., and X. Liu (2005a). Fuzzy approximate disturbance decoupling of MIMO nonlin-ear systems by backstepping and application to chemical processes, *IEEE Trans. Fuzzy Syst.*, 13(6): 832–847.

Chen, B., and X.P. Liu (2005b). Delay-dependent robust H-infinity control for T–S fuzzy sys-tems with time delay, *IEEE Trans. Fuzzy Syst.*, 13(4): 544–556.

Chen, B., X.P. Liu, S.C. Tong, and C. Lin (2008). Observer-based stabilization of T–S fuzzy systems with input delay, *IEEE Trans. Fuzzy Syst.*, 16(3): 652–663.

Chen, B.S., C.H. Lee, and Y.C. Chang (1996). H_∞ Tracking design of uncertain nonlinear SISO systems: Adaptive fuzzy approach, *IEEE Trans. Fuzzy Syst.*, 4(1): 32–43.

Chen, B.S., C.L. Tsai, and D.S. Chen (2003b). Robust H_∞ and mixed H_2/H_∞ filters for equaliza-tion designs of nonlinear communication systems: Fuzzy interpolation approach, *IEEE Trans. Fuzzy Syst.*, 11(3): 384–398.

Chen, B.S., C.S. Tseng, and H.J. Uang (1999). Robustness design of nonlinear dynamic systems via fuzzy linear control, *IEEE Trans. Fuzzy Syst.*, 7(5): 571–585.

Chen, B.S., C.S. Tseng, and H.J. Uang (2000). Mixed H$_2$/H$_\infty$ fuzzy output feedback control design for nonlinear dynamic systems: An LMI approach, *IEEE Trans. Fuzzy Syst.,* 8(3): 249–265.

Chen, B.S., Y.S. Yang, B.K. Lee, and T.H. Lee (2003c). Fuzzy adaptive predictive flow control of ATM network traffic, *IEEE Trans. Fuzzy Syst.*, 11(4): 568–581.

Chen, C.L., and M.H. Chang (1998). Optimal design of fuzzy sliding mode control: A comparative study, *Fuzzy Sets Syst.*, 93: 37–48.

Chen, C.L., G. Feng, and X.P. Guan (2005). Delay-dependent stability analysis and controller synthesis for discrete time T–S fuzzy systems with time delays, *IEEE Trans. Fuzzy Syst.,* 13(5): 630–643.

Chen, C.L., G. Feng, D. Sun, and Y. Zhu (2005). H-infinity output feedback control of discrete-time fuzzy systems with application to chaos control, *IEEE Trans. Fuzzy Syst.,* 13(4): 531–543.

Chen, G. (1996). Conventional and fuzzy PID controllers: An overview, *Int. J. Intell. Control Syst.*, 1: 235–246.

Chen, G., and H. Ying (1993). Stability analysis of nonlinear fuzzy PI control systems, *Proc. 3rd Int. Conf. on Fuzzy Logic Applications*, Houston, TX, pp.128–133.

Chen, J.Y. (2001). Rule regulation of fuzzy sliding mode controller design: Direct adaptive approach, *Fuzzy Sets Syst.*, 120(1): 159–168.

Chen, S.S., Y.C. Chang, S.F. Su, S.L. Chung, and T.T. Lee (2005). Robust static output-feedback stabilization for nonlinear discrete-time systems with time delay via fuzzy control approach, *IEEE Trans. Fuzzy Syst.*, 13(2): 263–272.

Cheng, C.M. (1998), "Multi-model fuzzy control for nonlinear systems," Ph.D. thesis, University of New South Wales, Australia.

Chiu, S. (1998). Using fuzzy logic in control applications: Beyond fuzzy PID control, *IEEE Control Syst. Mag.*, 18: 100–104.

Chiu, S., S. Chand, D. Moore, and A. Chaudhary (1991). Fuzzy logic for control of roll and moment for a flexible wing aircraft, *IEEE Control Syst. Mag.*, 11(1): 42–48.

Cho, K.H., C.W. Kim, and J.T. Lim (1993). On stability analysis of nonlinear plants with fuzzy logic controllers, *Proc. 5th IFSA World Congress*, Seoul, Korea, pp.1094–1097.

Choi, D.J., and P.G. Park (2003). H-infinity state-feedback controller design for discrete-time fuzzy systems using fuzzy weighting-dependent Lyapunov functions, *IEEE Trans. Fuzzy Syst.*, 11(2): 271–278.

Choi, H.H. (2007). LMI-Based nonlinear fuzzy observer-controller design for uncertain MIMO nonlinear systems, *IEEE Trans. Fuzzy Syst.*, 15(5): 956–971.

Cuesta, F., F. Gordillo, J. Aracil, and A. Ollero (1999). Stability analysis of nonlinear multivariable Takagi–Sugeno fuzzy control systems, *IEEE Trans. Fuzzy Syst.*, 7(5): 508–520.

Czogala, E., and W. Pedrycz (1981). On identification in fuzzy systems and its applications in control problems, *Fuzzy Sets Syst.*, 6: 73–83.

Da, F.P., and W.Z. Song (2003). Fuzzy neural networks for direct adaptive control, *IEEE Trans. Indust. Electron.*, 50(3): 507–513.

Delmotte, F., T.M. Guerra, and M. Ksantini (2007). Continuous Takagi–Sugeno's models: Reduction of the number of LMI conditions in various fuzzy control design techniques, *IEEE Trans. Fuzzy Syst.*, 15(3): 426–438.

Diao, Y.X., and K.M. Passino (2002). Adaptive neural/fuzzy control for interpolated nonlinear systems, *IEEE Trans. Fuzzy Syst.*, 10(5): 583–595.

Dong, J.X., Y.Y. Wang, and G.H. Yang (2009). Control synthesis of continuous-time T–S fuzzy systems with local nonlinear models, *IEEE Trans. Syst. Man Cybern. B*, 39(5): 1245–1258.

Dubois, D. and H. Prade (1980). *Fuzzy Sets and Systems: Theory and Applications.* Academic Press, Orlando, Florida.

Dutta, S. (1993). Fuzzy logic applications: Technological and strategic issues, *IEEE Trans. Eng. Manage.*, 40(3): 237–254.

Edgar, C.R., and B.E. Postlethwaite (2000). MIMO fuzzy internal model control, *Automatica*, 36(6): 867–877.

El Ghaoui, L., and S.I. Niculescu (2000). *Advances in Linear Matrix Inequalities Methods in Control*, SIAM: Philadelphia.

Elshafei, A.L. (2002). Adaptive fuzzy control of nonlinear systems via a variable-structure algorithm, *Proc. of the 2002 IEEE Inter. Symposium on Intelligent Control*, Vancouver, pp. 620–625.

Erbatur, K., and O. Kaynak (2001). Use of adaptive fuzzy systems in parameter tuning of sliding-mode controllers, *IEEE/ASME Trans. Mechatron.*, 6(4): 474–482.

Espada, A., and A. Barreiro (1999). Robust stability of fuzzy control systems based on conicity conditions, *Automatica*, 35(4): 643–654.

Essounbouli, N., A. Hamzaoui, and K. Benmahammed (2003). Adaptation algorithm for robust fuzzy controller of nonlinear uncertain systems, *Proc. 2003 IEEE Conf. on Control Applications*, Istanbul, pp. 386–391.

Fantuzzi, C., and R. Rovatti (1996). On the approximation capabilities of the homogeneous Takagi–Sugeno model, *Proc. 5th IEEE Int. Conf. on Fuzzy Systems*, New Orleans, pp. 1067–1072.

Farag, W.A., V.H. Quintana, and G. Lambert-Torres (1998). A genetic-based neuro-fuzzy approach for modeling and control of dynamical systems, *IEEE Trans. Neural Netw.*, 9(5): 756–767.

Farinwata, S.S., D. Pirovolou, and G.J. Vachtsevanos (1994). On input-output stability analysis of a fuzzy controller for a missile autopilot's yaw axis, *Proc. 3rd IEEE Int. Conf. on Fuzzy Systems*, Orlando, FL, pp. 930–935.

Feng, G. (1999). Analysis of a new algorithm for continuous time robust adaptive control, *IEEE Trans. Automat. Contr.*, 44: 1764–1768.

Feng, G. (2001). Approaches to quadratic stabilization of uncertain fuzzy dynamic systems, *IEEE Trans. Circuits Syst. I*, 48(6): 760–769.

Feng, G. (2002). An approach to adaptive control of fuzzy dynamic systems, *IEEE Trans. Fuzzy Syst.*, 10(2): 268–275.

Feng, G. (2003). Controller synthesis of fuzzy dynamic systems based on piecewise Lyapunov functions, *IEEE Trans. Fuzzy Syst.*, 11(5): 605–612.

Feng, G. (2004a). Stability analysis of discrete time fuzzy dynamic systems based on piecewise Lyapunov functions, *IEEE Trans. Fuzzy Syst.*, 12(1): 22–28.

Feng, G. (2004b). H-infinity controller design of fuzzy dynamic systems based on piecewise Lyapunov functions, *IEEE Trans. Syst. Man Cybern. B*, 34(1): 283–292.

Feng, G. (2005). Robust H-infinity filtering of fuzzy dynamic systems, *IEEE Trans. Aerospace Electron. Syst.*, 41(2): 658–670.

Feng, G. (2006). A survey on analysis and design of model-based fuzzy control systems, *IEEE Trans. Fuzzy Syst.*, 14(5): 676–697.

Feng, G., and J. Ma (2001). Quadratic stabilization of uncertain discrete-time fuzzy dynamic systems, *IEEE Trans. Circuits Syst. I*, 48(11): 1337–1343.

Feng, G., and D. Sun (2002). Generalized H_2 controller synthesis of fuzzy dynamic systems based on piecewise Lyapunov functions, *IEEE Trans. Circuits Syst. I*, 49(12): 1843–1850.

Feng, G., S.G. Cao, and N.W. Rees (2002). Stable adaptive control of fuzzy dynamic systems, *Fuzzy Sets Syst.*, 131: 217–224.

Feng, G., S.G. Cao, N.W. Rees, and C.K. Chak (1997). Design of fuzzy control systems with guaranteed stability, *Fuzzy Sets Syst.*, 85: 1–10.

Feng, G., C.L. Chen, D. Sun, and X.P. Guan (2005). H_∞ controller synthesis of fuzzy dynamic systems based on piecewise Lyapunov functions and bilinear matrix inequalities, *IEEE Trans. Fuzzy Syst.*, 13(1): 94–103.

Feng, G., M. Chen, D. Sun, and T.J. Zhang (2008). Approaches to robust filtering design of discrete time fuzzy dynamic systems, *IEEE Trans. Fuzzy Syst.*, 16(2): 331–340.

Feng, M., and C.J. Harris (2001a). Piecewise Lyapunov stability conditions of fuzzy systems, *IEEE Trans. Syst. Man Cybern. B*, 31: 259–262.

Feng, M., and C.J. Harris (2001b). Feedback stabilization of fuzzy systems via linear matrix inequalities, *Int. J. Syst. Sci.*, 32: 221–231.

Ferrari-Trecate, G., F.A. Cuzzola, D. Mignone, and M. Morari (2002). Analysis of discrete time piecewise affine and hybrid systems, *Automatica*, 38: 2139–2146.

Fischle, K., and D. Schroder (1999). An improved stable adaptive fuzzy control method, *IEEE Trans. Fuzzy Syst.*, 7: 27–40.

Flores, A., D. Saez, J. Araya, M. Berenguel, and A. Cipriano (2005). Fuzzy predictive control of a solar power plant, *IEEE Trans. Fuzzy Syst.*, 13(1): 58–68.

French, M., and E. Rogers (1998). Input/output stability theory for direct neuro-fuzzy controllers, *IEEE Trans. Fuzzy Syst.*, 6(3): 331–345.

Frey, C.W., and H.B. Kuntze (2001). A neuro-fuzzy supervisory control system for industrial batch processes, *IEEE Trans. Fuzzy Syst.*, 9(4): 570–577.

Furutani, E., M. Saeki, and M. Araki (1992). Shifted Popov criterion and stability analysis of fuzzy control systems, *Proc. 23rd IEEE Conf. Decision and Control*, Tucson, AZ, pp. 2790–2795.

Gahinet, P., A. Nemirovski, A. Laub, and M. Chilali (1995). *The LMI Control Toolbox*, The Mathworks Inc., Natick, MA.

Gao, Y., and M.J. Er (2003). Online adaptive fuzzy neural identification and control of a class of MIMO nonlinear systems, *IEEE Trans. Fuzzy Syst.*, 11(4): 462–477.

Glower, S., and J. Munighan (1997). Designing fuzzy controllers from a variable structures standpoint, *IEEE Trans. Fuzzy Syst.*, 5(1): 138–144.

Golea, N., A. Golea, and K. Benmahammed (2002). Fuzzy model reference adaptive control, *IEEE Trans. Fuzzy Syst.*, 10(4): 436–444.

Goodwin, G.C., and K.S. Sin (1984). *Adaptive Filtering, Prediction, and Control*, Prentice-Hall: Englewood Cliffs, NJ.

Guan, X.P., and C.L. Chen (2003). Adaptive fuzzy control for chaotic systems with H-infinity tracking performance, *Fuzzy Sets Syst.*, 139(1): 81–93.

Guan, X.P., and C.L. Chen (2004). Delay-dependent guaranteed cost control for T–S fuzzy systems with time delays, *IEEE Trans. Fuzzy Syst.*, 12(2): 236–249.

Guerra, T.M., and L. Vermeiren (2004). LMI-based relaxed nonquadratic stabilization conditions for nonlinear systems in the Takagi-Sugeno's form, *Automatica*, 40(5): 823–829.

Guesmi, T., H.H. Adballah, and A. Toumi (2004). Transient stability fuzzy control approach for power systems, *IEEE Int. Conf. on Industrial Technology*, Hammamet, Tunisia, pp. 1676–1681.

Guillemin, P. (1996). Fuzzy logic applied to motor control, *IEEE Trans. Indust. Appl.*, 32(1): 51–56.

Ha, Q.P., Q.H. Nguyen, D.C. Rye, and H.F. Durrant-Whyte (2001). Fuzzy sliding mode controllers with applications, *IEEE Trans. Indust. Electron.*, 48(1): 38–46.

Hagras, H.A. (2004). A hierarchical type-2 fuzzy logic control architecture for autonomous mobile robots, *IEEE Trans. Fuzzy Syst.*, 12(4): 524–539.

Han, H., C.Y. Su, and Y. Stepanenko (2001). Adaptive control of a class of nonlinear systems with nonlinearly parameterized fuzzy approximators, *IEEE Trans. Fuzzy Syst.*, 9(2): 315–323.

Han, Z.X., G. Feng, B.L. Walcott, and J. Ma (2000). Dynamic output feedback controller design for fuzzy systems, *IEEE Trans. Syst. Man Cybern. B*, 30(1): 204–210.

Haruki, T., and K. Kikuchi (1992). Video camera system using fuzzy logic, *IEEE Trans. Consumer Electron.*, 38(3): 624–634.

He, S.Z., S. Tan, C.C. Han, and P.Z. Wang (1993a). Control of dynamic processes using an online rule-adaptive fuzzy control system, *Fuzzy Sets Syst.*, 54: 11–22.

He, S.Z., S. Tan, F.L. Xu, and P.Z. Wang (1993b). Fuzzy self-tuning of PID controllers, *Fuzzy Sets Syst.*, 56: 37–46.

Henson, M.A., and D.E. Seborg (1997). *Nonlinear Process Control*. Prentice Hall: Upper Saddle River, NJ.

Hong, S.K., and R. Langari (2000a). Robust fuzzy control of a magnetic bearing system subject to harmonic disturbances, *IEEE Trans. Control Syst. Technol.*, 8(2): 366–371.

Hong, S.K., and R. Langari (2000b). An LMI-based H_∞ fuzzy control system design with TS framework, *Inf. Sci.*, 123(3–4): 163–179.

Horiuchi, J.I., and M. Kishimoto (2002). Application of fuzzy control to industrial bioprocesses in Japan, *Fuzzy Sets Syst.*, 128(1): 117–124.

Hsiao, F.H., C.W. Chen, Y.W. Liang, S.D. Xu, and W.L. Chiang (2005). T–S fuzzy controllers for nonlinear interconnected systems with multiple time delays, *IEEE Trans. Circuits Syst. I*, 52(9): 1883–1893.

Hu, B.G., G.K.I. Mann, and R.G. Gosine (1999). New methodology for analytical and optimal design of fuzzy PID controllers, *IEEE Trans. Fuzzy Syst.*, 7(5): 521–539.

Hu, B.G., G.K.I. Mann, and R.G. Gosine (2001). A systematic study of fuzzy PID controllers: Function based evaluation approach, *IEEE Trans. Fuzzy Syst.*, 9(5): 699–712.

Hu, L., and B. Huang (2005). Multirate robust digital control for fuzzy systems with periodic Lyapunov function, *IEEE Trans. Fuzzy Syst.*, 13(4): 436– 443.

Huang, S.J., and W.C. Lin (2003). Adaptive fuzzy controller with sliding surface for vehicle suspension control, *IEEE Trans. Fuzzy Syst.*, 11(4): 550–559.

Hwang, C.L. (2004). A novel Takagi–Sugeno-based robust adaptive fuzzy sliding-mode controller, *IEEE Trans. Fuzzy Syst.*, 12(5): 676–687.

Hwang, C.L., and C.Y. Kuo (2001). A stable adaptive fuzzy sliding-mode control for affine nonlinear systems with application to four-bar linkage systems, *IEEE Trans. Fuzzy Syst.*, 9(2): 238–252.

Hwang, C.L., and H.Y. Lin (2004). A fuzzy decentralized variable structure tracking control with optimal and improved robustness designs: Theory and applications, *IEEE Trans. Fuzzy Syst.*, 12(5): 615– 630.

Hwang, G.C., and S.C. Lin (1992). A stability approach to fuzzy control design for nonlinear systems, *Fuzzy Sets Syst.*, 48: 179–287.

Hwang, Y.R., and M. Tomizuka (1994). Fuzzy smoothing algorithms for variable structure systems, *IEEE Trans. Fuzzy Syst.,* 2(4): 277–284.

Ioannou, P.A., and J. Sun (1995). *Stable and Robust Adaptive Control*, Prentice-Hall: Englewood Cliffs, NJ.

Jang, J.S.R. (1993). ANFIS: Adaptive network-based fuzzy inference system, *IEEE Trans. Syst. Man Cybern.*, 23: 665–685.

Jang, J.S.R., and C.T. Sun (1995). Neuro-fuzzy modeling and control, *Proc. IEEE*, 83(3): 378–406.

Joh, J., Y.H. Chen, and R. Langari (1998). On the stability issues of linear Takagi–Sugeno fuzzy models, *IEEE Trans. Fuzzy Syst.*, 6(3): 402–410.

Johansen, T.A. (1994). Fuzzy model based control: Stability robustness and performance issues, *IEEE Trans. Fuzzy Syst.*, 2(1): 221–233.

Johansen, T.A., R. Shorten, and R. Murray-Smith (2000). On the interpretation and identification of dynamic Takagi–Sugeno models, *IEEE Trans. Fuzzy Syst.*, 8(3): 297–313.

Johansson, M., A. Rantzer, and K.E. Arzen (1999). Piecewise quadratic stability of fuzzy systems, *IEEE Trans. Fuzzy Syst.*, 7(6): 713–722.

Juang, C.F., and C.H. Hsu (2005). Temperature control by chip-implemented adaptive recurrent fuzzy controller designed by evolutionary algorithm, *IEEE Trans. Circuits Syst. I: Regular Papers*, 52(11): 2376–2384.

Kadmiry, B., and D. Driankov (2004). A fuzzy gain-scheduler for the attitude control of an unmanned helicopter, *IEEE Trans. Fuzzy Syst.*, 12(4): 502–515.

Kandel, A., O. Manor, Y. Klein, and S. Fluss (1999). ATM traffic management and congestion control using fuzzy logic, *IEEE Trans. Syst. Man Cybern. C*, 29(3): 474–480.

Kang, G., and W. Lee (1995). Design of fuzzy state controllers and observers, *Proc. Int. Joint Conf. 4th FUZZ-IEEE/2nd IFES*, Yokohama, pp. 1355–1360.

Kang, H. (1993). Stability and control of fuzzy dynamic systems via cell-state transitions in fuzzy hypercubes, *IEEE Trans. Fuzzy Syst.*, 1(4): 267–279.

Kaynak, O., K. Erbatur, and M. Ertugnrl (2001). The fusion of computationally intelligent methodologies and sliding-mode control-A survey, *IEEE Trans. Indust. Electron.*, 48(1): 4–17.

Keller, J.M., R.R. Yager, and H. Tahani (1992). Neural network implementation of fuzzy logic, *Fuzzy Sets Syst.*, 45: 1–12.

Kickert, W.J.M., and E.H. Mamdani (1978). Analysis of a fuzzy logic controller, *Fuzzy Sets Syst.*, 1: 29–44.

Kickert, W.J.M., and H.R. Van Nauta Lemke (1976). Application of a fuzzy logic controller in a warm water plant, *Automatica*, 12: 301–308.

Kiguchi, K., T. Tanaka, and T. Fukuda (2004). Neuro-fuzzy control of a robotic exoskeleton with EMG signals, *IEEE Trans. Fuzzy Syst.*, 12(4): 481–490.

Kim, E. (2004). Output feedback tracking control of robot manipulators with model uncertainty via adaptive fuzzy logic, *IEEE Trans. Fuzzy Syst.*, 12(3): 368–378.

Kim, E., and D. Kim (2001). Stability analysis and synthesis for an affine fuzzy system via LMI and ILMI: Discrete case, *IEEE Trans. Syst. Man Cybern. B*, 31(1): 132–140.

Kim, E., and S. Kim (2002). Stability analysis and synthesis for an affine fuzzy control system via LMI and ILMI: Continuous case, *IEEE Trans. Fuzzy Syst.*, 10(3): 391–400.

Kim, E., and H. Lee (2000). New approaches to relaxed quadratic stability condition of fuzzy control systems, *IEEE Trans. Fuzzy Syst.*, 8(5): 523–534.

Kim, E., and S. Lee (2005). Output feedback tracking control of MIMO systems using a fuzzy disturbance observer and its application to the speed control of a PM synchronous motor, *IEEE Trans. Fuzzy Syst.*, 13(6): 725–741.

Kim, S.H., and P.G. Park (2009). Observer-based relaxed H_∞ control for fuzzy systems using a multiple Lyapunov function, *IEEE Trans. Fuzzy Syst.*, 17(2): 477–484.

Kim, S.W., Y.W. Cho, and M. Park (1996). A multirule-base controller using the robust property of a fuzzy controller and its design method, *IEEE Trans. Fuzzy Syst.*, 4: 315–327.

Kim, W.C., S.C. Ahn, and W.H. Kwon (1995). Stability analysis and stabilization of fuzzy state space models, *Fuzzy Sets Syst.*, 71(1): 131–142.

King, P.J., and E.H. Mamdani (1977). The application of fuzzy control systems to industrial process, *Automatica*, 13: 235–242.

Kiriakidis, K. (2001). Robust stabilization of the Takagi–Sugeno fuzzy model via bilinear matrix inequalities, *IEEE Trans. Fuzzy Syst.*, 9(2): 269–277.

Kiriakos, K. (1998). Fuzzy model-based control of complex systems, *IEEE Trans. Fuzzy Syst.*, 6(4): 517–529.

Ko, H.S., and T. Niimura (2002). Power system stabilization using fuzzy-neural hybrid intelligent control, *Proc. of the 2002 IEEE Int. Symposium on Intelligent Control*, Vancouver, pp. 879–884.

Koo, T.J. (2001). Stable model reference adaptive fuzzy control of a class of nonlinear systems, *IEEE Trans. Fuzzy Syst.*, 9: 624–636.

Korba, P., R. Babuska, H.B. Verbruggen, and P.M. Frank (2003). Fuzzy gain scheduling: Controller and observer design based on Lyapunov method and convex optimization, *IEEE Trans. Fuzzy Syst.*, 11(3): 285–298.

Kordon, A., P.S. Dhurjati, Y.O. Fuentes, and B.A. Ogunnaike (1999). An intelligent parallel control system structure for plants with multiple operating regimes, *J. Process Control*, 9: 453–460.

Kornblum, R.J., and M. Tribus (1970). The use of Bayesian inference in the design of an end-point control system for the basic oxygen steel furnace, *IEEE Trans. Syst. Man Cybern.*, SMC–6:339–348.

Koska, B. (1992). *Neural Networks and Fuzzy Systems*, Prentice Hall: Englewood Cliffs, NJ.

Kothare, M.V., V. Balakrishnan, and M. Morari (1996). Robust constrained model predictive control using linear matrix inequalities, *Automatica*, 32(10): 1361–1379.

Krstic, M., I. Kanellakopoulos, and P. Kokotovic (1995). *Nonlinear and Adaptive Control Design*, John Wiley & Sons: New York.

Kumar, S. (2005). A review of smart volume controllers for consumer electronics, *IEEE Trans. Consumer Electron.*, 51(2): 600–605.

Kung, C.C., T.H. Chen, and C.H. Chen (2005). H_∞ state feedback controller design for T–S fuzzy systems based on piecewise Lyapunov function, *Proc. 14th IEEE Int. Conf. on Fuzzy Systems*, Reno, NV, pp. 708–713,

Kwok, H.F., D.A. Linkens, M. Mahfouf, and G.H. Mills (2004). SIVA: A hybrid knowledge-and-model-based advisory system for intensive care ventilators, *IEEE Trans. Inf. Technol. Biomed.*, 8(2): 161–172.

Lam, H.K., F.H.F. Leung, and P.K.S. Tam (2001). Nonlinear state feedback controller for non-linear systems: Stability analysis and design based on fuzzy plant model, *IEEE Trans. Fuzzy Syst.*, 9(4): 657–661.

Langari, R., and M. Tomizuka (1990). Stability of fuzzy linguistic control systems, *Proc. of IEEE Conf. Decision Control*, Honolulu, pp. 2185–2190.

Larkin, L.I. (1985). A fuzzy logic controller for aircraft flight control, In: *Industrial Applications of Fuzzy Control*, M. Sugeno, Ed., North Holland: Amsterdam, pp. 87–104.

Larsen, P.M. (1980). Industrial applications of fuzzy logic control, *Int. J. Man Mach. Stud.*, 12: 3–10.

Lazzerini, B., L.M. Reyneri, and M. Chiaberge (1999). A neuro-fuzzy approach to hybrid intelligent control, *IEEE Trans. Industry Appl.*, 35(2): 413–425.

Lee, C.C. (1990a). Fuzzy logic in control systems: Fuzzy logic controller—Part I, *IEEE Trans. Syst. Man Cybern.*, 20(2): 404–418.

Lee, C.C. (1990b). Fuzzy logic in control systems: Fuzzy logic controller—Part II, *IEEE Trans. Syst. Man Cybern.*, 20(2): 419–435.

Lee, H., and M. Tomizuka (2000). Robust adaptive control using a universal approximator for SISO nonlinear systems, *IEEE Trans. Fuzzy Syst.*, 8(1): 95–106.

Lee, H.J., J.B. Park, and G. Chen (2001). Robust fuzzy control of nonlinear systems with parametric uncertainties, *IEEE Trans. Fuzzy Syst.*, 9(2): 369–379.

Lee, K.R., J.H. Kim, and E.T. Jeung (2000). Output feedback robust H_∞ control of uncertain fuzzy dynamic systems with time-varying delay, *IEEE Trans. Fuzzy Syst.*, 8: 657–664.

Lee, S.H., and Z. Bien (1994). Design of expandable fuzzy inference processor, *IEEE Trans. Consumer Electron.*, 40(2): 171–175.

Lee, S.H., and J.T. Lim (2001). Multicast ABR service in ATM networks using a fuzzy-logic-based consolidation algorithm, *IEE Proc. Commun.*.148: 8–13.

Lee, Y.G., and S.H. Zak (2004). Uniformly ultimately bounded fuzzy adaptive tracking con-trollers for uncertain systems, *IEEE Trans. Fuzzy Syst.*, 12(6): 797–811.

Lee, Y.M., S.I. Jang, K.W. Chung, D.Y. Lee, W.C. Kim, and C.W. Lee (1994). A fuzzy-control processor for automatic focusing, *IEEE Trans. Consumer Electron.*, 40(2): 138–144.

Leephakpreeda, T. (1999). H_∞ stability robustness of fuzzy control systems, *Automatica*, 35(8): 1467–1470.

Leu, Y.G., W.Y. Wang, and T.T. Lee (2005). Observer-based direct adaptive fuzzy-neural con-trol for nonaffine nonlinear systems, *IEEE Trans. Neural Netw.*, 16(4): 853–861.

Leung, F.H.F., H.K. Lam, P.K.S. Tam, and Y.S. Lee (2003). Stable fuzzy controller design for uncertain nonlinear systems: Genetic algorithm approach, *Proc. the 12th IEEE Int. Conf. on Fuzzy Systems*, St. Louis, MO, pp. 500–505.

Lewis, F.L., and K. Liu (1996). Towards a paradigm for fuzzy logic control, *Automatica*, 32(2): 167–181.

Li, C.S., and C.Y. Lee (2003). Self-organizing neuro-fuzzy system for control of unknown plants, *IEEE Trans. Fuzzy Syst.*, 11(1): 135–150.

Li, C.S., C.Y. Lee, and K.H. Cheng (2004). Pseudoerror-based self-organizing neuro-fuzzy system, *IEEE Trans. Fuzzy Syst.*, 12(6): 812–819.

Li, H.X., L. Zhang, K.Y. Cai, and G.R. Chen (2005). An improved robust fuzzy-PID controller with optimal fuzzy reasoning, *IEEE Trans. Syst. Man Cybern. B*, 35(6): 1283–1294.

Li, T.H.S., and K.J. Lin (2004). Stabilization of singularly perturbed fuzzy systems, *IEEE Trans. Fuzzy Syst.*, 12(5): 579–595.

Li, T.H.S., S.J. Chang, and W. Tong (2004). Fuzzy target tracking control of autonomous mobile robots by using infrared sensors, *IEEE Trans. Fuzzy Syst.*, 12(4): 491–501.

Li, W., X.G. Chang, J. Farrell, and F.M. Wahl (2001). Design of an enhanced hybrid fuzzy P+ID controller for a mechanical manipulator, *IEEE Trans. Syst. Man Cybern. B*, 31(6): 938–945.

Lian, K.Y., and J.J. Liou (2006). Output tracking control for fuzzy systems via output feedback design, *IEEE Trans. Fuzzy Syst.*, 14(5): 628–639.

Lian, K.Y., C.S. Chiu, T.S. Chiang, and P. Liu (2001). LMI-based fuzzy chaotic synchronization and communications, *IEEE Trans. Fuzzy Syst.*, 9(4): 539–553.

Liang, Y.W., S.D. Xu, D.C. Liaw, and C.C. Chen (2008). A study of T–S model-based SMC scheme with application to robot control, *IEEE Trans. Indust. Electron.*, 55(11): 3964–3971.

Lin, C., Q.G. Wang, and T.H. Lee (2005). Stabilization of uncertain fuzzy time-delay systems via variable structure control approach, *IEEE Trans. Fuzzy Syst.*, 13(6): 787–798.

Lin, C.M., and C.F. Hsu (2003). Self-learning fuzzy sliding-mode control for antilock braking systems, *IEEE Trans. Control Syst. Technol.*, 11(2): 273–278.

Lin, C.M., and C.F. Hsu (2004). Supervisory recurrent fuzzy neural network control of wing rock for slender delta wings, *IEEE Trans. Fuzzy Syst.*, 12(5): 733–742.

Lin, C.T., and C.S.G. Lee (1994). Reinforcement structure/parameter learning for neural-network-based fuzzy logic control systems, *IEEE Trans. Fuzzy Syst.*, 2(1): 46–63.

Lin, F.J., and P.H. Shen (2006). Adaptive fuzzy-neural-network control for a DSP-based permanent magnet linear synchronous motor servo drive, *IEEE Trans. Fuzzy Syst.*, 14(4): 481–495.

Lin, S.C., and Y.Y. Chen (1997). Design of self learning fuzzy sliding mode controllers based on genetic algorithms, *Fuzzy Sets Syst.*, 86: 139–153.

Lin, W.S., and C.H. Tsai (2001). Self-organizing fuzzy control of multi-variable systems using learning vector quantization network, *Fuzzy Sets Syst.*, 124: 197–212.

Liu, B.D., C.Y. Chen, and J.Y. Tsao (2001). Design of adaptive fuzzy logic controller based on linguistic-hedge concepts and genetic algorithms, *IEEE Trans. Syst. Man Cybern. B*, 31(1): 32–53.

Liu, H.P., F.C. Sun, and Z.Q. Sun (2005). Stability analysis and synthesis of fuzzy singularly perturbed systems, *IEEE Trans. Fuzzy Syst.*, 13(2): 273–284.

Liu, X., and Q. Zhang (2003). New approaches to controller designs based on fuzzy observers for T–S fuzzy systems via LMI, *Automatica*, 39(9): 1571–1582.

Liu, X.J., F. Lara-Rosano, and C.W. Chan (2004). Model-reference adaptive control based on neurofuzzy networks, *IEEE Trans. Syst. Man Cybern. C*, 34(3): 302–309.

Lo, J.C., and Y.M. Chen (1999). Stability issues on Takagi–Sugeno fuzzy model: Parametric approach, *IEEE Trans. Fuzzy Syst.* 7(5): 597–607.

Lo, J.C., and Y.H. Kuo (1998). Decoupled fuzzy sliding mode control, *IEEE Trans. Fuzzy Syst.*, 6(3): 426–435.

Lo, J.C., and M.L. Lin (2004). Observer-based robust H-infinity control for fuzzy systems using two-step procedure, *IEEE Trans. Fuzzy Syst.*, 12(3): 350–359.

Lu, Y., and Y. Arkun (2000). Quasi-min-max MPC algorithms for LPV systems, *Automatica*, 36: 527–540.

Lu, Y., and Y. Arkun (2002). A scheduling quasi-min-max model predictive control algorithm for nonlinear systems, *J. Process Control*, 12: 589–604.

Ma, X., Z. Sun, and Y. He (1998). Analysis and design of fuzzy controller and fuzzy observer, *IEEE Trans. Fuzzy Syst.*, 6(1): 41–51.

Mamdani, E.H. (1974). Application of fuzzy algorithms for simple dynamic plant, *Proc. IEE*, 121: 1585–1588.

Mamdani, E.H. (1976). Advances in the linguistic synthesis of fuzzy controllers, *International Journal of Man-Machine Studies*, 8(6): 669-678.

Mamdani, E.H., and S. Assilian (1975). An experiment in linguistic synthesis with a fuzzy logic controller, *Int. J. Man Mach. Stud.*, 7: 1–13.

Mann, G.K.I., and R.G. Gosine (2005). Three-dimensional min-max-gravity based fuzzy PID inference analysis and tuning, *Fuzzy Sets Syst.*, 156: 300–323.

Mann, G.K.I., B.G. Hu, and R.G. Gosine (1999). Analysis of direct action fuzzy PID controller structures, *IEEE Trans. Syst. Man Cybern. B*, 29(3): 371–388.

Mann, G.K.I., B.G. Hu, and R.G. Gosine (2001). Two-level tuning of fuzzy PID controllers, *IEEE Trans. Syst. Man Cybern. B*, 31(2): 263–269.

Mannani, A., and H.A. Talebi (2003). A fuzzy Lyapunov-based control strategy for a macro-micro manipulator, *Proc. 2003 IEEE Conf. on Control Applications*, Istanbul, pp. 368–373.

Mar, J., and F.J. Lin (2001). An ANFIS controller for the car-following collision prevention system, *IEEE Trans. Vehicular Technol.*, 50(4): 1106–1113.

Meda-Campana, J.A., and B. Castillo-Toledo (2005). On the output regulation for TS fuzzy models using sliding modes, *Proc. American Control Conf.*, Portland, OR, pp. 4062–4067.

Melin, C., and B. Vidolov (1994). Passive two-rule-based fuzzy logic controllers: Analysis and application to stabilization, *Proc. Third IEEE Int. Conf. on Fuzzy Systems*, Orlando, FL, pp. 947–950.

Melin, P., and O. Castillo (2001). Intelligent control of complex electrochemical systems with a neuro-fuzzy-genetic approach, *IEEE Trans. Indust. Electron.*, 48(5): 951–955.

Mendel, J.M. (1995). Fuzzy logic systems for engineering: A tutorial, *Proc. IEEE*, 83: 345–377.

Middleton, R., G.C. Goodwin, D. Hill, and D. Mayne (1988). Design issues in adaptive control, *IEEE Trans. Autom. Control*, 33: 50–58.

Misir, D., H.A. Malki, and G. Chen (1996). Design and analysis of a fuzzy proportional-integral-derivative controller, *Fuzzy Sets Syst.*, 79: 297–314.

Mitra, S., and Y. Hayashi (2000). Neuro-fuzzy rule generation: Survey in soft computing framework, *IEEE Trans. Neural Netw.*, 11(3): 748–768.

Mizumoto, M. (1995). Realization of PID controls by fuzzy control methods, *Fuzzy Sets Syst.*, 70: 171–182.

Mollov, S., R. Babuska, J. Abonyi, and H.B. Verbruggen (2004). Effective optimization for fuzzy model predictive control, *IEEE Trans. Fuzzy Syst.*, 12(5): 661– 675.

Mudi, R.K., and N.R. Pal (1999). A robust self-tuning scheme for PI- and PD-type fuzzy controllers, *IEEE Trans. Fuzzy Syst.*, 7(1): 2–16.

Mudi, R.K., and N.R. Pal (2001). A note on fuzzy PI-type controllers with resetting action, *Fuzzy Sets Syst.*, 121(1): 149–159.

Munasinghe, S.R., M.S. Kim, and J.J. Lee (2005). Adaptive neurofuzzy controller to regulate UTSG water level in nuclear power plants, *IEEE Trans. Nuclear Sci.*, 52(1): 421–429.

Murakami, S., and M. Maeda (1985). Application of fuzzy controller to automobile speed control system, In: *Industrial Applications of Fuzzy Control*, M. Sugeno, Ed., North-Holland: Amsterdam, pp. 105–124.

Nakagaki, N., Y. Bando, T. Mori, S. Torikoshi, and S. Suzuki (1994). Wide aspect TV receiver with aspect detection and non-linear control for picture quality, *IEEE Trans. Consumer Electron.*, 40(3): 743–752.

Narendra, K.S., and A.M. Annaswamy (1989). *Stable Adaptive Systems*, Prentice Hall: Englewood Cliffs, NJ.

Nerenji, H.R., and P. Khedkar (1992). Learning and tuning fuzzy logic controllers through reinforcements, *IEEE Trans. Neural Netw.*, 3(5): 724–740.

Nguang, S.K., and W. Assawinchaichote (2003). H_∞ filtering for fuzzy dynamical systems with pole placement constraints, *IEEE Trans. Circuits Syst. 1*, 50(11): 1503–1508.

Niasar, A.H., H. Moghbeli, and R. Kazemi (2003). Yaw moment control via emotional adaptive neuro-fuzzy controller for independent rear wheel drives of an electric vehicle, *Proc. of 2003 IEEE Conf. on Control Applications*, Istanbul, pp. 380–385.

Nounou, H.N., and K.M. Passino (2004). Stable auto-tuning of adaptive fuzzy/neural controllers for nonlinear discrete-time systems, *IEEE Trans. Fuzzy Syst.*, 12(1): 70–83.

Ohtake, H., K. Tanaka, and H.O. Wang (2003). Piecewise nonlinear control, *Proc. 42nd IEEE Conference on Decision and Control*, Maui, HI, pp. 4735–4740.

dc Oliveira, M.C., J. Bernussou, and J.C. Geromel (1999). A new discrete time robust stability condition, *Syst. Control Lett.*, 36(2): 135–141.

Opitz, H.P. (1993). Fuzzy control and stability criteria, *Proc. EUFIT'93*, Aachen, Germany, pp. 130–136.

Ordonez, R., and P.M. Passino (1999). Stable multi-input multi-output adaptive fuzzy neural control, *IEEE Trans. Fuzzy Syst.*, 7(3): 345–353.

Ordonez, R., J. Zumberge, J.T. Spooner, and K.M. Passino (1997). Adaptive fuzzy control: Experiments and comparative analyses, *IEEE Trans. Fuzzy Syst.*, 5(2): 167–188.

Ostergaard, J.J. (1977). Fuzzy logic control of a heat exchanger process, In: *Fuzzy Automata and Decision Processes*, M.M. Gupta, G.N. Saridis, and B.R. Gaines, Eds., North-Holland: Amsterdam, pp. 285–320.

Ozkan, L., M.V. Kothare, and C. Georgakis (2000). Model predictive control of nonlinear systems using piecewise linear models, *Comput. Chem. Eng.*, 24: 793–799.

Pal, K., and N.R. Pal (1999). A neuro-fuzzy system for inferencing, *Int. J. Intell. Syst.*, 14: 1155–1182.

Pal, K., R.K. Mudi, and N.R. Pal (2002). A new scheme for fuzzy rule-based system identification and its application to self-tuning fuzzy controllers, *IEEE Trans. Syst. Man Cybern. B*, 32(4): 470–482.

Pal, K., N.R. Pal, and J.M. Keller (1998). Some neural net realizations of fuzzy reasoning, *Int. J. Intell. Syst.*, 13: 859–886.

Palm, R. (1992). Sliding mode fuzzy control, *Proc. 1st IEEE Int. Conf. on Fuzzy Systems*, San Diego, pp. 519–526.

Palm, R. (1993). Tuning of scaling factors in fuzzy controllers using correlation functions, *Proc. IEEE Int. Conf. on Fuzzy Systems*, San Francisco, pp. 691–696.

Palm, R. (1994). Robust control by fuzzy sliding mode, *Automatica*, 30: 1429–1437.

Palm, R., D. Driankov, and H. Hellendoorn (1996). *Model Based Fuzzy Control*, Springer-Verlag: New York.

Pappis, C.P., and E.H. Mamdani (1977). A fuzzy logic controller for a traffic junction, *IEEE Trans. Syst. Man Cybern.*, 7(10): 707–717.

Park, C.W., and Y.W. Cho (2004). T–S model based indirect adaptive fuzzy control using online parameter estimation, *IEEE Trans. Syst. Man Cybern. B*, 34(6): 2293–2302.

Park, Y.M., M.J. Tahk, and H.C. Bang (2004). Design and analysis of optimal controller for fuzzy systems with input constraint, *IEEE Trans. Fuzzy Syst.*, 12(6): 766–779.

Passino, K., and S. Yurkovich (1998). *Fuzzy Control*, Addison-Wesley: Reading, MA.

Pedrycs, W. (1993). *Fuzzy Control and Fuzzy Systems*, Research Studies Press: Somerset, UK.

Pfluger, N., J. Yen, and R. Langari (1992). A defuzzification strategy for a fuzzy logic control employing prohibitive information in command formulation, *Proc. of IEEE Internatonal Conference on Fuzzy Systems*, San Diego, pp. 717–723.

Pomares, H., I. Rojas, J. Gonzalez, M. Damas, B. Pino, and A. Prieto (2004). Online global learning in direct fuzzy controllers, *IEEE Trans. Fuzzy Syst.*, 12(2): 218–229.

Qi, R.Y., and M.A. Brdys (2005). Adaptive fuzzy modeling and control for discrete-time nonlinear uncertain systems, *Proc. of American Control Conf.*, Portland, OR, pp. 1108–1113.

Ray, K.S., and D.D. Majumder (1984). Application of circle criteria for stability analysis of linear SISO and MIMO system associated with fuzzy logic controller, *IEEE Trans. Syst. Man Cybern.*, SMC-14: 345–349.

Sala, A., T.M. Guerra, and R. Babuska (2005). Perspectives of fuzzy systems and control, *Fuzzy Sets Syst.*, 156: 432–444.

Sanner, R.M., and J.E. Slotine (1992). Gaussian networks for direct adaptive control, *IEEE Trans. Neural Netw.*, 3(6): 837–863.

Santibanez, V., R. Kelly, and M.A. Llama (2005). A novel global asymptotic stable set-point fuzzy controller with bounded torques for robot manipulators, *IEEE Trans. Fuzzy Syst.*, 13(3): 362–372.

Seker, H., M.O. Odetayo, D. Petrovic, and R.N.G. Naguib (2003). A fuzzy logic based-method for prognostic decision making in breast and prostate cancers, *IEEE Trans. Inf. Technol. Biomed.*, 7(2): 114–122.

Seng, T.L., M. Bin Khalid, and R. Yusof (1999). Tuning of a neuro-fuzzy controller by genetic algorithm, *IEEE Trans. Syst. Man Cybern. B*, 29(2): 226–236.

de Silva, C.W. (1995). *Intelligent Control: Fuzzy Logic Applications*, CRC Press: New York.

Singh, S. (1992). Stability analysis of discrete fuzzy control systems, *Proc. 2nd IEEE Int. Conf. on Fuzzy Systems*, San Diego, pp.527–534

Sio, K.C., and C.K. Lee (1998). Stability of fuzzy PID controllers, *IEEE Trans. Syst. Man Cybern. A*, 28(4): 490–495.

Skelton, R.E., T. Iwasaki, and K. Grigoriadis (1998). *A Unified Algebraic Approach to Linear Control Design*. Taylor & Francis: Bristol, PA.

Skoczowski, S., S. Domek, K. Pietrusewicz, and B. Broel-Plater (2005). A method for improving the robustness of PID control, *IEEE Trans. Indust. Electron.*, 52(6): 1669–1676.

Smith, M.L. (1994). Sensors, appliance control, and fuzzy logic, *IEEE Trans. Industry Appl.*, 30(2): 305–310.

Su, C.Y., M. Oya, and H. Hong (2003). Stable adaptive fuzzy control of nonlinear systems preceded by unknown backlash-like hysteresis, *IEEE Trans. Fuzzy Syst.*, 11(1): 1–8.

Su, J.P., T.M. Chen, and C.C. Wang (2001). Adaptive fuzzy sliding mode control with GA-based reaching laws, *Fuzzy Sets Syst.*, 120: 145–158.

Sugeno, M. (1985). *Industrial Applications of Fuzzy Control*, Elsevier Science: New York.

Sugeno, M. (1999). On stability of fuzzy systems expressed by fuzzy rules with singleton consequents, *IEEE Trans. Fuzzy Syst.*, 7(2): 201–224.

Sugeno, M. and G.T. Kang (1988). Structure identification of fuzzy model, *Fuzzy Sets and Systems*, 26(1): 15–33.

Sugeno, M., and M. Nishida (1985). Fuzzy control of model car, *Fuzzy Sets Syst.*, 16: 103–113.

Sugeno, M., and T. Yasukawa (1993). A fuzzy-logic-based approach to qualitative modeling, *IEEE Trans. Fuzzy Syst.*, 1(1): 7–31.

Sun, Y.L., and M.J. Er (2004). Hybrid fuzzy control of robotics systems, *IEEE Trans. Fuzzy Syst.*, 12(6): 755–765.

Takagi, H. (1992). Application of neural networks and fuzzy logic to consumer products, *Proc. 1992 Int. Conf. on Industrial Electronics, Control, Instrumentation, and Automation*, San Diego, pp. 1629–1633.

Takagi, H., N. Suzuki, T. Koda, and Y. Kojima (1992). Neural networks designed on approximate reasoning architecture and their applications, *IEEE Trans. Neural Netw.*, 3(5): 752–760.

Takagi, T., and M. Sugeno (1985). Fuzzy identification of systems and its applications to modeling and control, *IEEE Trans. Syst. Man Cybern.*, 15(1): 116–132.

Tanaka, K., and M. Sano (1994). A robust stabilization problem of fuzzy control systems and its application to backing up control of a truck-trailer, *IEEE Trans. Fuzzy Syst.*, 2: 119–134.

Tanaka, K., and M. Sugeno (1992). Stability analysis and design of fuzzy control systems, *Fuzzy Sets Syst.*, 12: 135–156.

Tanaka, K., and H.O. Wang (2001). *Fuzzy Control Systems Design and Analysis: A LMI Approach*, Wiley: New York.

Tanaka, K., T. Hori, and H.O. Wang (2003). A multiple Lyapunov function approach to stabilization of fuzzy control systems, *IEEE Trans. Fuzzy Syst.*, 11(4): 582–589.

Tanaka, K., T. Ikeda, and H.O. Wang (1998). Fuzzy regulators and fuzzy observers: Relaxed stability conditions and LMI-based designs, *IEEE Trans. Fuzzy Syst.*, 6(2): 250–265.

Tang, K.S., K.F. Man, G. Chen, and S. Kwong (2001). An optimal fuzzy PID controller, *IEEE Trans. Indust. Electron.*, 48: 757–765.

Tani, T., S. Murakoshi, and M. Umano (1996). Neuro-fuzzy hybrid control system of tank level in petroleum plant, *IEEE Trans. Fuzzy Syst.*, 4(3): 360–368.

Taniguchi, T., and M. Sugeno (2004). Stabilization of nonlinear systems based on piecewise Lyapunov functions, *Proc. 13th IEEE Int. Conf. on Fuzzy Systems*, Budapest, pp. 1607–1612.

Tao, C.W., and J.S. Taur (2005). Robust fuzzy control for a plant with fuzzy linear model, *IEEE Trans. Fuzzy Syst.*, 13(1): 30–41.

Tao, C.W., J.S. Taur, and M.L. Chan (2004). Adaptive fuzzy terminal sliding mode controller for linear systems with mismatched time-varying uncertainties, *IEEE Trans. Syst. Man Cybern. B*, 34(1): 255–262.

Teixeira, M.C.M., and S.H. Zak (1999). Stabilizing controller design for uncertain nonlinear systems using fuzzy models, *IEEE Trans. Fuzzy Syst.*, 7(2): 133–142.

Teixeira, M.C.M., E. Assuncao, and R.G. Avellar (2003). On relaxed LMI-based designs for fuzzy regulators and fuzzy observers, *IEEE Trans. Fuzzy Syst.*, 11(5): 613–623.

Teodorescu, H.N., L.C. Jain, and A. Kandel (1998). *Fuzzy and Neuro-Fuzzy Systems in Medicine*, CRC Press: Boca Raton, FL.

Tong, R.M. (1977). A control engineering review of fuzzy systems, *Automatica*, 13: 559–568.

Tong, R.M., M.B. Beck, and A. Latten (1980). Fuzzy control of the activated sludge wastewater treatment process, *Automatica*, 6: 695–701.

Tong, S.C., and H.X. Li (2003). Fuzzy adaptive sliding-mode control for MIMO nonlinear systems, *IEEE Trans. Fuzzy Syst.*, 11(3): 354–360.

Tsay, D.L., H.Y. Chung, and C.J. Lee (1999). The adaptive control of nonlinear systems using the Sugeno-type of fuzzy logic, *IEEE Trans. Fuzzy Syst.*, 7(2): 225–229.

Tseng, C.S., and B.S. Chen (2001). $H\infty$ fuzzy estimation for a class of nonlinear discrete-time dynamic systems, *IEEE Trans. Signal Process.*, 49(11): 2605–2619.

Tseng, C.S., B.S. Chen, and H.J. Uang (2001). Fuzzy tracking control design for nonlinear dynamic systems via T–S fuzzy model, *IEEE Trans. Fuzzy Syst.*, 9(3): 381–392.

Tsourdos, A., J.T. Economou, A.B. White, and P.C.K. Luk (2003). Control design for a mobile robot: A fuzzy LPV approach, *Proc. of 2003 IEEE Conf. on Control Applications*, Istanbul, pp. 552–557.

Tuan, H.D., P. Apkarian, T. Narikiyo, and M. Kanota (2004). New fuzzy control model and dynamic output feed back parallel distributed compensation, *IEEE Trans. Fuzzy Systems*, 12(1): 13–21.

Tuan, H.D., P. Apkarian, T. Narikiyo, and Y. Yamamoto (2001). Parameterized linear matrix inequality techniques in fuzzy control system design, *IEEE Trans. Fuzzy Syst.*, 9(2): 324–332.

Tzafestas, S.G., and K.C. Zikidis (2001). NeuroFAST: On-line neuro-fuzzy ART-based structure and parameter learning TSK model, *IEEE Trans. Syst. Man Cybern. B,* 31(5): 797–802.

Umbers, I.G., and P.J. King (1980). An analysis of human-decision making in cement kiln control and the implications for automation, *Int. J. Man Mach. Stud.*, 12: 11–23.

Uragami, M., M. Mizumoto, and K. Tanaka (1976). Fuzzy robot controls, *Cybernetics*, 6: 39–64.

Utkin, V.I. (1992). *Sliding Modes in Control Optimization*, Springer-Verlag: Berlin.

Velez-Diaz, D., and Y. Tang (2004). Adaptive robust fuzzy control of nonlinear systems, *IEEE Trans. Syst. Man Cybern. B*, 34(3): 1596–1601.

Wai, R.J., and L.J. Chang (2006). Stabilizing and tracking control of nonlinear dual-axis inverted-pendulum system using fuzzy neural network, *IEEE Trans. Fuzzy Syst.*, 14(1): 145–168.

Wai, R.J., and P.C. Chen (2004). Intelligent tracking control for robot manipulator including actuator dynamics via TSK-type fuzzy neural network, *IEEE Trans. Fuzzy Syst.*, 12(4): 552–560.

Wan, Z., and M. V. Kothare (2003). Efficient scheduled stabilizing model predictive control for constrained nonlinear systems, *Int. J. Robust Nonlin. Control*, 13, pp. 331–346.

Wang, H.O., K. Tanaka, and M.F. Griffin (1996). An approach to fuzzy control of nonlinear systems: Stability and design issues, *IEEE Trans. Fuzzy Syst.*, 4(1): 14–23.

Wang, J.S., and C.S.G. Lee (2003). Self-adaptive recurrent neuro-fuzzy control of an autonomous underwater vehicle, *IEEE Trans. Robot. Autom.*, 19(2): 283–295.

Wang, L., and G. Feng (2004). Piecewise H-infinity controller design of discrete time fuzzy systems, *IEEE Trans. Syst. Man Cybern. B*, 34(1): 682–686.

Wang, L., G. Feng, and T. Hesketh (2004). Piecewise generalized H2 controller synthesis of discrete time fuzzy systems, *IEE Proc. Control Theor. Appl. D*, 151(5): 554–560.

Wang, L.X. (1993). Stable adaptive fuzzy control of nonlinear systems, *IEEE Trans. Fuzzy Syst.*, 1(2): 146–155.

Wang, L.X. (1994). *Adaptive Fuzzy Systems and Control: Design and Stability Analysis*, Prentice-Hall: Englewood Cliffs, NJ.

Wang, L.X. (1997). *A Course in Fuzzy Systems and Control*, Prentice-Hall: London.

Wang, L.X., and J.M. Mendel (1992). Fuzzy basis functions, universal approximation, and orthogonal least squares learning, *IEEE Trans. Neural Netw.*, 3(5): 807–814.

Wang, W.J., and W.W. Lin (2005). Decentralized PDC for large-scale T–S fuzzy systems, *IEEE Trans. Fuzzy Syst.*, 13(6): 779–786.

Wang, W.J., and L. Luoh (2004). Stability and stabilization of fuzzy large-scale systems, *IEEE Trans. Fuzzy Syst.,* 12(3): 309–315.

Wang, W.J., and C.H. Sun (2005). Relaxed stability and stabilization conditions for a T–S fuzzy discrete system, *Fuzzy Sets Syst.*, 156(2): 208–225.

Wang, W.Y., C.Y. Cheng, and Y.G. Leu (2004). An online GA-based output-feedback direct adaptive fuzzy-neural controller for uncertain nonlinear systems, *IEEE Trans. Syst. Man Cybern. B*, 34(1): 334–345.

Wang, Y., Z.Q. Sun, and F.C. Sun (2004). Stability analysis and control of discrete-time fuzzy systems: A fuzzy Lyapunov function approach, *Proc. 5th Asian Control Conference*, Melbourne, pp. 1855–1860.

Wong, L.K., F.H.F. Leung, and P.K.S. Tam (2001). A fuzzy sliding controller for nonlinear systems, *IEEE Trans. Indust. Electron.,* 48(1): 32–37.

Woo, Z.W., H.Y. Chung, and J.J. Lin (2000). A PID type fuzzy controller with self-tuning scaling factors, *Fuzzy Sets Syst.*, 115: 321–326.

Wu, C.J., and A.H. Sung (1994). The application of fuzzy logic to JPEG, *IEEE Trans. Consumer Electron.*, 40(4): 976–984.

Wu, H.N., and H.Y. Zhang (2005). Reliable mixed L-2/H-infinity fuzzy static output feedback control for nonlinear systems with sensor faults, *Automatica*, 41(11): 1925–1932.

Wu, J.C., and T.S. Liu (1996). A sliding mode approach to fuzzy control design, *IEEE Trans. Control Syst. Technol.*, 4(2): 141–150.

Xiao, J., J.Z. Xiao, N. Xi, R.L. Tummala, and R. Mukherjee (2004). Fuzzy controller for wall-climbing microrobots, *IEEE Trans. Fuzzy Syst.*, 12(4): 466–480.

Xie, L. (1996). Output feedback control of systems with parameter uncertainty, *Int. J. Control*, 63: 741–750.

Xu, C., and Y.C. Shin (2005). Design of a multilevel fuzzy controller for nonlinear systems and stability analysis, *IEEE Trans. Fuzzy Syst.*, 13(6): 761–778.

Xu, J.X., C.C. Hang, and C. Liu (2000). Parallel structure and tuning of a fuzzy PID controller, *Automatica*, 36: 673–684.

Xu, S.Y., and J. Lam (2005). Robust H-infinity control for uncertain discrete-time-delay fuzzy systems via output feedback controllers, *IEEE Trans. Fuzzy Syst.*, 13(1): 82–93.

Yager, R.R. and D.P. Filev (1993). A simple adaptive defuzzification method, *IEEE Transactions on Fuzzy Systems*, 1(1): 69–78.

Yager, R.R., and D.P. Filev (1994). *Essentials of Fuzzy Modeling and Control*, Wiley: New York.

Yang, S.X., H. Li, M.Q.H. Meng, and P.X. Liu (2004). An embedded fuzzy controller for a behavior-based mobile robot with guaranteed performance, *IEEE Trans. Fuzzy Syst.*, 12(4): 436–446.

Yang, Y.S., and J.S. Ren (2003). Adaptive fuzzy robust tracking controller design via small gain approach and its application, *IEEE Trans. Fuzzy Syst.*, 11(6): 783–795.

Yang, Y.S., and C.J. Zhou (2005). Adaptive fuzzy H-infinity stabilization for strict-feedback canonical nonlinear systems via backstepping and small-gain approach, *IEEE Trans. Fuzzy Syst.*, 13(1): 104–114.

Yeh, Z.M. (1999). A systematic method for design of multivariable fuzzy logic control systems, *IEEE Trans. Fuzzy Syst.*, 7(6): 741–752.

Yi, Z., and P.A. Heng (2002). Stability of fuzzy control systems with bounded uncertain delays, *IEEE Trans. Fuzzy Syst.*, 10(1): 92–97.

Yin, T.K., and C.S.G. Lee (1995). Fuzzy model-reference adaptive control, *IEEE Trans. Syst. Man Cybern.*, 25(12): 1606–1615.

Ying, H. (1993). The simplest fuzzy controllers using different inference methods are different nonlinear proportional-integral controllers with variable gains, *Automatica*, 29: 1579–1589.

Ying, H. (1998). General SISO Takagi–Sugeno fuzzy systems with linear rule consequent are universal approximators, *IEEE Trans. Fuzzy Syst.*, 6(4): 582–587.

Yoneyama, J., M. Nishikawa, H. Katayama, and A. Ichikawa (2000). Output stabilization of Takagi–Sugeno fuzzy systems, *Fuzzy Sets Syst.*, 111: 253–266.

Yu, L.X., and Y.Q. Zhang (2005). Evolutionary fuzzy neural networks for hybrid financial prediction, *IEEE Trans. Syst. Man Cybern. C*, 35(2): 244–249.

Yu, W., and X.O. Li (2004). Fuzzy identification using fuzzy neural networks with stable learning algorithms, *IEEE Trans. Fuzzy Syst.*, 12(3): 411–420.

Yu, W.S., and C.J. Sun (2001). Fuzzy model based adaptive control for a class of nonlinear systems, *IEEE Trans. Fuzzy Syst.*, 9(3): 413–425.

Yuan, K., H.X. Li, and J.D. Cao (2008). Robust stabilization of the distributed parameter system with time delay via fuzzy control, *IEEE Trans. Fuzzy Syst.*, 16(3): 567–584.

Zadeh, L.A. (1965). Fuzzy sets, *Inf. Control*, 8: 338–353.

Zadeh, L.A. (1968). Fuzzy algorithm, *Inf. Control*, 12: 94–102.

Zadeh, L.A. (1971). Similarity relations and fuzzy orderings, *Inf. Sci.*, 3: 177–200.

Zadeh, L.A. (1973). Outline of a new approach to the analysis of complex systems and decision processes, *IEEE Trans. Syst. Man Cybern.* 3(1): 28–44.

Zadeh, L.A. (1975). The concept of a linguistic variable and its application to approximate reasoning, I, II, III, *Inf. Sci.*, 8: 199–251, 301–357; 9: 43–80.

Zeng, K., N.Y. Zhang, and W.L. Xu (2000). A comparative study on sufficient conditions for Takagi–Sugeno fuzzy systems as universal approximators, *IEEE Trans. Fuzzy Syst.*, 8(6): 773–780.

Zeng, X.J., and M.G. Singh (1994). Approximation theory of fuzzy systems-SISO case, *IEEE Trans. Fuzzy Syst.* 2(2): 162–176.

Zeng, X.J., and M.G. Singh (1995). Approximation theory of fuzzy systems-MIMO case, *IEEE Trans. Fuzzy Syst.* 3(2): 219–235.

Zeng, X.J., and M.G. Singh (1996). Approximation accuracy analysis of fuzzy systems as function approximators, *IEEE Trans. Fuzzy Syst.*, 4(1): 44–63.

Zhang, H.G., and L.L. Cai (2002). Nonlinear adaptive control using the Fourier integral and its application to CSTR systems, *IEEE Trans. Syst. Man Cybern. B*, 32(3): 367–372.

Zhang, H.G., L.L. Cai, and Z. Bien (2000). A fuzzy basis function vector-based multivariable adaptive controller for nonlinear systems, *IEEE Trans. Syst Man Cybern. B*, 30(1): 210–217.

Zhang, J. (2005). Modeling and optimal control of batch processes using recurrent neuro-fuzzy networks, *IEEE Trans. Fuzzy Syst.*, 13(4): 417–426.

Zhang, R.T., and Y.A. Phillis (1999). Fuzzy control of queueing systems with heterogeneous servers, *IEEE Trans. Fuzzy Syst.* 7(1): 17–26.

Zhang, T.J., G. Feng, and J.H. Lu (2007). Fuzzy constrained min-max model predictive control using piecewise Lyapunov functions, *IEEE Trans. Fuzzy Syst.*, 15(4): 686–698.

Zhao, Z.Y., M. Tomizuka, and S. Isaka (1993). Fuzzy gain-scheduling of PID controllers, *IEEE Trans. Syst. Man Cybern.* 23(5): 1392–1398.

Zheng, H., and K.Y. Zhu (2004). A fuzzy controller-based multiple-model adaptive control system for blood pressure control, *Proc. 8th Conf. Control, Automation, Robotics and Vision*, Kunming, China, pp. 1353–1358.

Zhou, S.S., G. Feng, and C.B. Feng (2005a). Robust control for a class of uncertain nonlinear systems: Adaptive fuzzy approach based on backstepping, *Fuzzy Sets Syst.*, 151(1): 1–20.

Zhou, S.S., G. Feng, J. Lam, and S.Y. Xu (2005b). Robust H-infinity control for discrete fuzzy systems via basis-dependent Lyapunov functions, *Inf. Sci.*, 174(3–4): 197–217.

Zhou, S.S., J. Lam, W.X. Zheng (2007). Control design for fuzzy systems based on relaxed nonquadratic stability and H_∞ performance conditions, *IEEE Trans. Fuzzy Syst.*, 15(2): 188–199.

Zimmermann, H.J. (1991). *Fuzzy Set Theory and Its Application*, 2nd edition, Kluwer Academic: Boston.

Zinober, A.S. (1994). *Variable Structure and Lyapunov Control*, Springer-Verlag: Berlin.

Zou, A.M., Z.G. Hou, and M. Tan (2008). Adaptive control of a class of nonlinear pure-feedback systems using fuzzy backstepping approach, *IEEE Trans. Fuzzy Syst.*, 16(4): 886–897.

Index

Page references in **bold** refer to tables.